LONDON MATHEMATICAL SOCIETY LECTURE NOTE SERIES

Managing Editor: Professor J.W.S. Cassels, Department of Pure Mathematics and Mathematical Statistics, University of Cambridge, 16 Mill Lane, Cambridge CB2 1SB, England

The books in the series listed below are available from booksellers, or, in case of difficulty, from Cambridge University Press.

London Mathematical Society Lecture Note Series. 187

Surveys in Combinatorics, 1993

Edited by

Keith Walker
University of Keele

CAMBRIDGE
UNIVERSITY PRESS

Published by the Press Syndicate of the University of Cambridge
The Pitt Building, Trumpington Street, Cambridge CB2 1RP
40 West 20th Street, New York, NY 10011-4211, USA
10 Stamford Road, Oakleigh, Melbourne 3166, Australia

First published 1993

Library of Congress cataloguing in publication data available

British Library cataloguing in publication data available

ISBN   0 521 44857 3

Transferred to digital printing 2005

Contents

Preface

The fourteenth British Combinatorial Conference is to be held here in July 1993 and, as is usual, the British Combinatorial Committee has invited nine distinguished combinatorial mathematicians to give survey talks on the latest developments in their own fields. This volume contains the papers which they have agreed to submit in advance of their talks.

The quality of these contributions and the interest shown are encouraging signs that the fourteenth conference will be as successful as its predecessors.

The sixtieth birthday of Crispin Nash-Williams falls in 1993, and to mark his tremendous contribution to the growth of combinatorics both in Britain and the world, one day of the conference is to be 'Nash-Williams day'. The talks by András Frank and Anthony Hilton on that day will be in his honour, as will be many of the contributed talks, and to him this volume is dedicated.

A special edition of Discrete Mathematics, edited by Douglas Woodall, will contain papers contributed to the conference.

I wish to thank the contributors and the referrees for their co-operation in meeting quite tight deadlines, Roger Astley of Cambridge University Press, and the London Mathematical Society and the Institute of Combinatorics and its Applications for their financial support.

Keith Walker
Keele, March 1993

Restricted colorings of graphs

Noga Alon *

Department of Mathematics

Raymond and Beverly Sackler Faculty of Exact Sciences

Tel Aviv University, Tel Aviv, Israel

and Bellcore, Morristown, NJ 07960, USA

Abstract

The problem of properly coloring the vertices (or edges) of a graph using for each vertex (or edge) a color from a prescribed list of permissible colors, received a considerable amount of attention. Here we describe the techniques applied in the study of this subject, which combine combinatorial, algebraic and probabilistic methods, and discuss several intriguing conjectures and open problems. This is mainly a survey of recent and less recent results in the area, but it contains several new results as well.

*Research supported in part by a United States Israel BSF Grant

1

1   Introduction

Graph coloring is arguably the most popular subject in graph theory. An interesting variant of the classical problem of coloring properly the vertices of a graph with the minimum possible number of colors arises when one imposes some restrictions on the colors available for every vertex. This variant received a considerable amount of attention that led to several fascinating conjectures and results, and its study combines interesting combinatorial techniques with powerful algebraic and probabilistic ideas. The subject, initiated independently by Vizing [51] and by Erdős, Rubin and Taylor [24], is usually known as the study of the *choosability* properties of a graph. In the present paper we survey some of the known recent and less recent results in this topic, focusing on the techniques involved and mentioning some of the related intriguing open problems. This is mostly a survey article, but it contains various new results as well.

A *vertex coloring* of a graph G is an assignment of a color to each vertex of G. The coloring is *proper* if adjacent vertices receive distinct colors. The *chromatic number* $\chi(G)$ of G is the minimum number of colors used in a proper vertex coloring of G. An *edge coloring* of G is, similarly, an assignment of a color to each edge of G. It is *proper* if adjacent edges receive distinct colors. The minimum number of colors in a proper edge-coloring of G is the *chromatic index* $\chi'(G)$ of G. This is clearly equal to the chromatic number of the line graph of G.

If $G = (V, E)$ is a (finite, directed or undirected) graph, and f is a function that assigns to each vertex v of G a positive integer $f(v)$, we say that G is *f-choosable* if, for every assignment of sets of integers $S(v) \subset Z$ to all the vertices $v \in V$, where $|S(v)| = f(v)$ for all v, there is a proper vertex coloring $c : V \mapsto Z$ so that $c(v) \in S(v)$ for all $v \in V$. The graph G is *k-choosable* if it is f-choosable for the constant function $f(v) \equiv k$. The *choice number* of G, denoted $ch(G)$, is the minimum integer k so that G is k-choosable. Obviously, this number is at least the classical chromatic number $\chi(G)$ of G. The choice number of the line graph of G, which we denote here by $ch'(G)$, is usually called the *list chromatic index* of G, and it is clearly at least the chromatic index $\chi'(G)$ of G.

As observed by various researchers ([51], [24], [1]), there are many graphs G for which the choice number $ch(G)$ is strictly larger than the chromatic number $\chi(G)$. A simple example demonstrating this fact is the complete bipartite graph $K_{3,3}$. If $\{u_1, u_2, u_3\}$ and $\{v_1, v_2, v_3\}$ are its two vertex-classes and $S(u_i) = S(v_i) = \{1, 2, 3\} \setminus \{i\}$, then there is no proper vertex coloring assigning to each vertex w a color from its class $S(w)$. Therefore, the choice number of this graph exceeds its chromatic number. In fact, it is easy to show that, for any $k \geq 2$, there are bipartite graphs whose choice number exceeds k; a general construction is given in the following section. The gap between the two parameters $ch(G)$ and $\chi(G)$ can thus be arbitrarily large. In view of this, the following conjecture, suggested independently by various researchers including Vizing, Albertson, Collins, Tucker and Gupta, which apparently appeared first in print in the paper of Bollobás and Harris ([15]), is somewhat surprising.

Conjecture 1.1 (The list coloring conjecture) *For every graph* G, $ch'(G) = \chi'(G)$.

This conjecture asserts that for *line graphs* there is no gap at all between the choice number and the chromatic number. Many of the most interesting results in the area are proofs of special cases of this conjecture, some of which are described in Sections 2, 3 and 4. The proof for the general case (if true) seems extremely difficult, and even some very special cases that have received a considerable amount of attention are still open.

The problem of determining the choice number of a given graph is difficult, even for small graphs with a simple structure. To see this, you may try to convince yourself that the complete bipartite graph $K_{5,8}$ is 3-choosable; a (lengthy) proof appears in [42]. More formally, it is shown in [24] that the problem of deciding if a given graph $G = (V, E)$ is f-choosable, for a given function $f : V \mapsto \{2, 3\}$, is Π_2^p-Complete. Therefore, if the complexity classes NP and $coNP$ differ, as is commonly believed, this problem is stricly harder than the problem of deciding if a given graph is k-colorable, which is, of course, NP-Complete. (See [27] for the definitions of the complexity classes above.) More results on the complexity of several variants of the choosability problem appear in [38], where it is also briefly shown how some of these variants arise naturally in the study of various scheduling problems.

The study of choice numbers combines combinatorial ideas with algebraic and probabilistic tools. In the following sections we discuss these methods and present the main results and open questions in the area. The paper is organized as follows. After describing, in Section 2, some basic and initial results, we discuss, in Section 3, an algebraic approach and some of its recent consequences. Various applications of probabilistic methods to choosability are considered in Section 4. A new result is obtained in Section 5 which presents a proof of the fact that the choice number of any simple graph with average degree d is at least $\Omega(\log d/ \log \log d)$. Thus, the choice number of a simple graph must grow with its average degree, unlike the chromatic number. The final Section 6 contains some concluding remarks and open problems, in addition to those mentioned in the previous sections.

2   Some basic results

One of the basic results in graph coloring is Brooks' theorem [17], that asserts that the chromatic number of every connected graph, which is not a complete graph or an odd cycle, does not exceed its maximum degree. The choosability version of this result has been proved, independently, by Vizing [51] (in a slightly weaker form) and by Erdős, Rubin and Taylor [24]. (See also [40]).

Theorem 2.1 ([51], [24]) *The choice number of any connected graph G, which is not complete or an odd cycle, does not exceed its maximum degree.*

Note that this suffices to prove the validity of the list coloring conjecture for simple graphs of maximum degree 3 whose chromatic index is not 3 (known as *class* 2 graphs), since by Vizing's theorem [50], the chromatic index of such graphs must be 4.

A graph is called *d-degenerate* if any subgraph of it contains a vertex of degree at most d. By a simple inductive argument, one can prove the following result.

Proposition 2.2 *The choice number of any d-degenerate graph is at most $d+1$.*

This simple fact implies, for example, that every planar graph is 6-choosable. It is not known if every planar graph is 5-choosable; this is conjectured to be the case in [24]- in fact, it may even be true that every planar graph is 4-choosable.

A characterization of all 2-choosable graphs is given in [24]. If G is a connected graph, the *core* of G is the graph obtained from G by repeatedly deleting vertices of degree 1 until there is no such vertex.

Theorem 2.3 ([24]) *A simple graph is 2-choosable if and only if the core of each of its connected components is either a single vertex, or an even cycle, or a graph consisting of two vertices with three even internally disjoint paths between them, where the length of at least two of the paths is exactly 2.*

Of course, one cannot hope for such a simple characterization of the class of all 3-choosable graphs, since, as observed by Gutner it follows easily from the complexity result mentioned in the introduction that the problem of deciding if a given graph is 3-choosable is NP-hard; in fact, as shown in [28], this problem is even Π_2^p-complete.

In Section 1 we saw an example of a graph with choice number that exceeds its chromatic number. Here is an obvious generalization of this construction. Let $H = (U, W)$ be a k-uniform hypergraph which is not 2-colorable; that is, every edge $w \in W$ has precisely k elements and for every 2-vertex coloring of H there is a monochromatic edge. If $|W| = n$ we claim that the complete bipartite graph $K_{n,n}$ is not k-choosable. Indeed, denote the vertices of H by $1, 2, \ldots$, and let $A = \{a_w : w \in W\}$ and $B = \{b_w : w \in W\}$ be the two vertex classes of $K_{n,n}$. For each $w \in W$, define $S(a_w) = S(b_w) = \{u \in U : u \in w\}$. One can easily check that there is a proper coloring c of the complete bipartite graph on $A \cup B$ assigning to each vertex a_w and b_w a color from its class $S(a_w) (= S(b_w))$ if and only if the hypergraph H is 2-colorable. Thus, by the choice of H, the choice number of $K_{n,n}$ is stricly bigger than k.

As shown by Erdős [23], for large values of k there are k-uniform hypergraphs with at most $n = (1 + o(1)) \frac{e \ln 2}{4} k^2 2^k$ edges which are not 2-colorable, showing that there are bipartite graphs with that many vertices on each side whose choice number exceeds k. We note that this estimate is nearly sharp, as a very simple probabilistic argument shows that, if $n < 2^{k-1}$, then $K_{n,n}$ is k-choosable. Indeed, given a list $S(v)$ of k colors for each vertex v in the two vertex classes A and B, let S be the set of all the colors used in the union of all the lists and let us choose a random partition (S_A, S_B) of S into two disjoint parts, where, for each $s \in S$ randomly and independently, s is chosen to be in S_A or in S_B with equal probability. The colors in S_A will be used to color vertices in A and those in S_B to color the vertices in S_B. For a fixed vertex a in A, the probability that its coloring will fail-that is, we will not be able to color it by a color from S_A- is precisely $1/2^k$, as this is the probability that all the colors in its class $S(a)$ were chosen to be in S_B. A similar estimate holds for the members of B, and hence the probability that there exists a vertex that will fail to receive a color is at most $|A \cup B|/2^k < 1$. This estimate can be slightly improved, using the method (or the result) of Beck ([10]), but the above simple argument suffices to demonstrate the relevance of probabilistic techniques in the study of choice numbers.

The *total chromatic number* of a graph G, denoted by $\chi''(G)$, is the minimum number of colors required to color all the vertices and edges of G, so that adjacent or incident elements receive distinct colors. The following conjecture is due to Behzad [11].

Conjecture 2.4 (The total coloring conjecture) *The total chromatic number of every simple graph G with maximum degree Δ is at most $\Delta + 2$.*

There are several papers dealing with this conjecture, and the following estimates are known. If G is a simple graph on n vertices with maximum degree Δ then, as shown by Hind [33], [34]:

$$\chi''(G) \leq \Delta + 1 + 2\lceil\sqrt{\Delta}\rceil,$$

and

$$\chi''(G) \leq \Delta + 1 + 2\lceil\frac{n}{\Delta}\rceil.$$

Chetwynd and Häggkvist [31] showed that, if $t! > n$, then

$$\chi''(G) \leq \Delta + 1 + t.$$

Note that, as observed by the authors of [15], the validity of Conjecture 1.1 (the list coloring conjecture) would imply that, for every simple graph with maximum degree Δ,

$$\chi''(G) \leq \chi'(G) + 2 \leq \Delta + 3.$$

Indeed, let $S = \{1, 2, \ldots, \chi'(G) + 2\}$ be our set of colors. Start with an arbitrary proper vertex coloring of G using these colors; this certainly exists, for example, by Brooks' Theorem and by the fact that $\chi'(G) \geq \Delta$. Now associate with each edge e of G a list $S(e)$ of all the colors in S except the ones appearing on its two ends. By the list coloring conjecture, there is a proper edge coloring of G using, for each edge e, a color from $S(e)$; this would give a proper total coloring of G. The fact that $|S| \leq \Delta + 3$ now follows from Vizing's theorem ([50]). It seems, however, that getting a $\Delta + O(1)$ upper estimate for the total chromatic number of a simple graph with maximum degree Δ should be much easier than getting a similar bound for the list chromatic index of such a graph.

3 An algebraic approach and its applications

An algebraic technique that, in various cases, supplies useful information on the choice numbers of given graphs, has been developed by M. Tarsi and the present author in [9]. In this section we describe this method and present some of its recent applications.

A subdigraph H of a directed graph D is called *Eulerian* if the indegree $d_H^-(v)$ of every vertex v of H is equal to its outdegree $d_H^+(v)$. Note that we do not assume that H is connected. H is *even* if it has an even number of edges, otherwise, it is *odd*. Let $EE(D)$ and $EO(D)$ denote the numbers of even and odd Eulerian subgraphs of D, respectively. (For convenience we agree that the empty subgraph is an even Eulerian subgraph.) The following result is proved in [9].

Theorem 3.1 *Let $D = (V, E)$ be a digraph, and define $f : V \mapsto Z$ by $f(v) = d_D^+(v) + 1$, where $d_D^+(v)$ is the outdegree of v. If $EE(D) \neq EO(D)$, then D is f-choosable.*

Note that the assertion of the theorem for the special case of acyclic di-graphs, which implies Proposition 2.2, can be proved by a simple inductive argument. The general case seems much more difficult. To prove this theorem, we need the following simple statement.

Lemma 3.2 *Let $P = P(x_1, x_2, \ldots, x_n)$ be a polynomial in n variables over the ring of integers Z. Suppose that the degree of P as a polynomial in x_i is at most d_i for $1 \leq i \leq n$, and let $S_i \subset Z$ be a set of $d_i + 1$ distinct integers. If $P(x_1, x_2, \ldots, x_n) = 0$ for all n-tuples $(x_1, \ldots, x_n) \in S_1 \times S_2 \times \ldots \times S_n$, then $P \equiv 0$.*

Proof We apply induction on n. For $n = 1$, the lemma is simply the assertion that a non-zero polynomial of degree d_1 in one variable can have at most d_1 distinct zeros. Assuming that the lemma holds for $n - 1$, we prove it for n ($n \geq 2$). Given a polynomial $P = P(x_1, \ldots, x_n)$ and sets S_i satisfying the hypotheses of the lemma, let us write P as a polynomial in x_n- that is,

$$P = \sum_{i=0}^{d_n} P_i(x_1, \ldots, x_{n-1})x_n^i,$$

where each P_i is a polynomial with x_j-degree bounded by d_j. For each fixed $(n - 1)$-tuple $(x_1, \ldots, x_{n-1}) \in S_1 \times S_2 \times \ldots \times S_{n-1}$, the polynomial in x_n obtained from P by substituting the values of $x_1, \ldots, x_{n-1}$ vanishes for all $x_n \in S_n$, and is thus identically 0. Thus $P_i(x_1, \ldots, x_{n-1}) = 0$ for all $(x_1, \ldots, x_{n-1}) \in S_1 \times \ldots \times S_{n-1}$. Hence, by the induction hypothesis, $P_i \equiv 0$ for all i, implying that $P \equiv 0$. This completes the induction and the proof of the lemma. $\square$

The *graph polynomial $f_G = f_G(x_1, x_2, \ldots, x_n)$* of a directed or undirected graph $G = (V, E)$ on a set $V = \{v_1, \ldots, v_n\}$ of n vertices is defined by $f_G(x_1, x_2, \ldots, x_n) = \Pi\{(x_i - x_j) : i < j , \{v_i, v_j\} \in E\}$. This polynomial has been studied by various researchers, starting already with Petersen [44] in 1891. See also, for example, [46], [39].

For $1 \leq i \leq n$, let $S_i \subset Z$ be a set of $d_i + 1$ distinct integers. For each i, $1 \leq i \leq n$, let $Q_i(x_i)$ be the polynomial $Q_i(x_i) = \Pi_{s \in S_i}(x_i - s)$. Let $\mathcal{I}$ be the ideal generated by the polynomials Q_i in the ring of polynomials $Z[x_1, \ldots, x_n]$. It is obvious that if $f_G(x_1, \ldots, x_n) \in \mathcal{I}$, then f_G vanishes on every common zero of all the polynomials Q_i. But this means that f_G

vanishes on every $(x_1, \ldots, x_n) \in S_1 \times S_2 \times \ldots \times S_n$; hence, for each assignment of values $x_i \in S_i$, there is an edge $v_i v_j$ of G with $x_i = x_j$. Therefore, there is no proper vertex coloring of G assigning to each vertex v_i a color from its set S_i. The following Nullstellensatz-type result asserts that the converse is also true.

Proposition 3.3 *Let $G = (V, E)$ be a graph on the set of vertices $V = \{v_1, \ldots, v_n\}$, and let S_i, $1 \leq i \leq n$, be sets of integers. Let $f_G = f_G(x_1, \ldots, x_n)$ be the graph polynomial of G, and let $Q_i(x_i)$ and $\mathcal{I}$ be as above. Then $f_G \in \mathcal{I}$ if and only if there is no proper vertex coloring c of G satisfying $c(v_i) \in S_i$, for all $1 \leq i \leq n$.*

Proof We have already seen that, if there is a coloring as above, then f_G is not in $\mathcal{I}$. It remains to show that if there is no such coloring, then $f_G \in \mathcal{I}$. The assumption that the required coloring does not exist is equivalent to the statement:

$$f_G(x_1, \ldots, x_n) = 0 \quad \text{for every } n\text{-tuple} \quad (x_1, \ldots, x_n) \in S_1 \times S_2 \times \ldots \times S_n. \tag{1}$$

For each i, $1 \leq i \leq n$, put

$$Q_i(x_i) = \Pi_{s \in S_i}(x_i - s) = x_i^{d_i+1} - \sum_{j=0}^{d_i} q_{ij} x_i^j.$$

Observe that,

$$\text{if } x_i \in S_i \text{ then } Q_i(x_i) = 0 \text{- that is, } x_i^{d_i+1} = \sum_{j=0}^{d_i} q_{ij} x_i^j. \tag{2}$$

Let $\overline{f_G}$ be the polynomial obtained by writing f_G as a linear combination of monomials and replacing, repeatedly, each occurrence of $x_i^{f_i}$ ($1 \leq i \leq n$), where $f_i > d_i$, by a linear combination of smaller powers of x_i, using the relations (2). The resulting polynomial $\overline{f_G}$ is clearly of degree at most d_i in x_i, for each $1 \leq i \leq n$, and satisfies $\overline{f_G} \equiv f_G \pmod{\mathcal{I}}$. Moreover, $\overline{f_G}(x_1, \ldots, x_n) = f_G(x_1, \ldots, x_n)$, for all $(x_1, \ldots, x_n) \in S_1 \times \ldots \times S_n$, since the relations (2) hold for these values of $x_1, \ldots, x_n$. Therefore, by (1), $\overline{f_G}(x_1, \ldots, x_n) = 0$ for every n-tuple $(x_1, \ldots, x_n) \in S_1 \times \ldots \times S_n$ and hence, by Lemma 3.2, $\overline{f_G} \equiv 0$. This implies that $f_G \in \mathcal{I}$, and completes the proof.

$\square$

The special case of the last proposition, for the case in which all the sets S_i are equal, implies that, for every fixed polynomial $Q(x)$ of one variable with k distinct integer roots, a graph G is not k-colorable if and only if the graph polynomial f_G lies in the ideal generated by the polynomials $Q(x_i)$. In fact, the assumption that the roots of $Q(x)$ are integral is not essential, as the proof works equally well in the ring of polynomials $K[x_1, \ldots, x_n]$ over any field K. See [9] for more details. This result is related to a theorem of Kleitman and Lovász ([41]) , who applied a method similar to that of [39], and showed that a graph $G = (V, E)$ is not k-colorable if and only if f_G lies in the ideal generated by the set of all graph polynomials of complete graphs on $k + 1$ vertices among those in V. As shown by De Loera in [19], the set of graph polynomials of complete $(k + 1)$-graphs, as well as the set of polynomials $Q(x_i)$ above, are both universal Gröbner bases for the ideals they generate. See [19] for more details.

It is not too difficult to see that the coefficients of the monomials that appear in the standard representation of f_G as a linear combination of monomials can be expressed in terms of the orientations of G. For each oriented edge $e = (v_i, v_j)$ of G, define its *weight* $w(e)$ by $w(e) = x_i$ if $i < j$, and $w(e) = -x_i$ if $i > j$. The weight $w(D)$ of an orientation D of G is defined to be the product $\Pi w(e)$, where e ranges over all oriented edges e of D. Clearly $f_G = \sum w(D)$, where D ranges over all orientations of G. This is simply because each term in the expansion of the product $f_G = \Pi\{(x_i - x_j) : i < j , \{v_i, v_j\} \in E\}$ corresponds to a choice of the orientation of the edge $\{v_i, v_j\}$ for each edge $\{v_i, v_j\}$ of G. Let us call an oriented edge (v_i, v_j) of G *decreasing* if $i > j$. An orientation D of G is called *even* if it has an even number of decreasing edges; otherwise, it is called *odd*.

For non-negative integers $d_1, d_2, \ldots, d_n$, let $DE(d_1, \ldots, d_n)$ and $DO(d_1, \ldots, d_n)$ denote, respectively, the sets of all even and odd orientations of G in which the outdegree of the vertex v_i is d_i, for $1 \leq i \leq n$. By the last paragraph, the following lemma holds.

Lemma 3.4 *In the above notation*

$$f_G(x_1, \ldots, x_n) = \sum_{d_1, \ldots, d_n \geq 0} (|DE(d_1, \ldots, d_n)| - |DO(d_1, \ldots, d_n)|)\Pi_{i=1}^n x_i^{d_i} . \quad \square$$

Consider, now, a fixed sequence $d_1, \ldots, d_n$ of nonnegative integers and let D_1 be a fixed orientation in $DE(d_1, \ldots, d_n) \cup DO(d_1, \ldots, d_n)$. For any orientation $D_2 \in DE(d_1, \ldots, d_n) \cup DO(d_1, \ldots, d_n)$, let $D_1 \oplus D_2$ denote the set of all oriented edges of D_1 whose orientation in D_2 is in the opposite direction. Since the outdegree of every vertex in D_1 is equal to its outdegree in D_2, it follows that $D_1 \oplus D_2$ is an Eulerian subgraph of D_1. Moreover, $D_1 \oplus D_2$ is even as an Eulerian subgraph if and only if D_1 and D_2 are both even or both odd. The mapping $D_2 \longrightarrow D_1 \oplus D_2$ is clearly a bijection between $DE(d_1, \ldots, d_n) \cup DO(d_1, \ldots, d_n)$ and the set of all Eulerian subgraphs of D_1. In the case D_1 is even, it maps even orientations to even (Eulerian) subgraphs, and odd orientations to odd subgraphs. Otherwise, it maps even orientations to odd subgraphs, and odd orientations to even subgraphs. In any case,

$$\Big| |DE(d_1, \ldots, d_n)| - |DO(d_1, \ldots, d_n)| \Big| = |EE(D_1) - EO(D_1)|$$

where $EE(D_1)$ and $EO(D_1)$ denote, as before, the numbers of even and odd Eulerian subgraphs of D_1, respectively. Combining this with Lemma 3.4, we obtain the following.

Corollary 3.5 *Let D be an orientation of an undirected graph $G = (V, E)$ on a set $V = \{v_1, \ldots, v_n\}$ of n vertices. For $1 \leq i \leq n$, let $d_i = d_D^+(v_i)$ be the outdegree of v_i in D. Then the absolute value of the coefficient of the monomial $\Pi_{i=1}^{n} x_i^{d_i}$ in the standard representation of $f_G = f_G(x_1, \ldots, x_n)$ as a linear combination of monomials, is $|EE(D) - EO(D)|$. In particular, if $EE(D) \neq EO(D)$, then this coefficient is not zero.* $\square$

Proof of Theorem 3.1 Let $D = (V, E)$ be a digraph on the set of vertices $V = \{v_1, \ldots, v_n\}$ and let $d_i = d_D^+(v_i)$ be the outdegree of v_i. Suppose that $EE(D) \neq EO(D)$. For $1 \leq i \leq n$, let $S_i \subset Z$ be a set of $d_i + 1$ distinct integers. We must show that there is a legal vertex-coloring $c : V \mapsto Z$ such that $c(v_i) \in S_i$, for all $1 \leq i \leq n$. Suppose that this is false and there is no such coloring. Then, by Proposition 3.3 and its proof, $\overline{f_D} \equiv 0$, where, as before, $\overline{f_D}$ is the reduction of the graph polynomial f_D using the relations (2). However, by Corollary 3.5, the coefficient of $\Pi_{i=1}^{n} x_i^{d_i}$ in f_D is nonzero, since, by assumption, $EE(D) \neq EO(D)$. Since the degree of each

x_i in this monomial is d_i, the relations (2) will not affect it. Moreover, as the polynomial f_D is homogeneous and each application of the relations (2) strictly reduces the degree, the process of replacing f_D by $\overline{f_D}$ will not create any new scalar multiples of $\Pi_{i=1}^{n} x_i^{d_i}$. Thus, the coefficient of $\Pi_{i=1}^{n} x_i^{d_i}$ in $\overline{f_D}$ is equal to its coefficient in f_D, and is not 0. This contradicts the fact that $\overline{f_D} \equiv 0$. Therefore, our assumption was false, and there is a legal coloring $c : V \mapsto Z$ satisfying $c(v_i) \in S_i$, for all $1 \leq i \leq n$. □

An interesting application of Theorem 3.1 has been obtained by Fleischner and Stiebitz in [26], solving a problem raised by Du, Hsu and Hwang in [21], as well as a strengthening of it suggested by Erdős.

Theorem 3.6 ([26]) *Let G be a graph on $3n$ vertices, whose set of edges is the disjoint union of a Hamilton cycle and n pairwise vertex-disjoint triangles. Then the choice number and the chromatic number of G are both 3.*

The proof is based on a subtle parity argument that shows that, if D is the digraph obtained from G by directing the Hamilton cycle as well as each of the triangles cyclically, then $EE(D) - EO(D) \equiv 2 (mod\ 4\ )$. The result thus follows from Theorem 3.1. We note that the result supplies no efficient algorithm for finding a proper 3-vertex coloring of such a graph, although the methods of [3], [25] or [5] do supply an efficient algorithm for the related (easier) problem of finding a proper 4-vertex coloring of a graph on $4n$ vertices, whose set of edges is the disjoint union of a Hamilton cycle and n pairwise vertex disjoint copies of K_4. Several extensions appear in [25], [5].

Another simple application of Theorem 3.1 is the following result, that solves an open problem from [24].

Theorem 3.7 ([9]) *The choice number of every planar bipartite graph is at most 3.*

This is tight, since $ch(K_{2,4}) = 3$.

Recall that the list coloring conjecture (Conjecture 1.1) asserts that $ch'(G) = \chi'(G)$ for every graph G. In order to try to apply Theorem 3.1 for tackling this problem, it is useful to find a more convenient expression for the difference $EE(D) - EO(D)$, where D is the appropriate orientation of a given line graph. Here is a brief derivation of such an expression for line graphs of regular graphs of class 1. Let $G = (V, E)$ be a d-regular graph satisfying $\chi'(G) = d$. Observe that the line graph $L(G)$ of G is $(2d - 2)$-regular, and hence has an Eulerian orientation D in which every outdegree is precisely $d-1$. Let $f_D(x_1, \ldots)$ denote the graph polynomial of D. Our objective is to compute the coefficient of the monomial Πx_i^{d-1} in the standard representation of f_D as a sum of monomials. Let us denote this coefficient by $C(D)$. Note that, by Corollary 3.5 and its proof, the absolute value of $C(D)$ is the absolute value of the difference between the number of even Eulerian orientations of $L(G)$ and the number of odd Eulerian orientations of it.

It is convenient to consider both this combinatorial interpretation and the interpretation as the appropriate coefficient. Starting with the latter, observe that the edges of the line graph $L(G)$ consist of $|V|$ edge disjoint cliques, each of size d. For every $v \in V$, there is a clique in $L(G)$ on all the d edges of G (which are vertices of $L(G)$) that are incident with v. Therefore, the graph polynomial f_D is a product of $|V|$ graph polynomials of complete graphs, each of size d. However, the graph polynomial of a complete graph is a Vandermonde determinant, and hence one can express f_D as a product of $|V|$ Vandermonde determinants. Interpreting the coefficient of the monomial Πx_i^{d-1} in this new expresion, we conclude that its absolute value is the absolute value of the difference between the number of even Eulerian orientations of $L(G)$ and the number of odd ones, where we count only (Eulerian) orientations in which each of the $|V|$ tournaments corresponding to the cliques around the vertices of G is acyclic. (It is not too difficult to show directly that the other orientations cancel each other, without using the interpretation as a product of Vandermonde determinants, but we omit the detailed argument using this direct approach.) Since an acyclic orientation of a tournament defines a permutation in the obvious way (or by expanding the Vandermonde determinants according to their definition), the last difference can be rewritten as follows.

For each vertex $v \in V$, let π_v be an arbitrary permutation of the edges of G incident with v. It is convenient to consider such a permutation as a bijection from the d edges above to the set $\{0, 1, \ldots, d-1\}$. Let SP denote the class of all sets of $|V|$ permutations $\Sigma = \{\sigma_v : v \in V\}$, where σ_v is a permutation of the edges of G incident with v so that for each edge $e = uv$, $\sigma_u(e) + \sigma_v(e) = d-1$. For each $\Sigma \in SP$, let $sign(\Sigma)$ denote the product of the signs of all $|V|$ permutations $\pi_v^{-1}(\sigma_v)$, $v \in V$. Then

$$|C(D)| = |\sum_{\Sigma \in SP} sign(\Sigma)|. \tag{3}$$

Next, associate with each $\Sigma = \{\sigma_v : v \subset V\}$ a partition $P = P(\Sigma)$ of the set of edges of G into $s = \lfloor (d+1)/2 \rfloor$ classes $P_0, P_1, \ldots, P_{s-1}$, by letting P_i denote the set of all edges $e = uv$ with $\sigma_u(e) = i$ (and $\sigma_v(e) = d-1-i$). Notice that for $0 \leq i < (d-1)/2$, P_i is a 2-factor of G, whereas for odd d, the last class $P_{(d-1)/2}$ is a perfect matching. If there is some 2-factor P_i that contains an odd cycle, then let $v_0, v_1, \ldots, v_r = v_0$ be the vertices of the first such cycle in the first such 2-factor, and suppose that $\sigma_{v_j}(v_j v_{j+1}) = i$ and $\sigma_{v_{j+1}}(v_j v_{j+1}) = d-1-i$, where the indices are reduced modulo r. Then we can define $\Sigma' \in SP$ as the collection of permutations σ'_v obtained from Σ, by defining $\sigma'_{v_j}(v_j v_{j+1}) = d-1-i$ and $\sigma'_{v_{j+1}}(v_j v_{j+1}) = i$, where, in each other place, each σ'_v coincides with the corresponding σ_v. One can easily check that this is a fixed-point-free involution on the members of SP of this type that switches the sign. Therefore, these members cancel each other in equation (3). We are thus left with the members $\Sigma \in SP$ for which every class in $P(\Sigma)$ is a 2-factor of even cycles (and one class is a perfect matching, in case of odd d). From each such partition $P(\Sigma)$, with l even cycles in all its 2-factors together, one can get 2^l proper edge colorings of G with d colors, by coloring the edges of each even cycle in P_i alternately with the colors i and $d-1-i$. It can easily be checked that this correspondence is a bijection between the remaining members of SP and the proper edge colorings. This yields the following interpretation of the coefficient $C(D)$. For every proper edge coloring c of G with the d-colors $\{0, \ldots, d-1\}$, the *sign of c*, denoted by $sign(c)$, is defined as the product of the signs of all the $|V|$ permutations $(\pi_v^{-1} c(e) : v \in e \in E(G))$. Let $EC(G)$ denote the set of all proper d edge colorings of G, and define

$$ec(G) = \sum_{c \in EC(G)} sign(c).$$

Proposition 3.8 *With the above notation*

$$|C(D)| = |ec(G)|. \quad \square$$

This proposition is described (very briefly, and only for the special case $G = K_{n,n}$) in [9]. The case $d = 3$ of it appears in [35] (see also [46] for the case of planar cubic graphs). Combining this proposition with Theorem 3.1, we conclude.

Corollary 3.9 *The list coloring conjecture holds for any d-regular graph G with chromatic index d that satisfies $ec(G) \neq 0$.* $\square$

Are there any interesting examples of graphs G as above, for which one can prove that $ec(G) \neq 0$? Any cubic graph with chromatic index 3, in which there is a perfect matching that appears in every proper 3-edge coloring, is such an example, and one can give infinitely many examples of this type. More interesting is the following result, observed independently by F. Jaeger and M. Tarsi (private communication), and by M. Ellingham and L. Goddyn [22].

Corollary 3.10 *For every 2-connected cubic planar graph G, $ch'(G) = 3$.*

Proof It is known ([49]; see also [35] for a short proof) that all the 3-edge colorings of a planar cubic graph have the same sign. On the other hand, the fact that every 2-connected cubic planar graph has chromatic index 3 is well known to be equivalent to the Four Color Theorem. The result thus follows from Corollary 3.9. $\square$

Note that the above result is a strengthening of the Four Color Theorem. The proof supplies no efficient procedure for finding a proper 3-edge coloring for a given 2-connected planar cubic graph with a list of 3 colors for each of its edges, that assigns to each edge a color from its list. As shown in [22], it is possible to extend the above proof to any d-regular planar multigraph with chromatic index d, establishing the following.

Theorem 3.11 ([22]) *The list chromatic index of any d-regular planar multigraph with chromatic index d is d.*

4   Probabilistic techniques

Probabilistic arguments have been applied by various researchers in the study of restricted colorings. We have already seen a simple example in Section 2. Here is another simple and elegant example, due to Chetwynd and Häggkvist ([31]), which deals with a variant of the restricted coloring problem, in which we color edges trying to avoid a forbidden color on each edge.

Proposition 4.1 ([31]) *Let $G = (V, E)$ be a graph with n vertices, maximum degree Δ, and chromatic index r $(\geq \Delta)$. For each edge $e \in E$, let $d(e) \in \{1, \ldots, r\}$ be a forbidden color for e. If t is the smallest integer satisfying $t! > n$, then there is a proper edge coloring $c : E \mapsto \{1, 2, \ldots, r + t\}$ satisfying $c(e) \neq d(e)$, for all $e \in E$.*

Proof Fix a proper edge coloring of G with the colors $1, 2, \ldots, r$, and consider a random permutation of these colors. For a fixed vertex v of G, the probability that there will be at least t edges incident with v that receive (after the permutation) their forbidden color, is at most

$$\binom{\Delta}{t} \frac{1}{r(r-1)\ldots(r-t+1)} \leq \frac{1}{t!} < 1/n.$$

Therefore, with positive probability, there is no vertex v in which there are at least t edges that received a forbidden color after the permutation. Let H be the subgraph of violations- that is, the subgraph of G consisiting of all edges that received their forbidden colors. We have seen that there is a permutation for which the maximum degree of H is at most $t-1$. Moreover, if there are any pairs of parallel edges in H, we can exchange the colors on such a pair and reduce the number of violations locally. Therefore, we may assume that H is simple and hence, by Vizing's Theorem [50], its edges can be properly colored with the additional t colors $r+1, \ldots, r+t$, supplying a proper edge coloring of G with the colors $1, \ldots, r+t$, as needed. $\square$

It is shown in [31] that the same reasoning can be applied to obtain an $r+t$ upper bound for the total chromatic number of any multigraph with n vertices and chromatic index r, provided that $t! > n$. A similar result has been proved by McDiarmid ([43]).

Returning to the list coloring conjecture, observe that, by Proposition 2.2, $ch'(G) \leq 2\Delta - 1$ for any graph G with maximum degree Δ; this can be improved to $2\Delta - 2$ for $\Delta > 2$, by Theorem 2.1. Probabilistic methods are particularly powerful, when one tries to obtain asymptotic results. For the case of simple graphs G with maximum degree Δ, a $(\frac{7}{4} + o(1))\Delta$ upper bound for $ch'(G)$ has been obtained by Bollobás and Hind [16], improving a $(\frac{11}{6} + o(1))\Delta$ upper bound proved by Bollobás and Harris [15]. (Here, and in the next sentence, the $o(1)$ term tends to 0 as Δ tends to infinity.) The final asymptotic result for this problem has been obtained recently by J. Kahn [36], who proved an asymptotically optimal $(1 + o(1))\Delta$ upper bound. His proof applies delicate probabilistic arguments, which are based on the technique developed by Rödl in [45]. For upper bounds for the maximum possible value of $ch'(G)$ for *multigraphs* with maximum degree Δ, see [18], [32].

Probabilistic arguments are applied in [6] to obtain a sharp estimate for the choice numbers of complete multipartite graphs with equal color classes. For two positive integers m and r, let K_{m*r} denote the complete r-partite graph with m vertices in each vertex class. For $r = 1$, K_{m*r} has no edges and hence, obviously, $ch(K_{m*1}) = 1$, for all m. Another trivial observation is the fact that $ch(K_{1*r}) = r$, for all r. In [24] it is shown that $ch(K_{2*r}) = r$ for all r. The following theorem determines, up to a constant factor, the choice number of K_{m*r} for all the remaining cases.

Theorem 4.2 ([6]) *There exist two positive constants c_1 and c_2 such that, for every $m \geq 2$ and for every $r \geq 2$,*

$$c_1 r \log m \leq ch(K_{m*r}) \leq c_2 r \log m.$$

A simple application of this theorem is the following.

Corollary 4.3 *There exists a positive constant b such that, for every n, there is an n-vertex graph G such that*

$$ch(G) + ch(G^c) \leq bn^{1/2}(\log n)^{1/2},$$

where G^c is the complement of G.

This settles a problem raised in [24], where the authors ask whether there exists a constant $\epsilon > 0$ so that, for all sufficiently large n and for every n-vertex graph G, $ch(G) + ch(G^c) > n^{1/2+\epsilon}$.

Another simple corollary of the above theorem deals with the choice numbers of random graphs. It is convenient to consider the common model $G_{n,1/2}$ (see, for example, [14]), in which the graph is obtained by taking each pair of the n labelled vertices $1, 2, \ldots, n$ to be an edge, randomly and independently, with probability $1/2$. (It is not too difficult to obtain similar results for other models of random graphs as well.) As proved by Bollobás in [13], almost surely (that is, with probability that tends to 1 as n tends to infinity), the random graph $G = G_{n,1/2}$ has chromatic number

$$(1 + o(1))n/2\log_2 n.$$

It is also known and easy- see for example, [14], [8], that almost surely G contains no independent set of size greater than $2\log_2 n$. Therefore, for $r = n/\log_2 n$ and $m = 2\log_2 n$, say, G has almost surely a proper coloring with r colors in which no color appears more than m times. It follows that G is almost surely a subgraph of K_{m*r} and hence, by Theorem 4.2, almost surely

$$ch(G) \le ch(K_{m*r}) = O(r\log m) = O(n\frac{\log\log n}{\log n}).$$

Hence, for almost all the graphs G on n vertices, $ch(G) = o(n)$ as n tends to infinity. This solves another problem raised in [24].

A sharper estimate has been obtained by Jeff Kahn (private communication), who determined the correct asymptotic behaviour of the choice number of the random graph. Here is the result, and its surprisingly simple proof.

Proposition 4.4 *The choice number of the random graph $G_{n,1/2}$ on n vertices is almost surely $(1 + o(1))n/2\log_2 n$.*

Proof As the choice number is at least as large as the chromatic number, the known estimate of the chromatic number of the random graph shows that the choice number of $G_{n,1/2}$ is almost surely at least $(1+o(1))n/2\log_2 n$. In [13], Bollobás shows that $G = G_{n,1/2}$ satisfies almost surely the following property: every set of at least $n/\log^2 n$ vertices of G contains an independent set of size $q = (1 + o(1))2\log_2 n$. Suppose, therefore, that G satisfies this property, and suppose that, for each vertex v of G, we are given a set $S(v)$ of at least $n/q + n/\log^2 n$ $(= (1+o(1))n/2\log_2 n)$ colors. As long as there is a color c that appears in at least $n/log^2 n$ sets $S(v)$, take an independent set of size at least q among the vertices whose color lists contain c, color them by c, delete them from the graph, and omit c from the color lists of the other vertices. When this process terminates, every vertex still has at least $n/\log^2 n$ colors in its list, and no color appears in more than $n/\log^2 n$ color lists. Therefore, by Hall's theorem, one can assign to each of the remaining vertices a color from its list so that no two vertices will get the same color. This completes the proof. $\square$

It is worth noting that, by applying martingales as in [47] (see also [8], pp. 84-86), one can show that, if E_n is the expectation of the choice number of $G_{n,1/2}$, then, for any $\lambda > 0$, only a fraction of at most $2e^{-\lambda^2/2}$ of the graphs on n labelled vertices have choice numbers that deviate from E_n by more than $\lambda\sqrt{n}$.

The following proposition establishes the upper bound for $ch(K_{m*r})$, asserted in Theorem 4.2. Although, as observed by J. Kahn, the proof of this proposition given in [6] can be simplified, we sketch the original proof here, since it applies an interesting splitting technique which has other applications as well (see, for example, [5]). To simplify notation, we omit all the floor and ceiling signs whenever these are not crucial. All the logarithms are to the natural base e, unless otherwise specified.

Proposition 4.5 *There exists a positive constant c such that, for all positive integers $m \geq 2$ and r, $ch(K_{m*r}) \leq cr\log m$.*

Proof Since rm is a trivial upper bound for $ch(K_{m*r})$ and since, for $c \geq 4$, say, $rm \leq cr\log m$ for all m satisfying $m \leq c$, we may assume that $m > c$ (where c will be chosen later). Let $V_1, V_2, \ldots V_r$ be the vertex classes of $K = K_{m*r}$, where $|V_i| = m$ for all i, and let $V = V_1 \cup \ldots \cup V_r$ be the set of all vertices of K. For each $v \in V$, let $S(v)$ be a set of at least $cr\log m$ distinct colors. We must show that there is a proper coloring of K, assigning to each vertex v a color from $S(v)$. Since $ch(K_{m*r})$ is a non-decreasing function of r, we may (and will) assume that r is a power of 2.

We consider two possible cases.

Case 1: $r \leq m$.

Let $S = \cup_{v \in V} S(v)$ be the set of all colors. Put $R = \{1, 2, \ldots, r\}$, and let $f : S \mapsto R$ be a random function, obtained by choosing the value of $f(c)$, randomly and independently for each color $c \in S$, according to a uniform distribution on R. The colors c for which $f(c) = i$ will be the ones to be used for coloring the vertices in V_i. To complete the proof for this case, it thus suffices to show that, with positive probability for every i $(1 \leq i \leq r)$, and for every vertex $v \in V_i$, there is at least one color $c \in S(v)$ such that $f(c) = i$.

Fix an i, and a vertex $v \in V_i$. The probability that there is no color $c \in S(v)$ such that $f(c) = i$ is clearly

$$(1 - \frac{1}{r})^{|S(v)|} \leq (1 - \frac{1}{r})^{cr \log m} \leq e^{-c \log m} \leq \frac{1}{m^c} < \frac{1}{rm},$$

where the last inequality follows from the fact that $r \leq m$ and $c \geq 4 > 2$. There are rm possible choices of i $(1 \leq i \leq r)$ and $v \in V_i$, and hence the probability that, for some i and some $v \in V_i$, there is no $c \in S(v)$ such that $f(c) = i$ is smaller than 1; this completes the proof in this case.

Case 2: $r > m$.

Here we apply a splitting trick, similar to the one used in [5]. As before, define $R = \{1, 2, \ldots, r\}$ and let $S = \cup_{v \in V} S(v)$ be the set of all colors. Put $R_1 = \{1, 2, \ldots, r/2\}$ and $R_2 = \{r/2 + 1, \ldots, r\}$. Let $f : S \mapsto \{1, 2\}$ be a random function obtained by choosing $f(c) \in \{1, 2\}$, for each $c \in S$ randomly and independently, according to a uniform distribution. The colors c for which $f(c) = 1$ will be used for coloring the vertices in $\cup_{i \in R_1} V_i$, whereas the colors c for which $f(c) = 2$ will be used for coloring the vertices in $\cup_{i \in R_2} V_i$.

For every vertex $v \in V$, put $S^0(v) = S(v)$, and define $S^1(v) = S^0(v) \cap f^{-1}(1)$ if v belongs to $\cup_{i \in R_1} V_i$, and $S^1(v) = S^0(v) \cap f^{-1}(2)$ if v belongs to $\cup_{i \in R_2} V_i$. Observe that in this manner the problem of finding a proper coloring of K in which the color of each vertex v is in $S(v) = S^0(v)$ has been decomposed into two independent problems. These are the problems of finding proper colorings of the two complete $r/2$-partite graphs on the vertex classes $\cup_{i \in R_1} V_i$ and $\cup_{i \in R_2} V_i$, by assigning to each vertex v a color from $S^1(v)$. Let $s_0 = cr \log m$ be the number of colors in each original list of colors assigned to a vertex. Using the standard tail estimates for binomial variables (see, for example, [8]), it is not too difficult to show that, for all sufficiently large c, with high probability,

$$|S^1(v)| \geq \frac{1}{2}s_0 - \frac{1}{2}s_0^{2/3}, \tag{4}$$

for all $v \in V$.

Let s_1 denote the minimum cardinality of a set $S^1(v)$, for $v \in V$. As shown above, we can ensure that

$$s_1 \geq \frac{1}{2}s_0 - \frac{1}{2}s_0^{2/3}.$$

We have thus reduced the problem of showing that the choice number of K_{m*r} is at most s_0 to that of showing that the choice number of $K_{m*(r/2)}$ is at most s_1.

Repeating the above decomposition technique, which we can repeat as long as $r/2^i > m$, we obtain, after j iterations, a sequence s_i, where $s_0 = cr \log m$ and

$$s_{i+1} \geq s_i/2 - s_i^{2/3}/2, \quad for \ 1 \leq i < j. \tag{5}$$

In order to show that the choice number of $K = K_{m*r}$ is at most s_0, it suffices to show that, for some i, the choice number of $K_{m*(r/2^i)}$ is at most s_i.

Let the number of iterations j be chosen so that j is the minimum integer satisfying $r/2^j \leq m$. Clearly, in this case, $r/2^j > m/2 \geq c/2$. A simple (but tedious) computation, which we omit, shows that

$$s_j \geq \frac{s_0}{2^{j+1}},$$

provided that c is sufficiently large.

To complete the proof of the proposition, observe that it suffices to show that the choice number of $K_{m*(r/2^j)}$ is at most s_j. However, $r/2^j \leq m$ and

$$s_j \geq s_0/2^{j+1} \geq \frac{c}{2}\frac{r}{2^j} \log m.$$

For a sufficiently large c, the result thus follows from Case 1. This completes the proof. $\square$

The lower bound in Theorem 4.2 can also be derived by probabilistic arguments, which we omit. Let us, however, present the simple derivation of Corollary 4.3 from the upper bound of this theorem.

Proof of Corollary 4.3 Define $m = \sqrt{n \log n}$ and $r = n/m = \frac{\sqrt{n}}{\sqrt{\log n}}$, and let G be the graph K_{m*r}. The complement G^c of G is a disjoint union of r cliques, each of size m, and thus $ch(G^c) = m = O(\sqrt{n \log n})$. By Theorem 4.2, $ch(G) = O(r \log m) = O(\sqrt{n \log n})$. Thus,

$$ch(G) + ch(G^c) = O(\sqrt{n \log n}),$$

as required. □

For a graph $G = (V, E)$, and for two integers $a \geq b \geq 1$, G is called $(a : b)$-*choosable* if, for every assignment of sets of colors $S(v) \subset Z$, each of cardinality a, for every vertex $v \in V$, there are subsets $T(v) \subset S(v)$, $|T(v)| = b$ for all v, so that, if u and v are adjacent, then $T(u)$ and $T(v)$ are disjoint. In particular, $(k : 1)$-choosability coincides with the previous definition of k-choosability. This definition is introduced in [24], where the authors raise the following question:

Suppose G is $(a : b)$-choosable, and suppose that $c/d > a/b$, where $a \geq b$ and $c \geq d$ are positive integers. Does it follow that G is $(c : d)$-choosable as well?

S. Gutner [28] showed that the answer is "no", by establishing the following result, proved by a simple probabilistic argument.

Proposition 4.6 ([28]) *For every two integers $n \geq k$ and for every $\epsilon > 0$, there is a $c_0 = c_0(n, k, \epsilon)$ such that the following holds: for every graph G on n vertices with chromatic number k, and for every two positive integers $c > c_0$ and d satisfying $c(1 - \epsilon) \geq kd$, G is $(c : d)$-choosable.*

Proof Let us fix a proper k-coloring of $G = (V, E)$, and let $V_1, \ldots, V_k$ be the k color classes. Given sets of colors $S(v) \subset Z$ of cardinality c for each vertex v of G, let S be the union of all the sets $S(v)$, and let us split S randomly into k pairwise disjoint subsets $S_1, \ldots, S_k$, where each $s \in S$ is chosen randomly and independently as a member of one of the sets S_i, according to a uniform distribution. The colors in S_i will be used for defining the sets $T(v)$, for $v \in V_i$. To complete the proof, it suffices to check that, if c is sufficiently large and $c(1 - \epsilon) \geq kd$, then with positive probability, $|S(v) \cap S_i| \geq d$ for every $1 \leq i \leq k$ and for every $v \in V_i$. However, for each such v, $|S(v) \cap S_i|$ is a binomial random variable with expectation c/k and variance $\frac{c}{k}(1 - \frac{1}{k}) < c/k$, and hence, by Chebyshev's Inequality, the probability that there is a vertex v in V_i for some i so that $|S(v) \cap S_i| < d$, is at most

$$n\frac{ck^2}{k\epsilon^2 c^2} < 1,$$

where the last inequality holds for all sufficently large c (as a function of ϵ, n and k). (Observe that this estimate can be improved by applying the more accurate known estimates for binomial distributions.) This completes the proof. $\square$

By the last proposition (with $n = 2m, k = 2, \epsilon = 1/3$ and $c = 3d$) for each integer m, the complete bipartite graph $K_{m,m}$ is, $(3d : d)$-choosable for all sufficiently large d, and yet it is not, say, $(100 : 1)$-choosable, if m is large enough, as mentioned in Section 2. As $100/1 > 3d/d$, this implies that the answer to the question preceding the last proposition is negative.

5 The minimum degree and choice numbers

Despite the close connection between choice numbers and chromatic numbers, these two invariants do not have the same properties. In this section, we prove a result that supplies an essential difference between the two. Let us call two graph invariants $\alpha(G)$ and $\beta(G)$ *related* if there are two functions $f(x)$ and $g(x)$, both tending to infinity as x tends to infinity, such that, for every (simple) graph G, $\beta(G) \geq f(\alpha(G))$ and $\alpha(G) \geq g(\beta(G))$. Roughly speaking, two invariants are related if they grow together, although one may grow much slower than the other. Examples of related parameters are the chromatic index of a graph and its maximum degree. Other examples of related parameters are the chromatic number of a graph and the minimum number of bipartite graphs required to cover all its edges. On the other hand, the chromatic number and the choice number of a graph are not related, as for any k there are graphs of chromatic number 2 whose choice number exceeds k.

The definition of a d-degenerate graph appears before Proposition 2.2.

For a simple graph G, let $d(G)$ denote the minimum integer d so that G is d-degenerate. Equivalently, $d(G)$ is the maximum integer d such that G has a subgraph with minimum degree d. The parameter $d(G)$ arizes naturally in the study of various problems and it is not diffciult to show that it is related, in the above sense, to the arboricity of G, which is the minimum number of forests whose union covers all edges of G. By Proposition 2.2, for any graph G, $d(G) \geq ch(G) - 1 \geq \chi(G) - 1$. On the other hand, for every k there are graphs with $\chi(G) = 2$ and $d(G) > k$ as shown by the family of complete bipartite graphs. Therefore the parameters $d(G)$ and $\chi(G)$ are *not* related. On the other hand, the parameters $ch(G)$ and $d(G)$, are related. One can show that the choice number of any simple graph with minimum (or average) degree d is at least $\Omega(\log d/\log\log d)$. This is proved in the following theorem; in its statement and proof, we make no attempt to optimize the constants.

Theorem 5.1 *Let G be a simple graph with average degree at least d. If s is an integer and*

$$d > 4\binom{s^4}{s}\log\left(2\binom{s^4}{s}\right),\tag{6}$$

then $ch(G) > s$.

Proof Let G be a simple graph with average degree at least d. We first show that, as is well known, every such G contains a bipartite graph, whose minimum degree is at least $d/4$. To see this, observe that G has an induced subgraph with minimum degree at least $d/2$, since one can repeatedly delete vertices of degree smaller than $d/2$ from G, as long as there are such vertices; since this process increases the average degree, it must terminate in a non-empty subgraph G' with minimum degree at least $d/2$. In G', one can take a spanning bipartite subgraph H with the maximum possible number of edges. If a vertex here has degree smaller than half of its degree in G', then shifting it to the other class of the bipartite graph will increase the number of edges and contradict the maximality in the choice of the bipartite graph. Now let $H = (V, E)$ be a bipartite subgraph of G, with minimum degree $\delta \geq d/4$ and with vertex classes A and B, where $|A| \geq |B|$. Let $S = \{1, 2, \ldots, s^4\}$ be our set of colors. Our objective is to show that there are subsets $S(v) \subset S$, where $|S(v)| = s$ for all $v \in V$, such that there is no proper coloring $c : V \mapsto S$ that assigns to every $v \in V$ a color $c(v) \in S(v)$. This will imply that

$$ch(G) \geq ch(H) > s,$$

and complete the proof.

The proof is probabilistic. For each vertex $b \in B$, let $S(b)$ be a random subset of cardinality s of S, chosen uniformly and independently among all the $\binom{s^4}{s}$ subsets of cardinality s of S. Call a vertex $a \in A$ *good* if, for every subset $C \subset S$ with $|C| = s$, there is a neighbor b of a in H such that $S(b) = C$. For a fixed $a \in A$, the probability that a is not good is at most

$$\binom{s^4}{s}\left(1 - \frac{1}{\binom{s^4}{s}}\right)^\delta \leq 1/2,$$

where the last inequality follows from the fact that $\delta \geq d/4$ and from assumption (6). Therefore, the expected number of good vertices $a \in A$ is at least $|A|/2$, and hence there is some choice of the s-subsets $S(b)$, $b \in B$ such that there are at least $|A|/2$ good vertices in A. Let us fix these subsets $S(b)$ and choose, for each $a \in A$, a subset $S(a) \subset S$, $|S(a)| = s$, randomly and independently according to a uniform distribution on the s-subsets of S. To complete the proof, we show that with positive probability there is no proper coloring $c : A \cup B \mapsto S$ of H assigning to each vertex $v \in A \cup B$ a color from its class $S(v)$.

There are $s^{|B|}$ possibilities for the restriction $c|_B$ of c to the vertices in B so that $c(b) \in S(b)$ for all b. Fix such a restriction, and let us estimate the probability that this restriction can be extended to a proper coloring c of the desired type. The crucial observation is that, if $a \in A$ is good, then the set of all colors assigned by $c|_B$ to the neighbors of a is a set that intersects every s-subset of S, since every s-subset is $S(b)$ for some neighbor b of a. Therefore, at least $s^4 - s + 1$ distinct colors are assigned by $c|_B$ to the neighbors of a. It is thus possible to choose a proper color for a from its set $S(a)$, only if $S(a)$ contains one of the set of at most $s - 1$ colors which differ from $c(b)$ for all the neighbors b of a. The probability that the randomly chosen set $S(a)$ satisfies this is at most

$$\frac{(s-1)\binom{s^4-1}{s-1}}{\binom{s^4}{s}} = \frac{s(s-1)}{s^4} < \frac{1}{s^2}.$$

Moreover, all these events for distinct good vertices $a \in A$ are mutually independent, by the independent choice of the sets $S(a)$. It follows that, if there are $g \; (\geq |A|/2)$ good vertices in A, the probability that a fixed partial coloring $c|_B$ can be extended to a full coloring $c : A \cup B \mapsto S$, assigning to each vertex a color from its class, is strictly less than

$$(1/s^2)^g \leq (1/s^2)^{|A|/2} \leq \frac{1}{s^{|B|}},$$

since $|A| \geq |B|$. As there are only $s^{|B|}$ possibilities for the partial coloring $c|_B$, and (as just shown) the probability that a fixed partial coloring would extend to a full coloring is strictly less than $1/s^{|B|}$, we conclude that with positive probability there is no coloring of the required type; this shows that $ch(H) > s$, completing the proof. $\square$

As shown in Section 2, the complete bipartite graphs $K_{d,d}$ supply an example with minimum degree d and choice number $(1 + o(1)) \log_2 d$, thus showing that Theorem 5.1 is nearly tight. Note also that there is a very simple polynomial time algorithm that finds, for a given input graph G, the value of the parameter $d(G)$- the minimum d so that G is d-degenerate. In view of Theorem 5.1, this supplies a polynomial algorithm that finds, for a given input graph G, a number s so that the choice number of G lies between s and $O(s^{4s} \log s)$. Although this is a very crude approximation, there is no known similar efficient approximation algorithm for the chromatic number of a graph. In fact, it would be very interesting to find any function $f : Z \mapsto Z$ and a polynomial time algorithm that finds, for a given input graph G, a number s so that the chromatic number of G is between s and $f(s)$. Although the problem of approximating the chromatic number of a graph received a considerable amount of attention, no such algorithm is known.

6   Concluding remarks and open problems

The most interesting open problem in the area is the list coloring conjecture (Conjecture 1.1), which is wide open, although it has been verified in several cases. It is easy to see that it is true for forests and for graphs with maximum degree 2. It also holds for graphs with no cycles of length bigger than 3, as shown in [28]. As mentioned in Section 2, the assertion of the conjecture holds for graphs with maximum degree 3 and edge chromatic index 4, by Theorem 2.1. Häggkvist [30] gave an interesting proof for all complete bipartite graphs $K_{r,n}$ with $r \leq \frac{2}{7}n$ and as discussed in Section 3 it holds for every planar d-regular multigraph with chromatic index d, by the result of [22] which applies the technique of [9]. It has also been proved for $K_{3,3}$ by H. Taylor (private communication); in [9], it is derived for $K_{4,4}$ and $K_{6,6}$ from Corollary 3.9. The last three examples are special cases of the case $G = K_{n,n}$ of the list coloring conjecture, which was formulated by J. Dinitz in 1979. Although Vizing had already raised the general conjecture in 1975, the special case of Dinitz became more popular, as it has the following appealing reformulation: given an arbitrary n by n array of n-sets, it is always possible to choose one element from each set, keeping the chosen elements distinct in every row and distinct in every column. (If all the n-sets are equal, then every n by n Latin square provides a good choice, and although it does seem that in any other case one has even more freedom, nobody has been able to transform this intuitive feeling into a rigorous proof.) As mentioned above, Dinitz's Conjecture is true for all $n \leq 4$ and for $n = 6$, but is not known for any other case, despite a considerable amount of effort by various researchers. Corollary 3.9 for this special case implies the following. Define the *weight* $w(L) \in \{-1, 1\}$ of an n by n Latin square to be the product of the signs of the $2n$ permutations appearing in its rows and its columns. If the sum $\sum w(L)$ is not 0, as L ranges over all n by n Latin squares, then the assertion of Dinitz's Conjecture holds for n. It is easy to see, however, that this sum is 0 for every odd $n \geq 3$, but it is conjectured in [9] to be non-zero for all even n (and this holds for $n = 2, 4, 6$).

The total coloring conjecture (Conjecture 2.4) is another problem that has received a considerable amount of attention. It would be nice to prove a $\Delta + O(1)$ upper bound for the total chromatic number of any simple graph with maximum degree Δ.

There are many known proofs in combinatorics that supply no efficient procedures for solving the corresponding algorithmic problems; see, for example, [4] for various representative examples. Some of the results described in Section 3 also have this flavour. Thus, for example, we know by Corollary 3.10 that $ch'(G) = 3$, for every 2-connected cubic planar graph G, but the proof supplies no polynomial time (deterministic or randomized) algorithms that produce, for a given such graph $G = (V, E)$ and given lists of colors $S(e)$, $e \in E$, each of size 3, a proper edge-coloring of G assigning to each edge a color from its list. Similarly, there is no known efficient procedure for solving the algorithmic problem suggested by Theorem 3.6. We note that, in contrast, one can give an efficient procedure for solving the algorithmic problem suggested by Theorem 3.7, based on Richardson's Theorem (see [12]), whose relevance to the problem has been pointed out by Bondy, Boppana and Siegel.

Another interesting problem is that of determining the largest possible choice number of a planar graph. This number is known to be at least 4 and at most 6, and in [24] it is conjectured that it is, in fact, 5.

The *Hadwiger number* $h(G)$ of a graph G is the maximum number h such that there are h pairwise vertex disjoint connected subgraphs of G with at least one edge of G between any pair of them. A well-known conjecture of Hadwiger [29] asserts that the Hadwiger number $h = h(G)$ of any graph is at least its chromatic number $\chi(G)$. This is known to be the case for all graphs with $h \leq 4$, where the case $h = 4$ is equivalent to the Four Color Theorem. Very recently, the case $h = 5$ has also been shown, by Robertson, Seymour and Thomas, to be equivalent to the Four Color Theorem. P. Seymour (private communication) has suggested that the stronger conjecture $h(G) \geq ch(G)$ may also hold. For graphs G with $h(G) \leq 3$, this can be deduced from Proposition 2.2. If it is true for $h(G) = 4$, then every planar graph has a choice number at most 4, contradicting the above mentioned conjecture of [24]. Needless to say, a proof of the general case seems beyond reach at present, but it may not be so difficult to find a counterexample, if one exists. We note that, by combining Proposition 2.2 with the known fact that the Hadwiger number of simple graphs with average degree at least d is at least $\Omega(d/\sqrt{\log d})$ (see [37], [48]), one easily concludes that, for every graph G, $h(G) \geq \Omega(\chi(G)/\sqrt{\log(\chi(G))})$.

Recall the definition of an $(a : b)$-choosable graph, given in the end of Section 4. In [24], the authors ask whether for any positive integers a, b

and m, any $(a : b)$-choosable graph is $(am : bm)$-choosable as well. This is proved in [28] for the special case $a = 2, b = 1$ and $m = 2$. The general case is still open, but we can prove the following partial result.

Proposition 6.1 *For any positive integers a, b and n, there is a positive integer $f = f(a, b, n)$ such that, for every integer m which is divisible by all integers smaller than f, any graph G on n vertices which is $(a : b)$-choosable, is $(am : bm)$-choosable as well.*

The proof applies the techniques of [7]. We omit the details.

Acknowledgement I would like to thank Mark Ellingham, Shai Gutner and Jeff Kahn for helpful comments.

References

[1] M. Aïder, *Réseaux d'interconnexion bipartis. Colorations généralisées dans les graphes*, Thése de 3ème cycle, Université Scientifique Technologique et Médicale de Grenoble, 1987.

[2] M. O. Albertson and D. M. Berman, *Cliques, colorings, and locally perfect graphs*, Congr. Numer. 39 (1983), 69-73.

[3] N. Alon, *The linear arboricity of graphs*, Israel J. Math. 62 (1988), 311-325.

[4] N. Alon, *Non-constructive proofs in Combinatorics*, Proc. of the International Congress of Mathematicians, Kyoto 1990, Japan, Springer Verlag, Tokyo (1991), 1421-1429.

[5] N. Alon, *The strong chromatic number of a graph*, Random Structures and Algorithms 3 (1992), 1-7.

[6] N. Alon, *Choice numbers of graphs; a probabilistic approach*, Combinatorics, Probability and Computing 1 (1992), 107-114.

[7] N. Alon, D. J. Kleitman, C. Pomerance, M. Saks and P. Seymour, *The smallest n-uniform hypergraph with positive discrepancy*, Combinatorica 7 (1987), 151-160.

[8] N. Alon and J. H. Spencer, **The Probabilistic Method**, Wiley, 1991.

[9] N. Alon and M. Tarsi, *Colorings and orientations of graphs*, Combinatorica 12 (1992), 125-134.

[10] J. Beck, *On 3-chromatic hypergraphs*, Discrete Math. 24 (1978), 127-
137.

[11] M. Behzad, *The total chromatic number of a graph; a survey*, in: Combinatorial Mathematics and its Applications (Proc. Conference Oxford
1969), D. J. A. Welsh, editor, Academic Press, New York, 1971, 1-9.

[12] C. Berge, **Graphs and Hypergraphs**, Dunod, Paris, 1970.

[13] B. Bollobás, *The chromatic number of random graphs*, Combinatorica
8 (1988), 49-55.

[14] B. Bollobás, **Random Graphs**, Academic Press, 1985.

[15] B. Bollobás and A. J. Harris, *List colorings of graphs*, Graphs and
Combinatorics 1 (1985), 115-127.

[16] B. Bollobás and H. R. Hind, *A new upper bound for the list chromatic
number*, Discr. Math. 74 (1989), 65-75.

[17] R. L. Brooks, *On coloring the nodes of a network*, Proc. Cambridge
Phil. Soc. 37 (1941), 194-197.

[18] A. Chetwynd and R. Häggkvist, *A note on list colorings*, J. Graph
Theory 13 (1989), 87-95.

[19] J. A. De Loera, *Gröbner bases for arrangements of linear subspaces
related to graphs*, to appear.

[20] Q. Donner, *On the number of list-colorings*, J. Graph Theory 16 (1992),
239-245.

[21] D. Z. Du, D. F. Hsu and F. K. Hwang, *The Hamiltonian property
of consecutive-d digraphs*, Mathematical and Computer Modelling, to
appear.

[22] M. N. Ellingham and L. Goddyn, *List edge colorings of some regular planar multigraphs*, to appear.

[23] P. Erdős, *On a combinatorial problem, II*, Acta Math. Acad. Sci. Hungar. 15 (1964), 445-447.

[24] P. Erdős, A. L. Rubin and H. Taylor, *Choosability in graphs*, Proc. West Coast Conf. on Combinatorics, Graph Theory and Computing, Congressus Numerantium XXVI, 1979, 125-157.

[25] M. Fellows, *Transversals of vertex partitions in graphs*, SIAM J. Discrete Math. 3 (1990), 206-215.

[26] H. Fleischner and M. Stiebitz, *A solution to a coloring problem of P. Erdős*, to appear.

[27] M. R. Garey and D. S. Johnson, *Computers and Intractability, A guide to the Theory of NP-Completeness*, W. H. Freeman and Company, New York, 1979.

[28] S. Gutner, M. Sc. thesis, Tel Aviv University, 1992.

[29] H. Hadwiger, *Über eine Klassifikation der Streckenkomplexe*, Vierteljschr. Naturforsch. Gessellsch. Zürich 88 (1943), 133-142.

[30] R. Häggkvist, *Towards a solution of the Dinitz problem ?*, Discrete Math. 75 (1989), 247-251.

[31] R. Häggkvist and A. Chetwynd, *Some upper bounds on the total and list chromatic numbers of multigraphs*, J. Graph Theory 16 (1992), 505-516.

[32] H. R. Hind, Ph. D. thesis, Peterhouse college, Cambridge.

[33] H. R. Hind, *An upper bound for the total chromatic number*, Graphs and Combinatorics 6 (1990), 153-159.

[34] H. R. Hind, *An upper bound for the total chromatic number of dense graphs*, J. Graph Theory 16 (1992), 197-203.

[35] F. Jaeger, *On the Penrose number of cubic diagrams*, Discrete Math. 74 (1989), 85-97.

[36] J. Kahn, *Asymptotically good list colorings*, in preparation.

[37] A. V. Kostochka, *A lower bound for the Hadwiger number of a graph as a function of the average degree of its vertices* (in Russian), Diskret. Analiz. Novosibirsk 38 (1982), 37-58.

[38] M. Kubale, *Some results concerning the complexity of restricted colorings of graphs*, Discrete Applied Math. 36 (1992), 35-46.

[39] S. Y. R. Li and W. C. W. Li, *Independence numbers of graphs and generators of ideals*, Combinatorica 1 (1981), 55-61.

[40] L. Lovász, **Combinatorial Problems and Exercises**, North Holland, Amsterdam, 1979, Problem 9.12.

[41] L. Lovász, *Bounding the independence number of a graph*, in: (A. Bachem, M. Grötschel and B. Korte, eds.), Bonn Workshop on Combinatorial Optimization, Annals of Discrete Mathematics 16 (1982), North Holland, Amsterdam.

[42] N. V. R. Mahadev, F. S. Roberts and P. Santhanakrishnan, *3-choosable complete bipartite graphs*, DIMACS Technical Report 91-62, 1991.

[43] C. McDiarmid, *Colorings of random graphs*, in:*Graph Colorings*, (R. Nelson and R. J. Wilson, eds.), Research Notes in Mathematics 218, Pitman(1990), 79-86.

[44] J. Petersen, *Die Theorie der regulären Graphs*, Acta Math. 15 (1891), 193-220.

[45] V. Rödl, *On a packing and covering problem*, European J. Combinatorics 5 (1985), 69-78.

[46] D. E. Scheim, *The number of edge 3-colorings of a planar cubic graph as a permanent*, Discrete Math. 8 (1974), 377-382.

[47] E. Shamir and J. H. Spencer, *Sharp concentration of the chromatic number in random graphs $G_{n,p}$*, Combinatorica 7 (1987), 121-130.

[48] A. G. Thomason, *An extremal function for contractions of graphs*, Math. Proc. Cambridge Philos. Soc. 95 (1984), 261-265.

[49] L. Vigneron, *Remarques sur les réseaux cubiques de classe 3 associés au probléme des quatre couleurs*, C. R. Acad. Sc. Paris, t. 223 (1946), 770-772.

[50] V. G. Vizing, *On an estimate on the chromatic class of a p-graph* (in Russian), Diskret. Analiz. 3 (1964), 25-30.

[51] V. G. Vizing, *Coloring the vertices of a graph in prescribed colors* (in Russian), Diskret. Analiz. No. 29, Metody Diskret. Anal. v. Teorii Kodov i Shem 101 (1976), 3-10.

Polynomials in Finite Geometries and Combinatorics

Aart Blokhuis *

Abstract

It is illustrated how elementary properties of polynomials can be used to attack extremal problems in finite and euclidean geometry, and in combinatorics. Also a new result, related to the problem of neighbourly cylinders is presented.

1. Introduction

In this paper I will present a number of problems and for the most part recent results in combinatorics in general and finite geometry in particular. The property these problems share is the fact that they can all be attacked using some kind of polynomial trick. Unfortunately I am not able to give a characterization of the kind of problem that can be solved with these methods, but I am convinced that seeing how the polynomials work in a number of cases should give an impression of the type of problems that might be attacked.

The starting point is usually a combinatorial problem of the following form: Given a set of points (or vectors, or sets) that satisfy some property, we want to say something about the size or the structure of this set. The approach is then to associate to this set a polynomial, or a collection of polynomials, and use properties of polynomials to obtain information on the size or structure of the set.

The setup of this paper is roughly as follows. We consider a particular property of polynomials and give examples where this property can be used to attack the problem.

In fact only very basic properties of polynomials will be used:
First of all, polynomials form a vector space. For a certain class of problems we can define a collection of polynomials forming an independent set

---

*Eindhoven University of Technology, P.O. Box 513, 5600 MB, Eindhoven, The Netherlands in a vector space whose dimension we can compute. This approach suggests itself in particular when the expected answer has a simple form in terms of binomial coefficients or powers of an integer. For a general survey on linear algebra methods in combinatorics, of which this is just a very special (although particularly nice) case, I refer to the book by László Babai and Péter Frankl [2]. Among the problems that can be attacked with this observation are the following. Equiangular lines, few-distance sets, sets with few intersections, Sperner capacities of certain graphs, Bollobás' theorem and generalizations. A very recent application of one particular intersection theorem of Frankl and Wilson has been the disproof of Borsuk's conjecture by Kahn and Kalai [31]. This will be the subject of the next section.

Another elementary but very useful property is the fact that a polynomial (defined over a commutative field) has no more roots than its degree. It provides us in fact with a double-edged sword. If we can find a polynomial vanishing at points corresponding to the objects of our problem then its degree will provide us with an upper bound for the size of our set. If on the other hand we encounter polynomials of bounded degree, but they already vanish more often, then we may infer that they vanish identically. The first 'edge' can be used to give surprisingly simple solutions of some problems in finite geometry. The classical example is the upper bound for the number of nuclei of a $q + 1$-set in $AG(2, q)$, the desarguesian affine plane of order q. Related to this are upper bounds for the size of quasi-odd sets and a recent generalization of the nuclei theorem that connects with the Brouwer-Schrijver and Jamison theorem on the size of a blocking set in $AG(2, q)$. This will be the subject of section 3.

A little bit in the same spirit as the previous is the problem of characterizing certain sets of points in finite projective and affine planes. The underlying principle is that properties of the set we are studying are reflected in properties of coefficients of some associated polynomial. It is here that the second edge of our sword will play a role. The examples in this case are quite technical, so we will not give proofs here, but just indicate the principles at work and mention some recent progress. Examples of this approach are the characterization of q-sets in $GF(q^2)$ such that all differences are squares, and Rédei's Theorem on the number of directions determined by the points on the graph of a function $GF(q) \rightarrow GF(q)$.

In the last section we present miscellaneous proofs involving polynomials. A basic property of the finite field $GF(p)$ is that $x^{p-1} = 1$ if and only if $x \neq 0$. This observation will be used to give a polynomial trick proof of a theorem of Erdős, Ginzberg and Ziv [25].

If we have a polynomial (in any finite number of arguments) of degree k defined over a finite field of size larger than $k + 1$ and we sum the value of this polynomial over all possible values of the argument(s), then we get zero, independently of the form of the polynomial. This observation lies at the basis of the impossibility of certain partitions of projective and affine space.

We finish the paper with a new result about equiangular planes, related to the well-known problem of touching cylinders.

2.   Polynomials form a vector space

In their beautiful paper [34] Lemmens and Seidel consider the following problem:

Let X be a collection of equiangular lines (or better, one-dimensional subspaces) in $\mathbf{R}^d$, that is a set of lines having the property that each pair determines the same angle. What can be said about the size of the set X in terms of the angle. In this paper they attribute the following *absolute bound* (valid for all possible angles) to Gerzon:

$$|X| \leq \binom{d+1}{2}.$$

In [22] sets of lines having a prescribed set of angles are studied and the results in [34] are generalized.

About one year later Koornwinder [32] came with a very short 'polynomial trick' proof of Delsarte, Goethals and Seidel's generalization of the Gerzon bound. The main idea is best illustrated by looking at the original case of equiangular lines.

For every line $l \in X$ fix a unit vector u on l. For each u obtained in this way define the homogeneous polynomial in d variables

$$F_u(x) = (u, x)^2 - \alpha^2(x, x),$$

where $\arccos(\alpha)$ equals the angle determined by two lines in X. The polynomials thus obtained are independent, since $F_u(v) = (1 - \alpha)\delta_{uv}$. Hence the size of X is bounded above by the dimension of the space of homogeneous polynomials of degree 2 in d variables, and this is $\frac{1}{2}d(d+1)$.

When I first learned about this proof (in 1979), I was very excited and I immediately tried to find other applications of the same idea, or maybe improvements. In fact I obtained my first result in mathematics using Koornwinder's trick plus a little idea of my own.

The problem in this case was to find an upper bound for a set X of points in R^d with the property that only two distances occur. In the special case that all points of X are on a sphere the upper bound $\frac{1}{2}d(d+3)$ was obtained by

Delsarte, Goethals and Seidel in [23]. This bound is essentially best possible, because equality occurs in dimensions 2, 6 and 22. Using Koornwinder's idea and the following polynomials

$$F_u(x) = (||x - u||^2 - a^2)(||x - u||^2 - b^2),$$

Larman, Rogers and Seidel in [33] obtained the bound $|X| \leq \frac{1}{2}(d+1)(d+4)$, and my little idea consisted of the observation that this system can be shown to remain independent if one adds the functions x_i, $1 \leq i \leq d$, and the constant function 1 to it [5]. This shows that in fact $|X| \leq \frac{1}{2}(d+1)(d+2)$.

Recently this idea of extending the independent family of polynomials has been used by Alon, Babai and Suzuki [1] to give (among other things) a simple proof of the following generalization of a famous intersection theorem of Ray-Chaudhuri and Wilson:

Let $K = \{k_1, \ldots, k_r\}$ and $L = \{l_1, \ldots, l_s\}$, and assume $k_i > s - r$ for all i. Let $\mathcal{F}$ be a family of subsets of an n-element set. Suppose that $|F| \in K$ for each $f \in \mathcal{F}$ and $|E \cap F| \in L$ for each pair of distinct sets $E, F \in \mathcal{F}$. Then

$$|\mathcal{F}| \leq \binom{n}{s} + \binom{n}{s-1} + \ldots + \binom{n}{s-r+1}.$$

I would just like to mention one further example due to Frankl and Wilson, because of a very recent, extremely nice application [27]:

Let $|X| = n = 4p$, p prime, and let $\mathcal{B}$ be a collection of subsets of X of size $2p$, such that no two elements of $\mathcal{B}$ intersect in precisely p points. Then

$$|\mathcal{B}| \leq 2\binom{4p-1}{p-1}.$$

The proof essentially consists of associating to every $B \in \mathcal{B}$ the polynomial

$$F_B(X) = (b_1 X_1 + b_2 X_2 + \ldots + b_{4p} X_{4p})^{p-1} - 1 \in \mathbf{F}_p[X_1, \ldots, X_{4p}].$$

Here $(b_1, \ldots, b_{4p})$ is the characteristic vector of the set B. Now $F_B(C) = 0$ unless $|B \cap C| \equiv 0 \bmod p$; so this means that $B = C$ or C is the complement of B, since $|B \cap C| = p$ does not occur. Now we may assume without loss of generality that our set $\mathcal{B}$ containing a given set also contains its complement, and restrict ourselves to the sets with $b_1 = 1$. Next in the expansion of our polynomial we may replace X_1 by 1, and reduce exponents larger than 1 to 1, since we are only looking at $(0, 1)$-vectors. Now everything happens in a $\binom{4p-1}{p-1}$-dimensional vector space and we are done.

This bound has been recently used by Kahn and Kalai [31] to construct a polytope in $\mathbf{R}^d$, where $d = \binom{4p}{2}$ that cannot be dissected in less than $1.1^{\sqrt{d}}$ parts of smaller diameter, thus giving a counterexample to Borsuk's conjecture from 1933, which asserts that a partition in $d+1$ parts of smaller diameter is always possible.

In the above it was more or less obvious how to define the polynomials (once one knows that a polynomial argument is available), but this is usually not the case.

The next example we consider comes from a problem in communication theory:

Let X be a subset of $\{0,1,2\}^n$ with the property that for every distinct pair of vectors $x = (x_i)$, $y = (y_i) \in X$, we have $x_j - y_j \equiv 1 \pmod 3$ for some index j.

Let $T(n)$ be the maximal size of such a set X. The problem is to determine

$$\lim_{n \to \infty} \frac{1}{n} \log T(n).$$

This problem, that occurs in [30], asks in fact for what the authors call the *Sperner capacity* of the cyclic triangle. It was solved in [20] where it is shown that this limit is 2. What is shown in fact is that $T(n)$ is always smaller than 2^{n-1}. This implies that this limit is 2, since we may always take for X the subset of $\{0,1\}^n$ consisting of the vectors having precisely $\lfloor n/2 \rfloor$ ones.

The following polynomial trick [6] shows that $T(n) \leq 2^n$. Identify $\{0,1,2\}$ with $GF(3)$, the field with three elements. For each vector $u \in GF(3)^n$ consider the polynomial

$$F_u(x) = (x_1 - u_1 - 1)(x_2 - u_2 - 1)...(x_n - u_n - 1).$$

We have $F_u(u) = (-1)^n$ is nonzero, for every $u \in GF(3)^n$, but if u, v are different elements from the subset X then $F_u(v) = 0$. So the polynomials F_u, $u \in S$, form an independent set in the vector space V of polynomials in the variables $x_1, ..., x_n$, of degree at most 1 in each separate variable. It follows that $|X| \leq \dim(V) = 2^n$.

As a final example of a somewhat different nature we consider the fundamental theorem of Bollobás in extremal set theory. For more background on this problem as well as a number of illustrative proofs of this result we refer to [2]. The basic version runs as follows:

Let A_i, and B_i, $i \in I$, be two collections of sets such that, for all $i, j \in I$ $|A_i| = r$, $|B_j| = s$ and $A_i \cap B_j = \emptyset$ if and only if $i = j$. Then

$$|I| \leq \binom{r+s}{r}.$$

This result can be proved using polynomials roughly as follows (for details and generalizations see [7]). Without loss of generality we may suppose that the sets A_i and B_j consist of real numbers. Now, to each set A_i associate a polynomial a_i of degree r and to each set B_j associate a polynomial b_j of degree s. The key observation in this case is that two sets have an empty intersection if and only if the resultant of the corresponding polynomials is nonzero. The resultant of a variable polynomial a of degree r and a fixed polynomial b of degree s is itself a homogeneous polynomial of degree s in the $r + 1$ coefficients of the polynomial a. The set of all such polynomials forms an $\binom{r+s}{s}$-dimensional vector space. The conditions in the theorem ensure again that these polynomials are independent, from which the bound follows.

There is a vector space version of Bollobás' theorem due to Lovász. If instead of subsets of a set we consider collections of subspaces A_i and B_i, $i \in I$, of a vector space, of dimensions r and s respectively such that A_i and B_j intersect nontrivially if and only if $i \neq j$, then again it follows that $|I| \leq \binom{r+s}{r}$.

The resultant proof above also has this flavour. Consider the vector space of polynomials of degree $< r + s$. Map the r-set A_i on the s-dimensional subspace consisting of all multiples of the associated polynomial a_i. In this way we obtain a natural representation of our sets as subspaces satisfying the conditions in Lovász' Theorem (note however that the roles of r and s are interchanged). Indeed two polynomials of degree r and s have a common multiple of degree $< r + s$ if and only if they have a common factor.

The reason I mention this interpretation is that it naturally leads to a slightly more direct proof of the threshold version of Bollobás' result by Füredi [28]. If the condition that A_i intersects B_j is replaced by the requirement that they should intersect in more than t points if and only if $i \neq j$, then the bound becomes $|I| \leq \binom{r+s-2t}{r-t}$. Füredi proves this by associating to the sets A_i and B_j subspaces of dimension r and s of a vector space over a sufficiently large field, having the property that the intersection is at least $(t+1)$-dimensional if and only if $i \neq j$. He then argues, that since the field is big enough there will be a subspace H of codimension t in general position, that is, having the expected intersection with everything in sight. Finally Lovász' theorem is applied to the $(r - t)$ and $(s - t)$-dimensional intersections of H with the A's and the B's.

Considering our sets as subspaces of dimension $r - t$ and $s - t$ of the vector space of polynomials of degree less then $r + s - t$, we may apply Lovász' theorem directly. Again the reason is that two polynomials of degree r and s respectively have a common multiple of degree less than $r + s - t$ if and only if they have a common factor of degree more than t.

3. No more zeros than the degree

In this section and in the following we will see how elementary properties of polynomials can be put to use in the investigation of certain combinatorial questions in finite projective planes.

Let $\Pi = PG(2,q)$ be the desarguesian projective plane of order q. Let B be a subset of Π of size $q + 1$. Since every point of Π is on $q + 1$ lines it could happen that every line through a certain point P contains exactly one point of B. In this case we call P a *nucleus* of B and the first problem we consider here is: How many nuclei can there be? If B is a line, then all other points of the plane are nuclei. So we assume that B is not a line.

This question (originating from Mazzocca) was motivated by the following result (proved for q even by Bruen and Thas [19], and for odd q by Segre and Korchmáros [40]):

Let B be a set of $q+1$ points not on a conic C in $PG(2,q)$ with the property that every two points of B are joined by an exterior line of C (that is, a line missing C). Then the points of B are all collinear.

In terms of nuclei this result can be reformulated as follows: The collection of nuclei of a set B that is not a line does not contain a conic.

Using a polynomial trick the following, surprising result was proved [14].

A $(q + 1)$-set of points B, that is not a line, has at most $q - 1$ nuclei.

Since an irreducible conic has $q + 1$ points we get the previous result as a corollary. There are two basic ingredients in the proof. The most important one is that a polynomial of degree d has at most d zeros. The second one is that the sum of all k-th roots of unity in $GF(q^2)$ is zero if $1 < k \mid q^2 - 1$.

The proof runs as follows: Since B is not a line, and has only $q + 1$ points, there must be a line l disjoint from it. Hence we may regard B as a set of points in the affine plane $AG(2,q) = PG(2,q) \setminus l$. Now identify the points of $AG(2,q)$ with the elements of $GF(q^2)$ in a suitable way and consider the polynomial

$$F(x) = \sum_{b \in B} (x - b)^{q-1}.$$

Let a be a nucleus of the set B. For $b \in B$ the value of $(a - b)^{q-1}$ is a $(q+1)$-st root of unity in $GF(q^2)$, and only depends on the direction of the line joining a and b, not on the position of b on that line. It follows that

$F(a) = 0$ if a is a nucleus, since in the sum defining F precisely all $(q+1)$-st roots of unity occur, and their sum is zero. Since the degree of F is $q - 1$, it follows that there are at most $q - 1$ nuclei.

Exactly the same polynomial is used in another result [11]. Let q be even. For a subset of $AG(2, q)$ it is impossible that every line intersects it in an odd number of points as an easy parity argument shows. The closest we can get to this is to require every non-zero intersection to be odd. Let us call such a set B a *quasi-odd set*. If we identify $AG(2, q)$ with $GF(q^2)$ as before and define $F(x)$ as above, then the points in B will be roots of F, this time however for a different reason. Every line through a point of B contains an even number of other points of B; so in the sum defining $F(x)$ they add up to zero, since q is even. Therefore we get the upper bound $q - 1$ for the size of a quasi-odd set. The surprising fact is that this bound is actually sharp. For example, the affine plane of order 8 contains Fano subplanes.

More generally we may construct quasi-odd sets of size $q - 1$ as follows. Let $L : GF(q) \to GF(q)$ be a function satisfying $L(x + y) = L(x) + L(y)$ for all x and y (equivalently, $L(x)$ is a so-called linearized polynomial, that is, a polynomial in $\langle x, x^2, x^4, \ldots, x^{2^{h-1}} \rangle$, where $q = 2^h$). Then

$$\{(x, L(x)/x) \in AG(2, q) \mid x \in GF(q)^*\}$$

is a quasi-odd set. In fact the number of points on a line is always of the form $2^k - 1$ for some k. To see this, consider the line with equation $ax + by + c = 0$. We have to determine the number of solutions of $ax + bL(x)/x + c = 0$ where $x \neq 0$. This is equivalent to $ax^2 + bL(x) + cx = 0$. As the left hand side of this equation is a linearized polynomial, if x and y are solutions, so also is $x + y$. Hence the total number of solutions (including $x = 0$) is a power of 2.

As a special case we may notice that (x, x^3) and $x \neq 0$ gives a nice embedding of the Steiner triple system corresponding to $PG(h - 1, 2)$ in $AG(2, q)$, where again $q = 2^h$.

As a side remark I should mention that there is an unexpected relation between quasi-odd sets and nuclei [11]: If B is a $(q + 1)$-set in $AG(2, q)$, with q even, with the maximal number $q - 1$ of nuclei, then these nuclei form a (maximal) quasi-odd set. It is not clear however whether this result really means something, since the only known example of a quasi-odd sets arising in this way is a set of $q - 1$ points on a line.

Recently an interesting generalization of the bound on the number of nuclei was obtained [8], again using the fact that a polynomial of degree d has at most d roots.

The idea is to extend the notion of nucleus to arbitrary sets. Let B be a subset of $AG(2, q)$. A point $a \notin B$ is called a *nucleus* of B if every line through a meets the set B. Again the question is: What can be said about the number of nuclei of B. First of all, in order to have nuclei, the cardinality of the set B obviously should be at least $q + 1$. On the other hand, if B is large enough, all points outside B can be nuclei. If this happens, then apparently every line meets B and B is a blocking set of the affine plane. Now a very surprising result by Jamison [29], and independently Brouwer & Schrijver [15] says that the size of a blocking set in $AG(2,q)$ is at least $2q - 1$. And it is very easy to construct examples of this size, for instance by taking the q points of a line together with one point on each parallel of this line. Now consider the set B consisting of a line l together with k points, each on a different parallel of l. This set has $q + k$ points and $k(q-1)$ nuclei. The nuclei are precisely the remaining points on the 'occupied' parallels. As it turns out $k(q - 1)$ is the maximal possible number of nuclei a $(q + k)$-set can have.

In order to prove this we try to construct a polynomial of degree $k(q - 1)$ having our nuclei as roots. The idea is again to look at the numbers $(a - b)^{q-1}$. When a is a nucleus and b runs through the set B, then we get again all different $(q + 1)$-st roots of unity, but now some of them may occur more than once. We somehow have to characterize this situation. The key observation is the following: If X is a sequence of $q + k$ elements from $GF(q^2)$, $k < q$, containing all $(q+1)$-st roots of unity, then $\sigma_k(X) = 0$ where σ_k stands for the k-th elementary symmetric polynomial. Using this the polynomial can be constructed in a straightforward way.

4. Blocking sets, squares and lacunary polynomials

One of my favourite open problems concerns the minimal size of a blocking set in projective planes of prime order. Let me give the definition. A set of points B in a projective plane is called a *blocking set* if every line intersects B as well as the complement of B. The main problem is to determine the minimal size of a blocking set and essentially there is only one result concerning this, Bruen's lower bound [16]: $|B| \geq q + \sqrt{q} + 1$, where q is the order of the plane. Equality implies that B is a Baer subplane.

I would be very disappointed if the following were not true. Let $\Pi = PG(2,q)$, the desarguesian projective plane.

1. If q is prime, then $|B| \geq \frac{3}{2}(q + 1)$.

2. If $q = q_1^m$ with $m > 1$ chosen minimal, then $|B| \geq q + \frac{q}{q_1} + 1$.

These bounds would be best possible (see for example [9]). The support for this conjecture is rather meagre. Part 1 is known to be true for $q = 3, 5, 7, 11$

and 13 [9]; part 2 of course reduces to Bruen's theorem if q is a square. Apart from that it is only known to be true for $q = 8$ and quite recently Klaus Metsch and I managed to show it for $q = 27$ [13].

For completeness, just let me mention a related problem. A double blocking set is a collection B, with the property that every line intersects B in at least two points. Is it true that a double blocking set in $PG(2,p)$, p prime, contains at least $3p$ points? Here the situation is even worse: It is only known (and a not too difficult exercise) for $p = 2, 3, 4, 5, 7$.

We now proceed to consider blocking sets of a special type. If B has size $q + N$ then a line can contain at most N points of B. If such a line exists the blocking set is called *of Rédei type*.

The reason for this terminology is the following result of L. Rédei [39], Satz 24.

Let X be a set of q points in $PG(2,q)$, where $q = p^n$, p prime, and let D be an exterior line, that is a line not intersecting X. Let N be the number of points on D that lie on a secant of X. Then, with $e = \lfloor n/2 \rfloor$,

$$N \ge \frac{q-1}{p^e + 1} + 1, \quad \text{and} \quad N \ge \frac{p+3}{2} \quad \text{if } q \text{ is prime,}$$

or $N = 1$ and the points of X are collinear.

The actual result by Rédei is a little bit more precise, it gives specific intervals for the possible values of N. Also it is phrased in a slightly different way, namely in terms of the number of difference quotients of a function $f : GF(q) \to GF(q)$. The above result is certainly not best possible if $n > 1$ is odd. My conjecture is that for e one can take instead of $\lfloor n/2 \rfloor$ the largest proper divisor of n. A step in this direction is contained by the following recent improvement on Rédei's theorem by Brouwer, Szőnyi and the author [10]. D_X denotes the set of points on D that are on a secant of X.

Let e (with $0 \le e \le n$) be the largest integer such that each line on a point of D_X meets X in a multiple of p^e points. Then we have one of the following:

1. *$e = 0$ and $(q + 3)/2 \le N \le q + 1$,*

2. *$1 \le e < n/3$, and $2 + (q - 1)/(p^e + 1) \le N \le (q - 1)/(p^e - 1)$,*

3. *$e = n/3$ and $N = p^{2e} + 1$ or $N = p^{2e} + p^e + 1$.*

4. *$e = n/2$ and $N = p^e + 1$,*

5. $e = n$ and $N = 1$.

Moreover $N \equiv 1 \pmod{p^e}$ and $(q, N) \neq (16, 7)$.

This is certainly not the place to give a complete proof of this result, but the principal idea, again involving polynomials, is easy to explain. We can view D as the line at infinity of the affine plane $AG(2, q)$ and we assume that the point on D corresponding to the vertical direction is not a secant of X. In that case the set X has exactly one point on every vertical line; so we may regard it as the graph of a function $f : GF(q) \to GF(q)$. One now considers the auxiliary polynomial in two variables:

$$H(x, y) := \prod_{w \in K}(x + yw - f(w)) = \sum_{j=0}^{q} h_j(y)x^j,$$

where $K = GF(q)$.

Every point on D that does not lie on a secant of X corresponds to an $a \in K$ with the property that $H(x, a) = x^q - x$. This gives us information on the coefficients $h_j(y)$ of $H(x, y)$. In particular, those h_j that are of small degree, will vanish identically, since they have to many zeros. This information combined with some special properties of so-called lacunary polynomials forms the body of the proof.

My first encounter with Rédei's theory on lacunary polynomials was around 1983. At that time I was looking at the following conjecture by van Lint and MacWilliams:

Let B be a set of size q in the field $GF(q^2)$, q odd, with the property that the difference of any two elements of B is a square. Assume that $0, 1 \in B$. Then $B = GF(q)$.

The special case that q is prime follows from the above result of Rédei, as was observed by van Lint and MacWilliams: if we identify $GF(q^2)$ in a suitable way with $AG(2, q)$ then half of the $q + 1$ directions correspond to square differences, but Rédei's result gives at least $(q + 3)/2$ in this case, unless the points of B are collinear.

The (very) basic idea of the proof of the conjecture for general q is this: Associate to the set B the polynomial

$$F(X) = \prod_{b \in B}(X - b).$$

Now the thing to prove becomes: $F(X) = X^q - X$. In other words, show that a lot of elementary symmetric functions in the b's vanish. How this is done can be read in [5], but why it actually works was (and is) not so clear to me. I know that also Chris Fisher was puzzled by it for a long time, but in a recent paper with A.A. Bruen [18] he gives a slightly more transparent proof based on what they call 'the Jamison Method'. This method consists of three steps:

1. Rephrase the theorem you want to prove as a relationship involving sets of points in an affine space.

2. Formulate the theorem in terms of polynomials over a finite field.

3. Calculate.

Without any further illustration this seems to be a rather vague method, but they demonstrate how it works in four cases: the flock theorem, the squares theorem, and two affine blocking set theorems.

The 2-dimensional case of Jamison's affine blocking set theorem [29] we already met in section 3. More generally this theorem says that any subset of the points of $AG(n,q)$ that intersects all hyperplanes contains at least $n(q-1)+1$ points.

The flock theorem, [26] and [38], says that the only way to partition the points of an elliptic quadric in $PG(3,q)$ in a set of $q-1$ conics and 2 points is by taking the conics in planes, all through the same line missing the quadric.

5. Miscellanea

In [25] Erdős, Ginzberg and Ziv solve the following problem: Let X be a sequence of $2n-1$ integers, show that among these numbers there are n whose sum is divisible by n. The essential difficulty lies in showing it for the case n is prime. Here we show a polynomial trick argument to deal with this problem. The property we use is that in $\mathrm{GF}(p)$, $x^{p-1}=1$ for all $x\neq 0$. So we let p be a prime, and we have a sequence X of $2p-1$ integers. Since we are only interested in whether a certain number is divisible by p, we may just look at the residues of our numbers mod p; so in fact we may consider X as a sequence in $\mathrm{GF}(p)$. Let

$$F(x_1, x_2, \ldots, x_{2p-1}) = \sum_I (x_{i_1} + x_{i_2} + \ldots + x_{i_p})^{p-1}.$$

Here $I = \{i_1, \ldots, i_p\}$ runs over all p-subsets of $\{1, 2, \ldots, 2p-1\}$. Let us compute the coefficient of a typical monomial $x_1^{e_1} x_2^{e_2} \ldots x_k^{e_k}$ of total degree $p-1$ (all $e_j > 0$). The number of p-sets I containing $1, \ldots, k$ equals

$$\binom{2p-1-k}{p-k};$$

so the coefficient will be a multiple of this, but this number is divisible by p, since $k \leq p - 1$, hence our polynomial vanishes identically.

Now let us compute F in the sequence X and let us assume that no subsequence of size p sums to 0 mod p. Then all terms occurring in the sum on the right will be 1, and the total sum amounts to

$$\binom{2p-1}{p} = 1 \pmod{p},$$

a contradiction. So in fact we have proved a little bit more, namely that the number of subsequences of X summing to 0 equals 1 mod p

A well-known partition puzzle asks whether it is possible to cover the 62 squares of a chessboard with two opposite corners removed with 32 domino tiles. The two opposite corners have the same colour and each domino covers a white and a black square; so this is impossible. This innocent looking argument has a whole range of variants. Here I will sketch a colouring argument involving (again) polynomials settling a partition problem in finite geometry (details can be found in [12]). The question is the following. Is it possible to partition the point set of $\Sigma = PG(3, q^2)$ in a Baer subspace, that is a $PG(3, q)$, and $q^4 - q$ lines? The reason that this is of interest, is that it is possible to show that any set of more than $q^4 - q$ mutually disjoint lines in $PG(2, q^2)$ is contained in a full spread of lines. For $q = 2$ a partition into a single Bear subspace and 14 lines of $PG(3, 4)$ is possible, as was shown in [36]; see also [21] for a classification of all possibilities. Rather surprisingly however, this turns out to be the only value for which it is possible.

It is not so hard to see that from a partition of the point set of $PG(3, q^2)$ like this one can obtain a partition of $AG(3, q^2) \setminus AG(3, q)$ into lines. Consider a plane π intersecting the Bear subspace in a Baer subplane, and look at the induced partition of the affine space $\Sigma \setminus \pi$. Now coordinatize $AG(3, q^2)$ in such a way that the $AG(3, q)$ we deleted has all coordinates in $GF(q)$. Now provide every point (x, y, z) with the weight $(xyz)^{q-1}$. As it turns out this gives total weight 0 on all lines, provided that $3(q - 1) < q^2 - 1$, and also total weight 0 of the entire space, but the total weight of $AG(3, q)$ is nonzero. Contradiction.

A problem dating back at least 30 years is the following: How large can a set of mutually touching congruent infinite circular cylinders be in $\mathbf{R}^d$. For $d = 3$ it is stated in [37] that 7 is possible, the best known upper bound seems to be the Ramsey number $R_3(4, 5)$ (this is the smallest number N with the property that if all triples from an N-set are coloured blue or red, then there is always a blue 4-set, that is a 4-set in which all triples are blue, or a red 5-set) [3]. For $d > 3$ it is not even known whether the number is finite. For more information on related problems concerning neighbourly

convex bodies we refer to the collection of problems in discrete geometry by
W. Moser and J. Pach [37], problem 55.

An alternative way of stating this problem is the following:

Let X be a set of equidistant lines in $\mathbf{R}^d$; how large can $|X|$ be? So the
question sounds a lot like the problems treated in section 2., about few-
distance sets and sets of lines with few angles. Note that a collection of lines
with few angles is the same as a few-distance set in projective space. Here
(real) projective space is provided with a metric in the natural way, that is
by using the angle between two one-dimensional subspaces corresponding
to two projective points to measure the distance.

Surprisingly the projective version of the cylinder problem can be attacked
using polynomials, although not as straightforwardly as usual, whereas the
euclidean version seems to be much harder.

Equidistant lines in (real) projective d-space correspond to equiangular 2-
dimensional subspaces in $\mathbf{R}^{d+1}$. I will use the shorter word plane for two-
dimensional subspace. The angle between two planes can be expressed in
terms of the largest eigenvalue of a certain matrix as follows [41].

Let $\pi_1 = \langle u, v \rangle$ and $\pi_2 = \langle s, t \rangle$ be two planes in $\mathbf{R}^{d+1}$, where u, v and s, t
are orthonormal bases of π_1 and π_2 respectively. Then the angle between π_1
and π_2 equals ϕ where $0 \le \phi \le \pi/2$ and $\cos^2 \phi = \lambda$, the largest eigenvalue
of AA^T for the matrix

$$A = \begin{pmatrix} (u,s) & (u,t) \\ (v,s) & (v,t) \end{pmatrix}.$$

Note that the angle of two planes is 0 if they intersect in more than just
the origin.

We now can prove the following.

Let Π be a set of equiangular planes in $\mathbf{R}^{d+1}$, and assume the common angle
is not zero. Then

$$|\Pi| \le \binom{2d+5}{4}.$$

Proof. The basic idea again is to associate to each of the planes a polynomial, and then show that the collection thus obtained is independent.
However it is a bit more involved then the examples we have met sofar.

Let $\pi_{u,v}$ and $\pi_{s,t}$ be two planes in $\mathbf{R}^{d+1}$, where the indices represent orthonormal bases. Denote by $f(u, v; s, t)$ the largest eigenvalue λ of the
matrix AA^T, with A as above. Then we may write

$$f(u, v; s, t) = g(u, v; s, t) + \sqrt{h(u, v; s, t)},$$

with $g = (a + d)/2$ and $h = ((a - d)^2 + 4bc)$, and $a = (u, s)^2 + (u, t)^2$, $b = c = (u, s)(v, s) + (u, t)(v, t)$ and $d = (v, s)^2 + (v, t)^2$.

In particular, $f(u, v; u, v) = g(u, v; u, v) = 1$ and $h(u, v; u, v) = 0$.

Now fix an orthonormal basis for each of the planes in Π and let $\lambda = 1 - \varepsilon$ for each pair in our collection.

So for $\{u, v\} \neq \{s, t\}$ (and $\pi_{u,v}, \pi_{s,t} \in \Pi$) we have $f = g + \sqrt{h} = 1 - \varepsilon$, from which it follows that
$$h - (1 - \varepsilon - g)^2 = 0.$$
On the other hand, if $\{u, v\} = \{s, t\}$, then obviously
$$h - (1 - \varepsilon - g)^2 = -\varepsilon^2.$$

Let $\Phi_{u,v}(x, y) = h(u, v; x, y) - (1 - \varepsilon - g(u, v; x, y))^2$. Then the system $\{\Phi_{u,v} \mid \pi_{u,v} \in \Pi\}$ is linearly independent in the space of polynomials of degree at most 4 in $x = (x_1, \ldots, x_{d+1})$ and $y = (y_1, \ldots, y_{d+1})$. Since we may restrict ourselves to those x and y with $(x, x) = (y, y) = 1$ we can in fact make the Φ's homogeneous of degree 4, and hence
$$|\Phi| \leq \binom{2d + 5}{4}.$$

References

[1] Alon N., Babai L. and Suzuki H., *Multilinear polynomials and Frankl - Ray-Chaudhuri - Wilson type intersection theorems.* J. Combin. Theory Ser. A 58 (1991), pp. 165–180.

[2] Babai L. and Frankl P., *Linear algebra methods in combinatorics I.* preliminary version, Department of Computer Science, University of Chicago, July 1988.

[3] Bezdek A., *Personal communication.*

[4] Blokhuis A., *A new upper bound for the cardinality of 2-distance sets in euclidean space.* Ann. Discrete Math. 20 (1984), pp. 65–66.

[5] Blokhuis A., *On subsets of $GF(q^2)$ with square differences.* Proc. Kon. Nederl. Akad. Wetensch. ser. A 87 = Indag. Math. 46 (1984), pp 369–372.

[6] Blokhuis A., *On the Sperner capacity of the cyclic triangle.* To appear in J. Algebraic Combin. (1993).

[7] Blokhuis A., *Solution of an extremal problem for sets using resultants of polynomials.* Combinatorica 10 (1990), pp. 393–396.

[8] Blokhuis A., *On nuclei and affine blocking sets*. To appear in J. Combin. Theory Ser. A (1993).

[9] Blokhuis A. and Brouwer A.E., *Blocking sets in Desarguesian projective planes*. Bull. London Math. Soc. 18 (1986) pp. 132–134.

[10] Blokhuis A., Brouwer A.E. and Szőnyi T., *The number of directions determined by a function f on a finite field*. Manuscript (1992).

[11] Blokhuis A. and Mazzocca F., *On maximal sets of nuclei in $PG(2,q)$ and quasi-odd sets in $AG(2,q)$*. in: Advances in finite geometries and designs, editors Hirschfeld J.W.P., Hughes D.R. and Thas J.A., Oxford University Press, 1991 pp. 27–34.

[12] Blokhuis A. and Metsch K., *On the number of lines in a maximal partial spread*. To appear in Des. Codes Cryptogr. (1993).

[13] Blokhuis A. and Metsch K., *On the size of a blocking set in $PG(2,27)$*. Manuscript (1992).

[14] Blokhuis A. and Wilbrink H.A., *A characterization of exterior lines of certain sets of points in $PG(2,q)$*. Geom. Dedicata 23 (1987), pp. 253–254.

[15] Brouwer A.E. and Schrijver A., *The blocking number of an affine space*. J. Combin. Theory Ser. A 24 (1978) pp. 251–253.

[16] Bruen A.A., *Blocking sets in finite projective planes*. SIAM J. Appl. Math. 21 (1971) pp. 380–392.

[17] Bruen A.A., *Nuclei of sets of $q+1$ points in $PG(2,q)$ and blocking sets of Rédei type*. J. Combin. Theory Ser. A 55 (1990) pp. 130–132.

[18] Bruen A.A. and Fisher J.C., *The Jamison method in Galois geometries*. Des. Codes Cryptogr. 1 (1991) pp. 199–205.

[19] Bruen A.A. and Thas J.A., *flocks, chains and configurations in finite geometries*. Atti Accad. Naz. Lincei Rend. 59 (1975) pp. 744–748.

[20] Calderbank A.R., Graham R.L., Shepp L.A., Frankl P. and Li W.-C.W., *The cyclic triangle problem*. To appear in J. Algebraic Combin. (1993).

[21] Dam E. van, *Classification of spreads of $PG(3,4) \setminus PG(3,2)$*. To appear in Des. Codes Cryptogr. (1993).

[22] Delsarte Ph., Goethals J.M. and Seidel J.J., *Bounds for systems of lines and Jacobi polynomials*. Philips Res. Rep. 30 (1975), pp. 91*–105*.

[23] Delsarte Ph., Goethals J.M. and Seidel J.J., *Spherical Codes and Designs.* Geom. Dedicata 6 (1977) pp. 363–388.

[24] Doyen J., *Lecture at Oberwolfach.* (1976).

[25] Erdős P., Ginzberg A. and Ziv A., *Theorem in additive number theory.* Bull. Res. Council Israel 10F (1961) pp. 41–43.

[26] Fisher J.C. and Thas J.A., *Flocks in $PG(3,q)$.* Math. Z. 169 (1979) pp. 1–11.

[27] Frankl P. and Wilson R.M., *Intersection theorems with geometric consequences.* Combinatorica 1 (1981) pp. 357–368.

[28] Füredi Z., *Geometrical solution of an intersection problem for two hypergraphs.* European J. Combin. 5 (1984), pp. 133–136.

[29] Jamison R., *Covering finite fields with cosets of subspaces.* J. Combin. Theory Ser. A 22 (1977), pp. 253–266.

[30] Gargano L., Körner J. and Vaccaro U., *Sperner theorems on directed graphs and qualitative independence.* To appear in J. Combin. Theory Ser. A (1993).

[31] Kalai G. and Kahn J., *A counter example to Borsuk's conjecture* To appear in Bull. Amer. Math. Soc. (1993).

[32] Koornwinder T.H., *A note on the absolute bound for systems of lines.* Proc. Kon. Nederl. Akad. Wetensch. Ser. A 79 = Indag. Math. 38 (1976), pp. 152–153.

[33] Larman D.G., Rogers C.A. and Seidel J.J., *On two-distance sets in euclidean space.* Bull. London Math. Soc. 9 (1977), pp. 261–267.

[34] Lemmens P.W.H. and Seidel J.J., *Equiangular lines.* J. Algebra 24 (1973), pp. 494–512.

[35] Lovász L. and Schrijver A., *Remarks on a theorem of Rédei.* Studia Sci. Math. Hungar. 16, (1981) pp. 449–454.

[36] Mesner D.M., *Sets of disjoint lines in $PG(3,q)$* Canad. J. Math. 19 (1967), pp. 273–280.

[37] Moser W. and Pach J., *100 Research Problems in Discrete Geometry.* (1986).

[38] Orr W.F., *The Miquelian inversive plane $IP(q)$ and the associated projective planes.* Ph.D. Thesis, Madison, (1973).

[39] Rédei L., *Lückenhafte Polynome über endlichen Körpern*. Birkhäuser Verlag, Basel (1970).

[40] Segre B. and Korchmáros G., *Una proprietà degli insiemi di punti di un piano di Galois caratterizzante quelli formati dai punti delle singole rette esterne ad una conica*. Atti Accad. Naz. Lincei Rend. *62 (1977)* pp. 1–7.

[41] Seidel J.J., *Angles and distances in n-dimensional euclidean and noneuclidean geometry I*. Proc. Kon. Nederl. Akad. Wetensch. Ser. A *58* = Indag. Math. *17 (1955)* pp. 329–541.

Models of Random Partial Orders

Graham Brightwell

Department of Statistical and Mathematical Sciences
London School of Economics and Political Science
Houghton St., London WC2A 2AE

§1. Introduction

The theory of random graphs has proved to be a success. It has provided
many interesting questions, provoking the development of new techniques
on the boundary between combinatorics and probability theory. Random
graphs have also proved tremendously valuable in applications outside
the theory: firstly in extremal questions, where bounds obtained by non-
constructive means are often the best known, and secondly in questions
coming from computer science, or, for instance, from biology, which turn
out to translate directly into natural questions about particular models
of random graphs. It is not appropriate to give an extensive bibliography
here, but the place to set off on a serious study of random graphs is the
monograph by Bollobás [7].

Another basic combinatorial structure is that of finite partially ordered sets, and it is very natural to ask whether an equally successful
theory of random partial orders can be developed. Regrettably, the simple answer will have to be "no". One reason for this is perhaps that partial
orders occur less often in applications than do graphs, but the more important problem is mathematical: the lack of "independence".

To explain this, it is convenient to recall first the most basic model
of random graphs. Let n be a positive integer, and p a real number
strictly between 0 and 1. Set $[n] = \{1, \ldots, n\}$. For each pair (i, j) of
integers from $[n]$ with $i < j$, let X_{ij} be a Bernoulli random variable with
$\Pr(X_{ij} = 1) = p$, with all the X_{ij} mutually independent. The random
graph $G_{n,p}$ is defined on the vertex set $[n]$ by setting i and j adjacent just
when $X_{ij} = 1$.

An alternative way to define the $G_{n,p}$ model is to let $\mathcal{G}(n,p)$ denote the set of all graphs on n vertices, equipped with a weight function so that the weight of a graph G with m edges is $p^m(1-p)^{N-m}$, where $N = \binom{n}{2}$. The sum of the weights is 1, so the weight function is also a probability function on the space $\mathcal{G}(n,p)$. A random graph $G_{n,p}$ is simply an element of this probability space. Note that, in $\mathcal{G}(n,1/2)$, all labelled n-vertex graphs are equiprobable.

More than anything else, it is the independence of the X_{ij} that makes the $G_{n,p}$ model pleasant to work with. Even in other random graph models, some form of "approximate" independence underlies the intuition and methods used. But independence for the related pairs in a model of random partial orders is a strong condition: the events $x < y$, $y < z$ and $x < z$ cannot be independent unless one of the first two has probability 0, or the third has probability 1.

Of course this is not the end of the story, but the fact remains that there is no model of random partial orders which will ever be as universally applicable as the $G_{n,p}$ model of random graphs. Yet there are already several examples of problems in the theory of partial orders where random methods have proved useful; we shall discuss these as we go along.

The main purpose of this article is to survey the current state of knowledge concerning what appear to be the major models of random orders. For each of these models, the emphasis so far has been on estimating the fundamental parameters of partial orders. Perhaps this has just laid the foundations for the subject, but there have already been interesting results and techniques, and a few surprises.

Before introducing any more random structures, we review some of the basic language of partial orders. A *partial order* (or *partially ordered set*, or *poset*, or simply *order*) is a pair $(X, <)$, where $<$ is a transitive, irreflexive relation on the set X. The set X will be finite except where otherwise stated. The partial order $(X, <)$ is a *linear order* or *total order* if, whenever x and y are distinct elements of X, either $x < y$ or $y < x$. A *suborder* (or, more properly, *induced suborder*) of $(X, <)$ is order $(Y, <')$ where $Y \subset X$ and $<' = < |_Y$. A *chain* in $(X, <)$ is a suborder of $(X, <)$ which is a total order. An *antichain* is a suborder of $(X, <)$ in which no pair is related. The *height* $H(X,<)$ of $(X,<)$ is the size of a longest chain, and the *width* $W(X,<)$ is the size of a largest antichain. A *linear extension* of $(X,<)$ is a total order $\prec$ on X such that $x \prec y$ whenever $x < y$. We denote the number of linear extensions of $(X,<)$ by $L(X,<)$. The *comparability graph* of $(X,<)$ is the graph defined on X by setting x and y adjacent if $x < y$ or $y < x$. The *maximum degree* of a partial order is the maximum degree of its comparability graph.

The *dimension* $\dim(X,<)$ of $(X,<)$ is the minimum number of total orders on X whose intersection (as relations) is $<$. There is a useful equivalent definition of dimension, which it is convenient to give here. Define the *co-ordinate order* $\preceq$ on $[0,1]^d$ $(d \geq 1)$ by setting $(x_1,\ldots,x_d) \preceq (y_1,\ldots,y_d)$ if $x_i \leq y_i$ for each i. The *dimension* of a partial order P is the minimum d such that P is isomorphic to a suborder of $[0,1]^d$ with the co-ordinate order. It is easy to see that the two definitions of dimension are equivalent.

The *linear sum* of two partial orders $(X,<)$ and $(Y,<')$ is the order $\prec$ formed on the disjoint union of X and Y by setting $a \prec b$ if either $a < b$, or $a <' b$, or $a \in X$ and $b \in Y$. Loosely, the linear sum is the order formed by "putting Y on top of X". The linear sum of three or more partial orders is defined similarly.

There are several books on partially ordered sets, but I don't know of one that covers the whole subject in any depth. Many concentrate on lattice theory, which is at the opposite end of the spectrum from the theory of random orders. Possibly the best place to start a study of partial orders is with the two conference volumes [36] and [37], both edited by Rival. Perhaps the most substantial work on the subject is Trotter's recent book [41], which concentrates on dimension theory.

This survey reflects my own personal preference, which is to regard the height, width, dimension, and number of linear extensions as the four most important parameters of partial orders. There are several other parameters that have attracted interest, such as the maximum degree, the number of related pairs, the *diameter* and the *jump number*. All of these have been studied for at least one model of random orders, but I shall not deal with them here.

Before we move on to study particular models of random orders, let us spend a little time with formalities. A *model of random partial orders* is a family of probability spaces $\mathcal{P}_n$, the elements of which are finite partial orders. Usually, but not always, the elements of $\mathcal{P}_n$ are partial orders with n elements. A random element of the space $\mathcal{P}_n$ will typically be denoted P_n. If f is a parameter of finite partial orders, then the value of $f(P_n)$ for an element P_n of $\mathcal{P}_n$ is a random variable F_n. We are interested in the behaviour of the random variable in the limit as $n \to \infty$. If $\mathcal{A}$ is a property of finite partial orders, we say that a random order P_n in a model $\mathcal{P}_n$ has property $\mathcal{A}$ *almost surely* if, as $n \to \infty$, the probability that a random element of $\mathcal{P}_n$ has $\mathcal{A}$ tends to 1. In this case, we also say that *almost every* partial order P_n in the model $\mathcal{P}_n$ has property $\mathcal{A}$.

Suppose $f(P)$ is a parameter of finite partial orders, and we have a model of random orders $\mathcal{P}_n$. It is usually overambitious to ask for the exact distribution of the corresponding random variable F_n, so we ask merely for partial information about it. The crudest type of information is simply the expected value $\mathbf{E}F_n$: this is sometimes, but not always, fairly easy to determine, at least approximately. A more useful type of result states that F_n almost surely lies in some range (depending on n). Note that it is not always the case that $\mathbf{E}F_n$ falls in the range: we shall see some examples later where the random variable is almost surely well below its expectation.

A sequence X_n of real-valued random variables *converges in distribution* to a random variable X (we write $X_n \xrightarrow{d} X$) if, for every x such that $\Pr(X \leq x)$ is continuous, $\Pr(X_n \leq x) \to \Pr(X \leq x)$ as $n \to \infty$. The information that a suitably normalised version of our random variable F_n converges in distribution to a random variable F gives us rather a strong handle on F_n. This type of result has been proved in relatively few cases for parameters of partial orders—hopefully that lies in the future.

A different type of information about the distribution of F_n is an upper estimate for the probability that F_n lies far from its mean. We aim for results of the form:

$$\Pr\big(|F_n - \mathbf{E}F_n| > \lambda\sigma(n)\big) \leq e^{-\lambda^2},$$

where $\sigma(n)$ is to be thought of as an upper bound for the standard deviation of F_n. A random variable for which such a result holds is said to be *sharply concentrated* about its mean. Often, this type of result can be proved even if little is known about $\mathbf{E}F_n$ or about the limit distribution of F_n: we shall see some examples later. The technique used in the proofs is inevitably a martingale inequality based on those of Azuma [4] or Hoeffding [27]. For further information about the use of these inequalities, see one of the survey articles by Bollobás [8,9] or McDiarmid [34].

We now introduce the most prominent models of random partial orders.

In the random graph model $G_{n,1/2}$, each labelled n-vertex graph has the same probability $2^{-\binom{n}{2}}$. A first shot at defining a model of n-vertex random partial orders is to follow this lead: take each partial order on the set $[n]$ to have the same probability. We shall call this the *uniform model* $\mathcal{U}_n$ for random partial orders, and we shall denote a random element of the space by U_n. In order to deal with this model, one needs to answer questions such as: about how many n-vertex partial orders are there? And what does a "typical" n-vertex partial order look like? These questions were answered by Kleitman and Rothschild in a 1975 paper [30]. This has turned out to be an influential paper: not only was it arguably the first paper on random partial orders, but the technique introduced there has become a paradigm for subsequent work on asymptotic enumeration of combinatorial structures.

In order to state their result, we introduce some notation. For disjoint sets $X_1, X_2, \ldots, X_k$ of vertices, let $\mathcal{A}(X_1, \ldots, X_k)$ denote the set of partial orders $<$ with vertex set $\cup_{i=1}^{k} X_i$ satisfying the following two conditions.

(1)   If $x \in X_i$, $y \in X_j$ and $x < y$, then $i < j$.

(2)   If $x \in X_i$, $y \in X_j$ and $i < j - 1$, then $x < y$.

Condition (1) implies that each X_i forms an antichain, and condition (2) specifies relations between non-consecutive sets, so the only freedom is that, for each $i < k$, the relations between X_i and X_{i+1} can be specified arbitrarily. Let $R = R(X_1, \ldots, X_k)$ be the set of pairs (x, y) such that $x \in X_i$ and $y \in X_{i+1}$ for some i; then a partial order $<$ in $\mathcal{A}(X_1, \ldots, X_k)$ is specified uniquely by the set $\{(x, y) \in R : x < y\}$. The number of partial orders in $\mathcal{A}(X_1, \ldots, X_k)$ is thus

$$2^{|R|} = 2^{\sum_{i=1}^{k-1} |X_i||X_{i+1}|}.$$

We can now state the result of Kleitman and Rothschild [30] as follows.

Theorem 1. *Let $\omega(n)$ be any function tending to infinity. Almost every n-vertex partial order U_n in the uniform model $\mathcal{U}_n$ lies in a class $\mathcal{A}(X_1, X_2, X_3)$, for some partition (X_1, X_2, X_3) of $[n]$ with $||X_2| - n/2| < \omega(n)$ and $||X_1| - n/4| < \omega(n)\sqrt{n}$.*   □

The bounds given in Theorem 1 on the sizes $|X_i|$ of the parts follow easily on counting the partial orders in the various $\mathcal{A}(X_1, X_2, X_3)$, so the significant portion of the proof is concerned with showing that the proportion of partial orders not lying in such a class is small. This is done by stating a number of properties which together imply that a partial order does lie in such a class, and showing, for each of the properties, that the proportion of partial orders not having the property is small. A consequence of Theorem 1 is the following estimate for the number $|\mathcal{U}_n|$ of partial orders on an n-vertex set: here i represents the number of elements in X_2.

$$|\mathcal{U}_n| = (1 + o(1)) \sum_{i=1}^{n} \binom{n}{i} 2^{i(n-i)+(n-i)}$$

$$= \begin{cases} C_e(1 + o(1))\binom{n}{n/2}2^{n^2/4+n/2} & n \text{ even} \\ C_o(1 + o(1))\binom{n}{(n-1)/2}2^{n^2/4+n/2} & n \text{ odd,} \end{cases}$$

where

$$C_e = \sum_{j=-\infty}^{\infty} 2^{-j^2-j} \simeq 2.5317402, \quad C_o = 2^{1/4} \sum_{j=-\infty}^{\infty} 2^{-j^2} \simeq 2.5317468.$$

(Yes, these two constants *are* slightly different.)

Theorem 1 has been widely viewed as a negative result. It says that almost every partial order has height three, and, even more, is *ranked*, that is, all the maximal chains have the same length (namely, three). Somehow this does not conform to the practising mathematician's view of what a "typical" partial order ought to look like. As regards extremal questions for partial orders, it is rare (but, as we shall see, not impossible) for a ranked partial order of height three to have any chance of providing a good bound.

Nevertheless there are things to be said in favour of $\mathcal{U}_n$. First, it has a certain status simply because it *is* the uniform model: a result about $\mathcal{U}_n$ really does apply to "most" partial orders. Second, as we now explain, the Kleitman-Rothschild Theorem allows us to prove results about the uniform model while working in a different model, where the relations in the random partial order *are* mutually independent.

For fixed disjoint sets $X_1, \ldots, X_k$, the set $\mathcal{A}(X_1, \ldots, X_k)$ can be made into a probability space by making each partial order in the set equally likely. Equivalently, we can take independent Bernoulli random variables $Y_{x,y}$ for each pair $(x, y) \in R$, with $\Pr(Y_{x,y} = 1) = 1/2$, and define a random partial order $A_{1/2}(X_1, \ldots, X_k)$ from $\mathcal{A}(X_1, \ldots, X_k)$ by setting $x < y$ whenever $(x, y) \in R$ and $Y_{x,y} = 1$. The Kleitman-Rothschild Theorem then tells us that, if almost every random order from $\mathcal{A}(X_1, X_2, X_3)$ (with suitable bounds on the $|X_i|$) has a certain property, then so does almost every random order from the uniform model.

More generally, for any $p = p(n)$ lying between 0 and 1, we can take the $Y_{x,y}$ to be independent Bernoulli random variables with $\Pr(Y_{x,y} = 1) = p$, and define the random partial order $A_p(X_1, \ldots, X_k)$ as above.

Experience so far suggests that the interesting structure in the $\mathcal{A}(X_1, \ldots, X_k)$ model already exists when $k = 2$ and the two sets are exactly the same size. A *random bipartite order* is a random order $A_p(X, Y)$, where X and Y are disjoint sets each of size n. Alternatively, a random bipartite order is formed by taking a random bipartite graph with vertex sets X and Y and probability p of an edge, and setting $x < y$ iff $x \in X$, $y \in Y$, and xy is an edge of the bipartite graph. The major result concerning random bipartite orders is due to Erdős, Kierstead and Trotter [22], who obtained good estimates for the dimension of almost every random bipartite order, in the process proving the existence of partial orders whose dimension is much larger than their maximum degree. Random bipartite orders are discussed further in §2.

There is another natural way to use a random graph to define a random partial order. Given any graph G on $[n]$, we define the associated partial order $<_G$ by $x <_G y$ if there is an increasing sequence $x = x_1, x_2, \ldots, x_k = y$ of numbers from $[n]$ such that $x_i x_{i+1}$ is an edge of G for each i. In other words, for $i < j$ in $[n]$, we interpret an edge between i and j as a relation $i \prec_G j$, and take $<_G$ to be the transitive closure of $\prec_G$. Any model $\mathcal{G}_n$ of random graphs G_n on $[n]$ now gives rise to a model of random orders $<_{G_n}$.

We define the space $\mathcal{P}_{n,p}$ of *random graph orders* to be the space derived from $\mathcal{G}(n,p)$ in the above way. This method of constructing random orders has occurred to a number of people independently, but relatively little has been written concerning the model. The earliest paper I am aware of that deals with $\mathcal{P}_{n,p}$ is a 1984 paper of Barak and Erdős [6]. That paper is entitled "On the maximal number of strongly independent vertices in a random acyclic directed graph"; in our terminology this parameter is just the width of a random graph order. Random graph orders were introduced as such by Albert and Frieze in a 1989 paper [2].

One good point of this model is that every n-vertex partial order does occur with some probability, but a typical random graph order looks very different from a typical random order in the uniform model. It is not too hard to see that random graph orders, at least for constant values of p, are extremely "tall and thin": the height of $\mathcal{P}_{n,p}$ is almost surely at least cn for some constant $c = c(p)$. We shall see later that the structure of a typical $\mathcal{P}_{n,p}$, for p constant, is that of a linear sum of many small partial orders. This rather prevents a random graph order from having any "global" structure, which makes it unlikely to be of use in answering extremal questions for partial orders.

Those familiar with the $\mathcal{G}(n,p)$ model of random graphs will not have been too surprised to hear that $\mathcal{P}_{n,p}$ for p constant is relatively uninteresting. It is likely that future work will concentrate on the case where p is a decreasing function of n. Random graph orders are discussed in detail in §3.

Probably the most important model of random orders so far developed is the model $\mathcal{P}_k(n)$ of *random k-dimensional orders*. A random k-dimensional order is constructed by taking k linear orders on the set

$[n]$, uniformly and independently at random from the set of all $n!$ linear orders, and forming the intersection of the k orders. Evidently such an order has dimension at most k. (More formally, the set $\mathcal{P}_k(n)$ of partial orders on $[n]$ of dimension at most k is made into a probability space by assigning each order $P \in \mathcal{P}_k(n)$ a probability proportional to the number of k-tuples of linear orders on $[n]$ whose intersection is P.)

As there are two equivalent definitions of dimension, so there are two equivalent definitions of random k-dimensional orders. The alternative definition, based on the co-ordinate order in the cube, is perhaps more appealing. Consider the cube $[0,1]^k$, equipped with the standard uniform product measure, and the co-ordinate order $\preceq$. Construct a random order by taking n points labelled $1, \ldots, n$ independently at random in $[0,1]^k$, according to the measure, and taking the order induced by $\preceq$ on these points. The k co-ordinate projections give independent random linear orders on the (labels of the) n points, and the random order constructed is just the intersection of these orders. Thus the two models of random k-dimensional orders are equivalent. We shall use the notation $P_k(n)$ for a random k-dimensional order in either definition.

Random k-dimensional orders were introduced as an object of study in a 1985 paper of Winkler [45], but they had in fact occurred in various different guises earlier. The case $k = 2$ is of particular interest. In this case, one may as well assume that one of the two linear orders is the standard order on $[n]$, and the other is just a random permutation of $[n]$. The height of $P_2(n)$ is the length of the longest increasing subsequence of the random permutation, while the width is the length of the longest decreasing subsequence. These two parameters are clearly identically distributed, and there has been great interest in estimating them. This problem is often known as Ulam's problem, since it was apparently first mentioned in [42].

Notice that we do recover independence, in a completely different way, in the random k-dimensional order model. Here it is not the relations that are independent, but the positions of the points in the cube. It is sometimes more convenient still to use a slightly different model, with yet another form of independence.

Given the cube $[0,1]^k$ with the standard measure, let X be a set of points in the cube generated by a Poisson process of density n, and form a random order by taking the order induced on X by the co-ordinate order of the cube. We use $P'_k(n)$ to denote a random order in this model. What we lose is that the number of points in the partial order is unlikely to be exactly n: what we gain is that the set of points in one region of the cube is independent of the set of points in another disjoint region. Asymptotic results about $P'_k(n)$ can usually be converted to asymptotic results about $P_k(n)$ without too much difficulty: the relationship between the two models is similar to that between $\mathcal{G}(n,p)$ and the model $\mathcal{G}(n,M)$ $(M = \lfloor p\binom{n}{2} \rfloor)$ of random graphs where each n-vertex graph with M edges is equally likely.

Random k-dimensional orders are treated in §4. There is an earlier survey article on random k-dimensional orders by Winkler [47], which emphasises different aspects of the theory.

A random order $P_k(n)$ or $P'_k(n)$ inherits some of the structure of the cube $[0,1]^k$ in which it is embedded. If we have any other infinite partial order, with a structure we wish to mimic in a discrete setting, we can hope to carry out the same procedure. Bollobás and Brightwell [11] investigated a class of spaces equipped with a measure and a (compatible) partial order called *box spaces*. They found that the height of a random partial order defined in a box space by taking a Poisson distribution of density n (or by taking n points independently at random) always behaves similarly to that of a random k-dimensional order. We shall discuss this further in §5.

One particular topic not discussed in this survey is that of 0-1 laws in models of random partial orders. This is a topic which has attracted a fair amount of interest in the wake of progress made on 0-1 laws for random graph models. See the papers of Compton [20] for the uniform model, Łuczak [33] for random graph orders, and Winkler [45,46,48] and Spencer [38] for random k-dimensional orders. The reason for the omission of this subject is partly lack of space, partly lack of expertise, and mostly the existence of the excellent recent surveys by Compton [21] and Winkler [50].

§2. Random bipartite orders

As we mentioned in the introduction, most of the interesting behaviour of random $\mathcal{A}(X_1, \ldots, X_k)$ orders occurs in random bipartite orders, so for this section we shall focus our attention on that case.

Let X and Y be disjoint sets of size n, let $p = p(n)$ lie between 0 and 1, and consider a random bipartite order $A_p(X, Y)$ such that, for each pair (x, y) with $x \in X$ and $y \in Y$, $x < y$ in the order with probability p.

Estimating the height of a random bipartite order is of course trivial. Estimating the width is almost equally trivial, at least provided $np - \log n \to \infty$, when almost surely the underlying random bipartite graph has a complete matching (see Bollobás [7, Chapter VII.3]). For this section, we concentrate on estimates for the dimension and the number of linear extensions of a random bipartite order.

There are many upper bounds for the dimension of a partial order in terms of other parameters. One fundamental and well-known bound is that $\dim(X, <) \leq W(X, <)$ (see, for instance, Trotter [41]). Another, due to Füredi and Kahn [24], is that $\dim(X, <) \leq 50\Delta \log^2 \Delta$, where Δ is the maximum degree of $(X, <)$, However, until recently it was not known whether there are orders $(X, <)$ with $\dim(X, <) > \Delta + 1$. (The partial order S_d formed by the 1-element and $(d-1)$-element subsets of a d-element set, ordered by inclusion, has dimension d and maximum degree $d - 1$.) Erdős, Kierstead and Trotter [22] showed that almost every $A_p(X, Y)$, for a suitable $p(n)$, has this property. This is probably the best example so far of the use of random orders in an extremal problem. We state the general version of their result, and then mention some specific cases.

Theorem 2. *For every $\epsilon > 0$, there exists $\delta > 0$ such that, if $\log^{1+\epsilon} n/n < p < 1 - n^{-1+\epsilon}$, then almost surely*

$$\dim(A_p(X, Y)) > (\delta pn \log pn)/(1 + \delta p \log pn). \qquad \square$$

The conclusion of Theorem 2 takes on a different character depending on whether $p \log pn$ is greater or less than 1. If $p \log n \leq 1$, the lower bound on the dimension is of the form $\delta pn \log pn$. On the other hand, if $p \log n \geq 1$ (in particular, if p is constant), then we have the lower bound in the following result.

Theorem 3. *For every* $\epsilon > 0$, *there is a constant* c *such that, if* $1/\log n < p < 1 - n^{-1+\epsilon}$, *then almost surely*

$$n(1 - \log(1/p)/3\log n) > \dim(A_p(X,Y)) > n(1 - c/p\log n). \qquad \Box$$

For p constant, both upper and lower bounds in Theorem 3 are of the form $n(1 - c/\log n) = n(1 - o(1))$. For $p = 2c/\log n$, the dimension of almost every $A_p(X,Y)$ is at least $n/2$, while the maximum degree is at most, say, $2np = 4cn/\log n$. In other words, for this choice of p, almost every $A_p(X,Y)$ has dimension at least $\alpha \Delta \log \Delta$, for some constant $\alpha > 0$. This is to be compared with the Füredi-Kahn upper bound of $50\Delta \log^2 \Delta$ for the dimension.

There is not the space here to say much about the proofs of Theorems 2 and 3. The interested reader can find the proof either in the original paper of Erdős, Kierstead and Trotter [22], or in Trotter's book [41]. Both sources also contain a proof of the following result. Recall that U_n denotes a random n-vertex partial order in the uniform model.

Theorem 4. *There exist constants* $c_1, c_2 > 0$ *such that, almost surely,*

$$\frac{n}{4}\left(1 - \frac{c_1}{\log n}\right) < \dim(U_n) < \frac{n}{4}\left(1 - \frac{c_2}{\log n}\right). \qquad \Box$$

The number of linear extensions of a random bipartite order, for constant p, was studied by Brightwell [17]. Intuitively, we would expect that most linear extensions of $A_p(X,Y)$ consist of the elements of X in some order, with the elements of Y in some order above them, except perhaps for a small patch in the middle where a few elements of X appear above a few elements of Y. Indeed, there are $(n!)^2$ linear extensions where no elements of X come above elements of Y, and on average another $(1-p)(n!)^2$ where just the top element of X comes above the bottom element of Y. This leads us to suspect that the number of linear extensions is almost surely about $C(p)(n!)^2$, and indeed this is the case.

Let us evaluate the *expected* number of linear extensions of $A_p(X,Y)$. A linear order on the set $X \cup Y$ can be considered as a triple consisting of: an arbitrary linear order $\prec_X$ on X, an arbitrary linear order $\prec_Y$ on Y, and a choice α of which n of the $2n$ positions in the linear order will be occupied by the elements of X. For such a triple, let $l = l(\alpha)$ denote the number of *reversals*, i.e., pairs of elements $x \in X$, $y \in Y$, with x coming above y in the proposed linear order. Note that the number of reversals is independent of $\prec_X$ and $\prec_Y$.

The probability that a triple $(\prec_X, \prec_Y, \alpha)$ is a linear extension of the random bipartite order $A_p(X, Y)$ is exactly $(1-p)^l$, the probability that none of the l reversed pairs are related in the random order. Hence the expected number of linear extensions of $A_p(X, Y)$ is

$$(n!)^2 \sum_\alpha (1-p)^{l(\alpha)} = (n!)^2 \sum_{l=0}^{\infty} c_l (1-p)^l,$$

where the first sum is over all the $\binom{2n}{n}$ choices α of n places in the linear order, and c_l is the number of choices α with l reversals. It is not hard to see that this coefficient c_l is the number of partitions of the integer l in which no part is greater than n. For $l \leq n$, this is just the partition function $\rho(l)$.

For constant p, the sum above converges quickly, and the expected number of linear extensions of $A_p(X, Y)$ is thus asymptotically

$$(n!)^2 \sum_{l=0}^{\infty} \rho(l)(1-p)^l = (n!)^2 \prod_{i=1}^{\infty} (1 - (1-p)^i)^{-1},$$

using the standard formula for the generating function of the partition function. We denote the product $\prod_{i=1}^{\infty}(1 - (1-p)^i)$ by $\eta(p)$: we shall see that it also plays an important role in the theory of random graph orders. As an example, $\eta(1/2)^{-1} \simeq 3.463$, so $\mathbf{E}A_{1/2}(X, Y) = (1 + o(1))3.463 \ldots (n!)^2$.

This calculation does not imply that almost every random bipartite order has about $(n!)^2 \eta(p)^{-1}$ linear extensions, but the following result of Brightwell [17] states that this is indeed the case.

Theorem 5. *Let $0 < p < 1$ be a constant. Then there is a constant $c = c(p)$ such that*

$$(n!)^2 \eta(p)^{-1} \left(1 - \frac{c \log^3 n}{n}\right) \leq L(A_p(X, Y)) \leq (n!)^2 \eta(p)^{-1} \left(1 + \frac{c \log^3 n}{n}\right)$$

almost surely.   □

In fact, the proof of Theorem 5 in [17] can be extended to show that $L(A_p(X,Y)) = (n!)^2 \eta(p)^{-1}(1 + o(1))$ almost surely, provided that $pn^{1/7}/\log^{4/7} n \to \infty$. The exponent $1/7$ here is almost certainly not best possible, but we cannot expect such a result to hold for p as small as, say, $\log n/n$.

The proof of Theorem 5 can be extended to show that, provided the $|X_i|$ are not too disparate in size, the number of linear extensions of a random order $A_p(X_1, \ldots, X_k)$ is almost surely $|X_1|! \ldots |X_k|! \eta(p)^{1-k}(1 + o(1))$. This shows that, for any $\omega(n) \to \infty$, the number $L(U_n)$ of linear extensions of an n-element random order in the uniform model almost surely lies between $(n/2)! (n/4)!^2/\omega(n)$ and $(n/2)! (n/4)!^2 \omega(n)$. However, as we shall now show, the average number of linear extensions of an n-element random order (i.e., $\mathbf{E}L(U_n)$) is significantly larger than this.

We double count the set of pairs $(P, \prec)$, where P is a partial order on $[n]$ and $\prec$ is a linear extension of P. On the one hand, this is the total number of partial orders on $[n]$ times the average number of linear extensions. On the other hand, it is $n!$ times the number of partial orders having the standard linear order $1 \prec 2 \prec \ldots \prec n$ as a linear extension. A lower bound on this last quantity thus gives a lower bound on the average number of linear extensions.

For convenience, we assume that n is even. Choose any m with $1 \le m \le n/2$. Then set $X_1 = \{1, \ldots, m-1\}$, $X_2 = \{m, \ldots, m+n/2-1\}$, $X_3 = \{m+n/2, \ldots, n\}$. Each order in $\mathcal{A}(X_1, X_2, X_3)$ has the standard order as a linear extension, and almost all of them arise from just one choice of m. Hence there are at least $(n/2)2^{n^2/4}(1 + o(1))$ orders having the standard order on $[n]$ as a linear extension.

Thus we have, using the above together with the Kleitman-Rothschild estimate for the number of n-vertex partial orders,

$$C_e(1 + o(1)) \binom{n}{n/2} 2^{n^2/4 + n/2} \mathbf{E}L(U_n) \ge n!(n/2)2^{n^2/4}(1 + o(1)),$$

and so

$$\mathbf{E}L(U_n) \ge \frac{(n/2)!^2 n(1 + o(1))}{2C_e 2^{n/2}} \ge (1 + o(1))\frac{1}{C_e\sqrt{\pi}}(n/2)! (n/4)!^2 \sqrt{n}.$$

One way of interpreting this argument is to see that contributions to the sum come equally from all the classes $\mathcal{A}(X_1, X_2, X_3)$ with $|X_2| = n/2$, not just those with $||X_1| - n/4| \leq \omega(n)\sqrt{n}$. My suspicion is that the formula above gives the correct order of magnitude for $\mathbf{E}L(U_n)$, although the constant can certainly be improved.

§3. Random graph orders

For this section, we shall consider a random order $P_{n,p} \in \mathcal{P}_{n,p}$, generated by taking G to be a random graph in $\mathcal{G}(n,p)$, and setting $P_{n,p}$ equal to the order $<_G$ on $[n]$. As mentioned in the introduction, I feel that the most interesting ranges of $p = p(n)$ will turn out to have $p(n) \to 0$ as $n \to \infty$. Some results about random graph orders $P_{n,p}$ with $p(n) \to 0$ can be found in Bollobás and Brightwell [14].

For this survey, we restrict ourselves to the case where the edge probability p is constant. Given this restriction, it is possible (and convenient) to consider the random order to be defined on the entire set of integers, and to view $P_{n,p}$ as the restriction of this infinite random order to $[n]$.

Thus we consider the random graph $G_{\mathbf{Z},p}$ defined on $\mathbf{Z}$ by taking each pair of vertices to be adjacent independently with probability p. We then form $P_{\mathbf{Z},p}$ by treating each edge ij with $i < j$ in $\mathbf{Z}$ as a relation $i \prec j$, and taking the transitive closure.

A vertex x of $P_{\mathbf{Z},p}$ is called a *post* if every other vertex of the partial order is related to it. If x is a post, then the rest of the random partial order is divided into the set $\{j \in \mathbf{Z} : j < x\}$, all elements of which are below x in $P_{\mathbf{Z},p}$, and the set $\{j \in \mathbf{Z} : j > x\}$, all elements of which are above x in $P_{\mathbf{Z},p}$. A result of Alon, Bollobás, Brightwell and Janson [3], which we shall prove shortly, states that there are almost surely infinitely many posts in $P_{\mathbf{Z},p}$. This means that the infinite random order is almost surely the linear sum of many finite random orders.

It is easy enough to find the probability that a particular vertex x is a post. Indeed, for each $k \geq 1$, the probability that $x <_G (x+k)$, given that x is less than all of $x+1, \ldots, x+k-1$, is $1-(1-p)^k$. Thus the probability that every integer above x is related to x is $\prod_{i=1}^{\infty}(1-(1-p)^i) = \eta(p)$, the function introduced in the previous section. The event that every integer below x is related to x also has probability $\eta(p)$, and is independent of the behaviour above x, so the probability that x is a post is $\eta(p)^2$.

Of course, the events that various vertices are posts are far from independent. (For instance, the probability that two successive integers are both posts is easily seen to be $p\eta(p)^2$, so the probability that $x + 1$ is a post given that x is a post is just p.) Thus this does not suffice to prove that there almost surely are posts in $P_{\mathbf{Z},p}$. For that, we need the following lemma from Alon, Bollobás, Brightwell and Janson [3].

Lemma 6. *For every* $0 < p < 1$ *there is a constant* $C = C(p) > 1$ *such that, for every sufficiently large* k, *the probability that none of the* k *elements* $2k, 4k, 6k, \ldots 2k^2$ *are posts in* $G_{\mathbf{Z},p}$ *is at most* C^{-k}.

Sketch of proof.   Observe that the event "none of the k elements $2k, 4k, 6k, \ldots 2k^2$ are posts in $P_{\mathbf{Z},p}$" is equivalent to the event "for each of the elements $2jk$ $(j = 1, \ldots, k)$, there is an element n_j of $\mathbf{Z}$ incomparable with $2jk$ in $P_{\mathbf{Z},p}$". Given this latter event, we have either: (i) for each j, there is an element n_j in the interval $[(2j - 1)k, (2j + 1)k]$ incomparable with $2jk$ in the partial order, or (ii) for some j, every element in the interval $[(2j - 1)k, (2j + 1)k]$ is comparable with $2jk$ in the partial order, but there is an element n_j outside this interval incomparable with $2jk$. It is easy to check that both of these events have suitably small probability. □

Theorem 7. *Almost surely, the posts in* $P_{\mathbf{Z},p}$ *form a two-way infinite sequence.*

Proof.   It is obvious that the events $\{i$ is the first post$\}$ have the same probability; since these events are disjoint, the probability has to be 0. Hence the sequence of posts almost surely has no first element. Similarly, there is almost surely no last element, so the sequence of posts is either empty or two-way infinite.

But Lemma 6 tells us that the probability that there is no post is less than C^{-k} for every k, and hence is 0. Therefore the sequence of posts is almost surely two-way infinite.   □

It is now relatively straightforward to prove more about the structure of the partial order. Let $\ldots, U_{-1}, U_0, U_1, \ldots$ denote the random variables giving the positions of the infinite sequence of posts, with (to be definite) U_0 being the first post at or to the right of 0. We call the posets P_j induced on the intervals $(U_j, U_{j+1}]$ the *factors* of the partial order. Then it can be shown that the various distances $U_{j+1} - U_j$ are mutually independent, identically distributed, random variables (except that, since we have chosen our labelling scheme to force $0 \in (U_{-1}, U_0]$, the distribution of $U_0 - U_{-1}$ is slightly different). Moreover, we have $\mathbf{E}(U_{j+1} - U_j)^r < \infty$ for every $r < \infty$ and every j. Furthermore, for any finite partial order P and any integer j, the event that the factor P_j is isomorphic to P is independent of the nature of any other factors. Thus $\Pr(P_0 \cong P)$ and $\Pr(P_{-1} \cong P)$ are probability distributions on the set of all finite partial orders P, and the infinite random partial order can be generated by taking a random partial order for P_{-1} according to the second distribution, random orders for P_i ($i \neq -1$) according to the first, and forming the linear sum.

The passage from the infinite random order to the finite order $P_{n,p}$ is effected by truncating the infinite order, and suitably adjusting the initial and final factors.

Say that a parameter f of partial orders is *additive* if, whenever P is the linear sum of P_1 and P_2, we have $f(P) = f(P_1) + f(P_2)$. A trivial example of an additive parameter is $f(P) = c|P|$ for any constant c: others are the height of the partial order, and the logarithm of the number of linear extensions.

For a random graph order $P_{n,p}$, and an additive parameter f, the random variable $f(P_{n,p})$ is the sum of the independent random variables $f(P_j)$, where the P_j are the factors of the random order. This leads us to suspect that $f(P_{n,p})$ might typically have an asymptotically normal distribution. The situation is however somewhat complicated by the fact that the number of factors is itself a random variable that is *not* independent of the P_j. Also, if $f(P) = c|P|$ for every P, then $f(P_{n,p})$ is, of course, identically equal to cn. Nevertheless, if we exclude this case and impose a moment condition on f, then our intuition is borne out, as shown by the next result. This is essentially from Alon, Bollobás, Brightwell and

Janson [3], although only the special case where f is the logarithm of the number of linear extensions is discussed there. The result is basically a consequence of the theory of stopped random walks: see Gut [25, Chapter 4, Theorem 2.3].

Theorem 8. *Let p be a constant with $0 < p < 1$. Let f be an additive parameter of partial orders P that is not proportional to $|P|$. Let Y and Z be the random variables given by $Y = f(P_0)$ and $Z = f(P_{-1})$. Suppose that $\mathbf{E}(Y^r)$ and $\mathbf{E}(Z^r)$ are finite for all $r \in \mathbf{N}$. Then there are constants $\mu = \mu(p) > 0$ and $\sigma = \sigma(p) > 0$ such that $\mathbf{E}f(P_{n,p})/n \to \mu$ and $\mathrm{Var}f(P_{n,p})/n \to \sigma^2$. Furthermore*

$$\frac{f(P_{n,p}) - \mu n}{\sigma\sqrt{n}} \xrightarrow{d} N(0,1),$$

with convergence of all moments. □

It seems rather hard to find explicit formulae for the expectations $\mu(p)$, for natural additive parameters, but bounds have been found. For $f(P)$ the logarithm of the number of linear extensions, Alon, Bollobás, Brightwell and Janson [3] gave formulae for upper and lower bounds for $\mu(p)$, giving in particular that $.507 < \mu(1/2) < .625$. For $f(P)$ the height, Albert and Frieze [2] proved that $.565 < \mu(1/2) < .610$, and their method can be used to find upper and lower bounds for general $\mu(p)$.

Theorem 8 tells us that the height and the logarithm of the number of linear extensions of $P_{n,p}$ converge in distribution to normal random variables. This does not suffice to give us good bounds for the probability in the tail of the distribution. However, the "method of bounded differences" involving the use of martingale inequalities, can be used to show that, if $f(P_{n,p})$ is either of the two random variables, then

$$\Pr(|f(P_{n,p}) - \mathbf{E}f(P_{n,p})| > \lambda g(p)\sqrt{n}\log n) \leq e^{-c\lambda^2},$$

for some absolute constant c, some function $g(p)$, and λ in a suitable range. For the height, see Albert and Frieze [2]: for the logarithm of the number of linear extensions, see Alon, Bollobás, Brightwell and Janson [3].

For the case of the number of linear extensions, there is one somewhat amusing, and at first sight rather startling, consequence of the above results. The results imply that $\log(L(P_{n,p}))$ is sharply concentrated about its mean $\mu(n,p) = \mu(p)n(1 + o(1))$. Furthermore, if $\omega(n) \to \infty$, then almost surely $L(P_{n,p}) < e^{\mu(p)n + \omega(n)\sqrt{n}}$. However, the expected number of linear extensions is fairly easy to calculate, and turns out to be asymptotic to $\eta(p)p^{-n}$ as $n \to \infty$. For all constant p, it follows from the known bounds on $\mu(p)$ that $e^{\mu(p)} < p^{-1}$. For instance, if $p = 1/2$, the expected number of linear extensions is about 2^n, while the actual number of linear extensions can be shown to be almost surely at most $(1.87)^n$. (In truth, one ought not to be too surprised at this behaviour: the same phenomenon occurs with the product of a fixed number of non-constant independent random variables.)

The situation for the width of $P_{n,p}$ is totally different. Note that the width of a linear sum is the maximum of the widths of the factors. As an antichain in the random order is an independent set in the underlying random graph, the width is almost surely at most $c(p) \log n$, for some $c(p)$ (see, e.g., Bollobás [7, Chapter 11]). As there are almost surely $\Omega(n)$ factors, it is not too surprising to learn that the width is very close to being determined almost surely. The following theorem, due to Barak and Erdős [6] in the case $p = 1/2$, is founded on the result that all largest antichains are almost surely made up of "almost" consecutive elements of $[n]$.

Theorem 9. *Suppose p is a constant with $0 < p < 1$, and $q = 1 - p$.*

$$\text{If } \sqrt{\log n}\left(w - \sqrt{\frac{2 \log n}{\log(1/q)}} - \frac{1}{2}\right) \to \infty, \text{then almost surely}$$

$$W(P_{n,p}) < w.$$

$$\text{If } \sqrt{\log n}\left(w - \sqrt{\frac{2 \log n}{\log(1/q)}} - \frac{1}{2}\right) \to -\infty, \text{then almost surely}$$

$$W(P_{n,p}) \geq w.$$

$\square$

Thus, unless n and p are such that the value $\sqrt{2 \log n / \log(1/q)} + 1/2$ is very close to an integer, the width of $P_{n,p}$ is almost surely equal to the next integer below this value.

The dimension of $P_{n,p}$ is again the maximum of the dimensions of its factors, so one would expect that it too is almost determined. However, this is an open problem. As the dimension of a partial order is at most the width, Theorem 9 gives an upper bound of $\sqrt{2\log n/\log(1/q)} + 1/2 + \epsilon$ for $\dim(P_{n,p})$. Albert and Frieze [2] observed (in the case $p = 1/2$) that this bound is of the right order: the dimension is at least $(1 - \epsilon)\sqrt{\log n/\log(1/pq)}$. They conjecture (again in the case $p = 1/2$) that this latter estimate is correct. However, the following result from Bollobás and Brightwell [14] disproves this. The proof gives the flavour of the subject, as well as being an application of Theorem 3 on the dimension of a random bipartite order.

Theorem 10. *Suppose p is a constant with $0 < p < 1$, $q = 1 - p$, and ϵ is any positive constant. Then*

$$\dim(P_{n,p}) \geq (1 - \epsilon)\sqrt{\frac{\log n}{\log(1/q)}}.$$

Proof. Set $m = (1 - \epsilon/2)\sqrt{\log n/\log(1/q)}$, and $l = \lfloor n/2m \rfloor$. For $i = 1,\ldots,l$, let A_i be the event that the sets $\{2mi + 1, 2mi + 2,\ldots,2mi + m\}$ and $\{2mi + m + 1,\ldots,2mi + 2m\}$ both form antichains in $P_{n,p}$. The events A_i are mutually independent and all have probability $q^{m(m-1)} \geq n^{-1+\epsilon/2}$. Thus, almost surely, at least one of the events A_i occurs.

Let i be the least integer such that A_i occurs, and consider the order induced by $P_{n,p}$ on the set $\{2mi + 1,\ldots,2mi + 2m\}$. This is clearly distributed as a random bipartite order $A_p(X, Y)$, where $|X| = |Y| = m$. By Theorem 3, this order almost surely has dimension at least $(1 - \epsilon/2)m$, as required. $\square$

Taking $p = q = 1/2$ as an example, the bounds given by Albert and Frieze [2] are

$$(1 - \epsilon)\sqrt{\frac{\log n}{2\log 2}} \leq \dim(P_{n,1/2}) \leq (1 + \epsilon)\sqrt{\frac{2\log n}{\log 2}}.$$

Theorem 10 improves the upper bound to $(1 - \epsilon)\sqrt{\log n/\log 2}$: an argument to improve the lower bound to $(1 + \epsilon)\sqrt{4\log n/3\log 2}$ is given in

Bollobás and Brightwell [14]. It would be of some interest to determine the correct value of the constant: my guess is that the bound in Theorem 10 is correct up to lower order terms.

To conclude this section, here is a possible motivation for studying random graph orders with lower edge probabilities. One of the major areas of application for partial orders is that of comparison sorting: one has a (large) set of objects, which is known to have an underlying total order, and one is asked to find this order by means of comparing pairs of elements (i.e., asking whether $x < y$ or $y < x$). The structure one has after making some comparisons is then just a partial order on the set of objects.

Historically, computer scientists have been primarily interested in sorting sequentially (i.e., the answer to each question is known before the next has to be formulated), but there is increasing interest in parallel sorting algorithms, where a batch of questions is asked in one go. In particular, one might well wish to ask batches of, say, $O(n)$ questions, and guarantee to find the order in $O(\log n)$ rounds. Ajtai, Komlós and Szemerédi [1] were the first to show that this is possible, and their proof makes use of random graphs. One might wish to permit even more questions per round, and use even fewer rounds: in this situation, one possible way of approaching the first round is to take a random graph on the set of objects, with the appropriate number of edges, and compare each pair of adjacent elements. The chief virtue of such an approach is that the worst-case performance is likely to be good. However, it is important also to study the average-case performance of such an algorithm, and this amounts to studying a random graph order with an appropriate probability p.

For example, Bollobás and Brightwell [10] proved the following result. For a graph G on a set X of n vertices, and a linear ordering $\prec$ of X, let $t(G, \prec)$ be the number of relations in the partial order $<$ on X defined by treating each edge xy of G with $x \prec y$ as a relation $x < y$, and taking the transitive closure. If $p(n) = \omega(n) \log n \log \log n / (n \log \log \log n)$, where $\omega(n) \to \infty$, then almost every $G = G_{n,p}$ has $t(G, \prec) = \binom{n}{2} - o(n^2)$—in other words, most of the relations of $\prec$ can be deduced from comparisons corresponding to edges of G— for *every* linear order $\prec$. This is best possible: if $p = c \log n \log \log n / (n \log \log \log n)$ for c constant, then for almost every $G = G_{n,p}$ there is some underlying order $\prec$ such that $t(G, \prec) \leq \binom{n}{2} - \delta(c)n^2$. For further details, and other results, see Bollobás and Brightwell [10]. However, if we take a random $G = G_{n,p}$ with $p = \omega(n) \log n / n$ and an independent *random* linear order $\prec$, then almost surely $t(G, \prec) = \binom{n}{2} - o(n^2)$. Indeed, this result is exactly the result that, for such a value of p, the number of relations in a random graph order $P_{n,p}$ is almost surely $\binom{n}{2} - o(n^2)$. (This is not too hard to prove.)

§4. Random k-dimensional orders

For this section, let $k \in \mathbf{N}$ be fixed, and consider the random k-dimensional order $P_k(n)$. Bollobás and Brightwell [13] have recently initiated the study of $P_k(n)$ where k is an increasing function of n, but there is not the space here to discuss their results, which have a rather different flavour from those for k constant.

As already mentioned, the problem of determining the height (or equivalently the width) of $P_2(n)$ was much studied even before random k-dimensional orders were considered as such. Ulam [40] was apparently the first to ask for the expected length of a longest increasing subsequence in a random permutation of $[n]$. As we have seen, this is the same as the expected height of $P_2(n)$. Baer and Brock [5] conjectured that the expectation is about $2\sqrt{n}$, based on experimental data. The first theoretical result was due to Hammersley [24], who showed that the height of $P_2(n)$, divided by $\sqrt{n}$, tends to some constant c in probability. Hammersley [24] and Kingman [27] gave bounds for c. Then Logan and Shepp [30] showed that $c \geq 2$, and the proof was completed when Veršik and Kerov [42] proved that $c \leq 2$. Neither of these last two proofs is at all elementary: a combinatorial proof that $c \leq 2$ was later given by Pilpel [33].

Steele [37] was the first to consider the height of $P_k(n)$ for general k, in a different guise. He conjectured that $H(P_k(n))n^{-2^{1-k}}$ tends to a constant c_k for every k. However, Winkler [43], in the first paper to consider random partial orders in their own right, showed that $H(P_k(n))$ almost surely lies between $a_k n^{1/k}$ and $e n^{1/k}$, for some constant $a_k > 0$. Bollobás and Winkler [14] proved the following result. Some parts of the proof build on the earlier work of Hammersley and Kingman for the $k = 2$ case.

Theorem 11. *There are constants c_k such that $H(P_k(n))n^{-1/k} \to c_k$ in probability, with $c_k < e$ for every k, and $c_k \to e$ as $k \to \infty$.*

Proof. We give a brief sketch only. For convenience, we deal with the Poisson model $P'_k(n)$: the result can be converted into the required one without too much effort.

Fix $\epsilon > 0$. Let $c_k = \sup_n (\mathbf{E}H(P'_k(n))/n^{1/k})$ (if this supremum is finite), and choose an n_0 such that $\mathbf{E}H(P'_k(n_0)) \geq (c_k - \epsilon)n_0^{1/k}$. Now take any $n \geq n_1$, where n_1 is "much" larger than n_0, and consider various disjoint subcubes lying along the diagonal of $[0,1]^k$, each of volume n_0/n. To be more precise, set $d = (n_0/n)^{1/k}$—the jth such subcube has lowest point $((j-1)/d, \ldots, (j-1)/d)$ and highest point $(j/d, \ldots, j/d)$. The height of the longest chain in $[0,1]^k$ is at least the sum of the heights of the longest chains in each of these $\lfloor (n/n_0)^{1/k} \rfloor$ subcubes. But these heights are distributed as $H(P'_k(n_0))$. By the Central Limit Theorem, for a suitably large n_1, the sum of these heights is almost surely at least $\lfloor (n/n_0)^{1/k} \rfloor (c_k - 2\epsilon)n_0^{1/k} \geq (c_k - 3\epsilon)n^{1/k}$. Hence $H(P'_k(n))n^{-1/k} \to c_k$ in probability.

The fact that the supremum c_k is finite, and indeed $c_k \leq e$, follows simply by counting the expected number of chains of length $(e + \epsilon)n^{1/k}$ in $P'_k(n)$. The strict inequality $c_k < e$ requires a slightly more subtle argument. The lower bound for c_k showing that $c_k \to e$ is obtained by constructing a chain "greedily", choosing at each stage an element of the Poisson process above the previous element so as to minimise the sum of the co-ordinates. $\qquad\square$

The determination of the constants c_k for $k > 2$ is still open, and there are no particularly appealing conjectures. It is not even known whether the c_k form an increasing sequence. Work by Richard Silverstein, communicated to me by Peter Winkler, suggests a value for c_3 of about 2.35. Silverstein's experiments used up to about 10^5 points, but even with that many points the suggested expected height is still only about 109, and the second decimal place is not presented with any great confidence.

None of the results mentioned above provides any information about the rate of convergence of $\mathbf{E}H(P_k(n))$, or any estimates for the probability of deviations of the height from $c_k n^{1/k}$. Frieze [23] used martingale methods to show that $H(P_2(n))$ was almost surely within $n^{1/3+\epsilon}$ of its mean. Bollobás and Brightwell [12] improved the exponent to $1/4 + \epsilon$ in the $k = 2$ case, and extended the result to the case of general k. The proof again involves martingale methods.

Theorem 12. *For every integer $k \geq 2$, there is a constant C_k such that, for n sufficiently large,*

$$\Pr\left(|H(P_k(n))| - c_k n^{1/k}| > \frac{\lambda C_k n^{1/2k} \log^{3/2} n}{\log\log n}\right) \leq e^{-\lambda^2}$$

for every λ with $2 < \lambda < n^{1/2k}/\log\log n$. $\qquad\qquad\square$

Theorem 12 can be used to give some lower bounds on the rate of convergence of $H(P_k(n))n^{-1/k}$ to c_k. The result also shows that the variance of $H(P_k(n))$ is at most $n^{1/k} \log^3 n$. Steele [40] showed that the variance of $H(P_2(n))$ is at most $Cn^{1/2}$. It seems natural to conjecture that the variance is about $n^{1/k}$, but no non-trivial lower bounds whatsoever are known.

Other parameters of random k-dimensional partial orders have been rather less studied. Winkler noted in [45] that, for every finite partial order P of dimension at most k, $P_k(n)$ almost surely contains a copy of P. One implication is that the dimension of $P_k(n)$ is almost surely equal to k. In [45], Winkler also provided bounds for the width of $P_k(n)$, and asked about the number of linear extensions. Brightwell [18] proved that the width almost surely lies between $(\sqrt{k}/2 - C)n^{1-1/k}$ and $4kn^{1-1/k}$, and that the number of linear extensions lies between $(e^{-2}n^{1-1/k})^n$ and $(2kn^{1-1/k})^n$. It is also shown in [18] that both the width and the logarithm of the number of linear extensions are sharply concentrated about their means.

One last remark about the model $P_k(n)$ seems in order. If we are going to restrict ourselves to considering only partial orders of dimension k, one might well ask whether we can use the uniform model: give every

k-dimensional partial order on n points the same probability. For $k = 2$, Winkler [49] proved that almost every 2-dimensional order P in this uniform model is the intersection of a unique pair of linear extensions, except that, if x and y are to be consecutive elements in both linear orders, but incomparable, then one may freely choose which linear extension has x above y. The conclusion is that the uniform model and the $\mathcal{P}_2(n)$ model are very closely related. Indeed, for any property $\mathcal{A}$ of partial orders, almost every $P_2(n)$ has $\mathcal{A}$ iff almost every 2-dimensional order in the uniform model has $\mathcal{A}$. One would not expect the situation for $k > 2$ to be quite this benign, but nevertheless one would expect that random k-dimensional orders in the uniform model look very similar to random $P_k(n)$ orders.

Finally, an application. A partial order is said to be *ranked* if all its maximal chains have the same length. Since an n-element partial order has either a chain or an antichain of size $\sqrt{n}$, there is always a ranked suborder of size at least $\sqrt{n}$. But is this best possible? Linial, Saks and Shor [31] improved the lower bound to $\sqrt{2n}$ and also proved that almost every $P_2(n)$ has no ranked suborder of size greater than $4e\sqrt{n}$.

§5. Box spaces

The random k-dimensional order can be formed by taking n points at random in $[0,1]^k$, and taking the order induced on these points. But there is nothing special about the k-dimensional cube: one can make the same definition for any space X which is simultaneously a partial order and a probability space. In order to say anything at all about the resulting random order, it is reasonable to suppose that, for all points x of X, the sets $\{z \in V : x \leq z\}$ and $\{z \in V : x \geq z\}$ are measurable: this implies that the *intervals* $\langle x, y \rangle \equiv \{z : x \leq z \leq y\}$ are measurable. Even so, it is difficult to imagine that anything can be said in general about random orders defined in this way. Bollobás and Brightwell [11] investigated the situation when the following extra conditions are imposed on V: (i) V has a greatest and a least element, (ii) every two intervals $\langle x, y \rangle$ and $\langle u, v \rangle$, both having positive probability, are isomorphic up to a scale factor, (iii) there are some intervals of non-zero probability. (The isomorphism in

(ii) has to preserve the order and, up to the scale factor, the probability measure). They called such spaces *box spaces*. One may check that $[0,1]^k$ is a box space, but there are (at least a few) other examples, which we shall come to shortly.

We state the main result of Bollobás and Brightwell [11] as follows. Let $H_X(n)$ denote the height of the random partial order defined by taking n points at random in the box space X.

Theorem 13. *For every box-space X, there are constants $d \in [1, \infty]$, $m \in [1 + 1/\Gamma(1/d), 4)$ such that $n^{-1/d} H_X(n) \to m$ in probability.* $\qquad\square$

If $d = \infty$, the conclusion should be interpreted to mean that, with probability 1, $H_X(n) = 1$. The constant d is called the *dimension*. The proof of Theorem 13 follows that of Theorem 11, once one has established the existence of a "dimension". Thus the behaviour of $[0,1]^k$ is typical of box spaces in this respect at least. Furthermore the analogue of Theorem 12 (giving stronger convergence results) holds for random orders in general box-spaces as well as for $P_k(n)$.

Another example of a box space is given by considering $\mathbf{R}^{k-1} \times \mathbf{R}$, with the usual measure, and the partial order given by $(\underline{x}, t) \le (\underline{y}, t')$ if $|\underline{x} - \underline{y}| \le t' - t$, where $|\cdot|$ is the Euclidean metric. If $\langle (\underline{x}, t), (\underline{y}, t') \rangle$ is an interval in this space of measure 1, then the interval is a box space. The easiest way to see this is to note that, if $\underline{x}$ is interpreted as a "position" and t as a "time", then this is the causality order on the standard Lorentzian space-time manifold. The required isomorphisms up to a scale-factor are Lorentz transformations combined with expansions. We denote this box-space by M_k: its dimension is indeed k.

Theorem 13 thus applies to M_k, and this result has implications for a model of discrete space-time. Take a space-time manifold, which (at least locally) is Lorentzian, take points in it according to a Poisson process with some (large) fixed density, and look at just the order induced on this set of points. Is the macroscopic structure of the manifold captured by this discrete partial order? This model and question were proposed by Bombelli, Lee, Meyer and Sorkin [16]. One version of the question is to ask whether the distance between two points in the manifold is captured by some parameter of the partial order. If the two points x and y, are timelike (i.e., $x < y$ or $y < x$ in the causality order), then a natural suggestion for the definition of distance in the random order is simply the height of the order restricted to the interval $\langle x, y \rangle$. Theorem 13 says that, over large scales, this height is asymptotically proportional to $(\text{Volume}\langle x, y \rangle)^{1/k}$, which is indeed the required manifold distance. This application, and other related ones, are explored by Brightwell and Gregory [19]. Such discrete models of space-time, preserving the manifold structure on large scales, might be of interest to physicists exactly because the small-scale structure is *not* that of a manifold, so that there is at least the possibility of defining a quantum theory where the behaviour depends dramatically on the scale.

As with random k-dimensional orders, not every finite partial order occurs as a suborder of M_k. Those that do are called $(k-1)$-sphere orders, as they can be represented by a set of balls in $(k-1)$-dimensional space, ordered by inclusion. 2-sphere orders are usually called *circle orders*, and these have attracted a fair amount of attention, see for instance the survey article by Urrutia [43]. Thus the random order in M_3 can be regarded as a "random circle order". One could imagine definitions of "random square orders" for instance, but such random orders do not seem to be obtainable from a box space.

Another model of random orders is that of *random interval orders*. Here one takes n pairs of points (x, y) with $x < y$ uniformly at random in $[0, 1]$, and set $(x, y) \leq (u, v)$ if $y \leq u$. The orders that can arise are called interval orders. We define the space M_2 by taking $\{(x, y) \in [0, 1]^2 : x < y\}$ with the usual measure, and setting $(x, y) \leq (u, v)$ if $y \leq u$. Adding a maximum and a minimum point to M_2 and normalising the measure makes it into a box space of dimension 2, and it is easy to see that random interval orders are just random orders in M_2. Therefore the height of a random interval order is roughly proportional to $\sqrt{n}$—Justicz, Scheinerman and Winkler [28] (and, independently, Bollobás and Brightwell [11]) showed that the height is asymptotic to $2\sqrt{n}/\sqrt{\pi}$. More on random interval orders, including a proof of the delightful fact that the probability of an isolated element in the comparability graph is always exactly 2/3, is to be found in Justicz, Scheinerman and Winkler [28].

There are other ways of generating new box spaces from old, but the class of known examples is still very small. All so far discovered can be derived from the Lorentzian space-time manifolds of various dimensions in one way or another (the cube $[0,1]^k$ for example is the Cartesian product of copies of the 1-dimensional manifold). It would be of great interest to find other examples.

References.

[1] M.Ajtai, J.Komlós and E.Szemerédi, Sorting in $C \log n$ parallel steps, *Combinatorica* **3** (1983) 1–19.

[2] M.Albert and A.Frieze, Random graph orders, *Order* **6** (1989) 19–30.

[3] N.Alon, B.Bollobás, G.Brightwell and S.Janson, Linear extensions of a random partial order, to appear in *Ann. Appl. Prob.*

[4] K.Azuma, Weighted sums of certain dependent random variables, *Tôhoku Math. Journal* **19** (1967) 357–367.

[5] R.M.Baer and P.Brock, Natural sorting over permutation spaces, *Math. Comp.* **22** (1968) 385–510.

[6] A.Barak and P.Erdős, On the maximal number of strongly independent vertices in a random acyclic directed graph, *SIAM J. Algebraic and Disc. Meths.* **5** (1984) 508–514.

[7] B.Bollobás, *Random Graphs*, Academic Press, London, 1985, xv+447pp.

[8] B.Bollobás, Martingales, isoperimetric inequalities and random graphs, in *Combinatorics*, A.Hajnal, L.Lovász and V.T.Sós Eds., Colloq. Math. Sci. Janos Bolyai **52**, North Holland, 1988 pp.113–139.

[9] B.Bollobás, Sharp concentration of measure phenomena in random graphs, in *Random Graphs '87*, M.Karonski, J.Jaworski and A.Rucinski Eds., John Wiley and Sons, New York, 1990, pp.1–15.

[10] B.Bollobás and G.Brightwell, Graphs whose every transitive orientation contains almost every relation, *Israel J. Math.* **59** (1987) 112–128.

[11] B.Bollobás and G.Brightwell, Box-spaces and random partial orders, *Trans. Amer. Math. Soc.* **324** (1991) 59–72.

[12] B.Bollobás and G.Brightwell, The height of a random partial order: concentration of measure, *Ann. Appl. Prob.* **2** (1992) 1009–1018.

[13] B.Bollobás and G.Brightwell, Random high dimensional orders, submitted.

[14] B.Bollobás and G.Brightwell, Random graph orders with small edge-probability, in preparation.

[15] B.Bollobás and P.M.Winkler, The longest chain among random points in Euclidean space, *Proc. Amer. Math. Soc.* **103** (1988) 347–353.

[16] L.Bombelli, J.Lee, D.Meyer and R.D.Sorkin, Spacetime as a causal set, *Phys. Rev. Lett.* **59** (1987) 521.

[17] G.Brightwell, Linear extensions of random orders, to appear in *Discrete Maths*.

[18] G.Brightwell, Random k-dimensional orders: width and number of linear extensions, to appear in *Order*.

[19] G.Brightwell and R.Gregory, Structure of random discrete spacetime, *Phys. Rev. Lett.* **66** (1991) 260–263.

[20] K.J.Compton, The computational complexity of asymptotic problems. I: Partial orders, *Inform. and Computation* **78** (1988) 103–123.

[21] K.J.Compton, 0-1 laws in logic and combinatorics, in *Algorithms and Order*, I.Rival Ed., NATO ASI Series, Kluwer Academic Publishers, Dordrecht, 1989, 353–383.

[22] P.Erdős, H.Kierstead and W.T.Trotter, The dimension of random ordered sets, *Random Structures and Algorithms* **2** (1991) 253–275.

[23] A.M.Frieze, On the length of the longest monotone subsequence in a random permutation, *Ann. Appl. Prob.* **1** (1991) 301–305.

[24] Z.Füredi and J.Kahn, On the dimensions of ordered sets of bounded degree, *Order* **3** (1986) 17–20.

[25] A.Gut, *Stopped Random Walks*, Springer-Verlag, New York, 1988, ix+199pp.

[26] J.M.Hammersley, A few seedlings of research, *Proc. 6th Berkeley Symp. Math. Stat. Prob.*, U. of California Press (1972) 345–394.

[27] W.Hoeffding, Probability inequalities for sums of bounded random variables, *J. Amer. Stat. Assoc.* **27** (1963) 13–30.

[28] J.Justicz, E.Scheinerman and P.M.Winkler, Random intervals, *Amer. Math. Monthly* **97** #10 (December 1990) 881–889.

[29] J.F.C.Kingman, Subadditive ergodic theory, *Ann. Prob.* **1** (1973)

883–909.

[30] D.J.Kleitman and B.L.Rothschild, Asymptotic enumeration of partial orders on a finite set, *Trans. Amer. Math. Soc.* **205** (1975) 205–220.

[31] N.Linial, M.Saks and P.Shor, Largest induced suborders satisfying the chain condition, *Order* **2** (1985) 265–268.

[32] B.F.Logan and L.A.Shepp, A variational problem for Young tableaux, *Advances in Mathematics* **26** (1977) 206–222.

[33] T.Łuczak, First order properties of random posets, *Order* **8** (1991) 291–297.

[34] C.J.H.McDiarmid, On the method of bounded differences, in *Surveys in Combinatorics 1989, Invited Papers at the 12th British Combinatorial Conference*, J.Siemons Ed., Cambridge University Press, 1989, pp.148–188.

[35] S.Pilpel, Descending subsequences of random permutations, *IBM Research Report #52283* (1986).

[36] I.Rival (Ed.) *Proceedings of the Symposium on Ordered Sets*, Reidel Publishing, Dordrecht, 1982.

[37] I.Rival (Ed.) *Graphs and Order*, Reidel Publishing, Dordrecht, 1985.

[38] J.Spencer, Nonconvergence in the theory of random orders, *Order* **7** (1991) 341–348.

[39] J.M.Steele, Limit properties of random variables associated with a partial ordering of $\mathbf{R}^d$, *Ann. Prob.* **5** (1977) 395–403.

[40] J.M.Steele, Talk given at LMS Symposium on Random Methods in Combinatorics, Durham 1991.

[41] W.T.Trotter, *Combinatorics and Partially Ordered Sets: Dimension Theory*, The Johns Hopkins University Press, Baltimore, 1992, xiv+307pp.

[42] S.M.Ulam, Monte Carlo calculations in problems of mathematical physics, in *Modern Mathematics for the Engineer*, E.F.Beckenbach Ed., McGraw Hill, N.Y. (1961).

[43] J.Urrutia, Partial orders and Euclidean geometry, in *Algorithms and Order*, I.Rival Ed., Kluwer Academic Publishers, 1989, pp.387–434.

[44] A.M.Veršik and S.V.Kerov, Asymptotics of the Plancherel measure of the symmetric group and the limiting form of Young tableaux, *Dokl.*

Akad. Nauk. SSSR **233** (1977) 1024–1028.

[45] P.Winkler, Random orders, *Order* **1** (1985) 317–335.

[46] P.Winkler, Connectedness and diameter for random orders of fixed dimension, *Order* **2** (1985) 165–171.

[47] P.Winkler, Recent results in the theory of random orders, in *Applications of Discrete Mathematics*, R.D.Ringeisen and F.S.Roberts Eds., SIAM Publications, Philadelphia, 1988, pp.59–64.

[48] P.Winkler, A counterexample in the theory of random orders, *Order* **5** (1989) 363–368.

[49] P.Winkler, Random orders of dimension 2, *Order* **7** (1991), 329–339.

[50] P.Winkler, Random structures and 0-1 laws, in *Finite and Infinite Combinatorics of Sets and Logic* (R.Woodrow Ed.), NATO Advanced Science Institutes Series, Kluwer Academic Publishers, to appear.

APPLICATIONS OF SUBMODULAR FUNCTIONS

ANDRÁS FRANK*

1993

Dedicated to C.St.J.A. Nash-Williams on the occasion of his 60th birthday.

ABSTRACT Submodular functions and related polyhedra play an increasing role in combinatorial optimization. The present survey-type paper is intended to provide a brief account of this theory along with several applications in graph theory.

1. INTRODUCTION

In 1960 C.St.J.A. Nash-Williams generalized the following easy but pretty result of H.E. Robbins [1939]: *the edges of an undirected graph G can be oriented so that the resulting directed graph $D := \vec{G}$ is strongly connected if and only if G is 2-edge-connected.*

To formulate the generalization let us call a directed graph [undirected graph] k**-edge-connected** if there are k edge-disjoint directed (undirected)) paths from each node to each other.)

WEAK ORIENTATION THEOREM 1.1 [Nash-Williams, 1960] *The edges of an undirected graph G can be oriented so that the resulting directed graph is k-edge-connected if and only if G is $2k$-edge-connected.*

The neccessity of the condition is straightforward and the main difficulty lies in proving its sufficiency. Actually, Nash-Williams proved a stronger result. To formulate it, we need the following notation. Given a

---

* Research Institute for Discrete Mathematics, University of Bonn, Nassestr. 2, Bonn-1, Germany, D-5300. On leave from the Department of Computer Science, Eötvös University, Múzeum krt. 6-8, Budapest, Hungary, H-1088

directed or undirected graph G, let $\lambda(x, y; G)$ denote **local edge-conn-ectivity** from x to y, that is, the maximum number of edge-disjoint paths from x to y.

STRONG ORIENTATION THEOREM 1.2 [Nash-Williams, 1960]
Every undirected graph $G = (V, E)$ has an orientation $D := \vec{G}$ so that

$$\lambda(x, y; D) = \lfloor \lambda(x, y; G)/2 \rfloor \quad \text{holds for every pair of nodes } x, y \in V. \quad (1.1)$$

In addition, the orientation can be chosen such a way that the difference between the in- and out-degree of each node is at most 1.

Nash-Williams calls an orientation satisfying (1.1) **well-balanced.** If graph G is $2k$-edge-connected, then $\lambda(x, y; G) \geq 2k$ for every pair of nodes x, y and hence Theorem 1.2 implies (the non-trivial part of) Theorem 1.1.

Nash-Williams' original proof (for an outline, see also [Nash-Williams, 1969]) used a very sophisticated inductive argument. What he actually proved was a theorem (Theorem 4.1 below), interesting for its own sake, on the existence of a certain pairing of the odd-degree nodes. This pairing theorem easily implies the strong orientation theorem. In the following quotation Theorems 1 and 2 refer to the strong orientation theorem and to this node-pairing theorem. In [1969] Nash-Williams writes:

> *The comparatively complicated nature of the foregoing proof, ... as contrasted with the comparatively simple and natural character of Theorems 1 and 2 might suggest that conceivably the most simple, natural, and insightful proof of those theorems has not yet been found. ... The only (very vague) suggestion I can offer in this direction is to observe that... the proof of Theorem 2 seemed to me to have a somewhat matroid-like flavor. ...I have sometimes wondered whether there might be a way of using matroids, or something like matroids, to give a better and more illuminating proof of our two theorems.*

As far as the weak orientation theorem is concerned Nash-Williams' anticipation was perfectly correct: there are simple proofs for this result using submodular functions and we understand pretty well how this theorem is related to polymatroids and submodular flows, structures with "somewhat matroid-like" features.

One purpose of the present paper is to explain this relationship and its usefulness in detail. A consequence is that several variations of the weak orientation problem, such as the minimum cost and/or degree-constrained k-edge-connected orientation, become tractable. Using this general background, similar questions can be answered if the starting graph is a mixed graph and we are allowed to orient the undirected edges. We shall also point out how orientation problems may help in solving apparently different combinatorial problems.

As far as the strong orientation theorem is concerned it is sad to say that, despite the thirty years passed, not much more new is known. In addition to Nash-Williams' original proof, only W. Mader [1978] offered a different proof. Mader's proof, however, can hardly be considered simpler than the original one. It uses a deep and difficult "splitting" theorem (which was actually the main goal of Mader's paper) and, on top of it, another rather sophisticated argument is required. An encouragement in this direction is that recently a simpler proof of Mader's splitting theorem has been found [Frank, 1992b] that uses submodularity. In Section 4, following in part the approach of Mader's proof, we outline a simplified proof for Nash-Williams' pairing theorem. But I keep feeling that there must be an even more illuminating proof which finally will lead to methods to solve the minimum cost and/or the degree-constrained well-balanced orientation problem.

Orientation problems form one class of applications of submodular functions. Our second purpose is to exhibit the main concepts, results and ideas concerning submodular functions. Beside orientations, we will consider three other areas where techniques using submodular functions have proved extremely useful.

Results concerning the splitting-off operation are exhibited in Section 2. Splitting-off is a general reduction technique that finds applications in the three other applications. Section 5 offers an overview of results on packing and covering with arborescences and trees. Augmentation problems consist of finding a minimum cardinality or cost of new edges whose addition makes a given graph satisfy certain prescribed connectivity properties. This is the content of Section 6. Disjoint paths problems are also a rich source of applications of submodular functions. In a survey paper [Frank, 1990] many of these have been exhibited. Therefore in Section 7 we restrict ourselves only to outline some recent applications.

Section 8 comprises the elements of the theory of submodular functions. Finally, in Section 9 we exhibit a relationship of the general theory with the concrete problems discussed in earlier sections.

We conclude this introductory section by a list of definitions and notation. Given two elements s, t and a subset X of a ground-set U, we say that X is an $s\bar{t}$-set if $s \in X, t \notin X$. X **separates** s **from** t (or s and t) if $|X \cap \{s, t\}| = 1$. A family $\mathcal{F} = \{X_1, \ldots, X_t\}$ of pairwise disjoint subsets of U is called a **sub-partition** of U. If their union is U, $\mathcal{F}$ is a **partition** of U. We do not distinguish between a one-element set and its only element. For a set X and an element x, $X + x$ denotes the union of X and $\{x\}$. Two subsets $X, Y \subseteq U$ are **intersecting** if none of $X \cap Y, X - Y, Y - X$ is empty. If in addition $X \cup Y \neq U$, then X, Y are called **crossing.** X, Y are **co-disjoint** if $X \cup Y \neq U$.

Let $G = (U, E)$ be an undirected graph. $d_G(X, Y)$ denotes the number of undirected edges between $X - Y$ and $Y - X$. $\bar{d}_G(X, Y) := d_G(X, U - Y)(= d_G(U - X, Y)$. $d_G(X)$ stands for $d_G(X, U - X)$. Observe that $\bar{d}_G(X, Y) = \bar{d}_G(U - X, U - Y)$. When it does not cause ambiguity we leave out the subscript. For a directed graph $D = (U, A)$, $\varrho_D(X)$ denotes the number of edges entering X, $\delta_D(X) := \varrho_D(U - X)$ and $\beta_D(X) := \min(\varrho_D(X), \delta_D(X))$. Note that $\beta_D(X) = \beta_D(U - X)$. $d_D(X, Y)$ denotes the number of edges with one end in $X - Y$ and one end in $Y - X$.

By a **mixed graph** $N = (V, E \cup A)$ we mean a graph composed from an undirected graph (V, E) and from a directed graph (V, A).

PROPOSITION 1.3 For $X, Y \subseteq U$,

$$d_G(X) + d_G(Y) = d_G(X \cap Y) + d_G(X \cup Y) + 2d_G(X, Y), \qquad (1.2A)$$

$$d_G(X) + d_G(Y) = d_G(X - Y) + d_G(Y - X) + 2\bar{d}_G(X, Y), \qquad (1.2B)$$

$$\varrho_D(X) + \varrho_D(Y) = \varrho_D(X \cup Y) + \varrho_D(X \cap Y) + d_D(X, Y). \qquad (1.3)$$

An **arborescence** F is a directed tree in which every node but one has in-degree 1 and the exceptional node, called the **root**, is of in-degree 0. (Equivalently, there is a directed path from the root to every other node of F.) By a **branching** we mean a directed forest where each in-degree is at most 1.

For a graph $G = (V, E)$ and $X \subseteq V$, $E_G(X)$ (respectively, $S_G(X)$) denotes the set of edges of G with both end-points (at least one end-point) in X. We call an undirected graph **Eulerian** if each degree is even. A directed graph is **di-Eulerian** if the in-degree of each node is equal to the out-degree.

For a function $m : V \to \mathbf{R}$ we use the notation $m(X) := \sum(m(x) : x \in X)$. For a digraph $D = (V, A)$ and a vector $x : A \to \mathbf{R}$ we write $\varrho_x(X) := \sum(x(a) : a \in A, a \text{ enters } X)$.

2. SPLITTING-OFF

Let M be a mixed graph. **Splitting off** a pair of edges $e = us, f = st$ of M means that we replace e and f by a new edge ut. The resulting mixed graph will be denoted by M^{ef}. This operation is defined only if both e and f are undirected (respectively, directed) and then the newly added edge ut is considered undirected (directed). Accordingly, we speak of undirected or directed splittings.

When a splitting off operation is performed, the local edge-connectivity never increases. The content of the splitting off theorems is that under certain conditions there is an appropriate pair $\{e = us, f = st\}$ of edges whose splitting preserves all local or global edge-connectivity between nodes distinct from s.

These theorems prove to be extremely powerful in attacking connectivity problems. Before exhibiting some of the known splitting off theorems we illustrate their use with a simple example.

THEOREM 2.1 [Lovász, 1974, 1979] *Suppose that in an undirected graph $G = (V + s, E)$*

$$d(X) \geq K = 2k \text{ for every } \emptyset \neq X \subset V \qquad (2.1)$$

where s is a given node of even degree. Then for every edge $f = st$ there is an edge $e = su$ so that $\{e, f\}$ can be split off without violating (2.1).

Proof. Lovász proved this result for any integer K but we need it only for even K and the proof in this case is simpler. Call a set $\emptyset \neq X \subset V$ **dangerous** if $d(X) \leq K+1$. A pair of edges $\{e, f\}$ is said to be **splittable** if their splitting off does not destroy (2.1). This is equivalent to saying that (*) there is no dangerous set X separating u, t from s.

CLAIM *The union of two crossing dangerous $t\bar{s}$-sets is dangerous.*

Proof. Since $f = st \in E$, by (1.2) we have $(K+1) + (K+1) \geq d(X) + d(Y) = d(X-Y) + d(Y-X) + 2\bar{d}(X,Y) \geq K + K + 2$. Hence $K+1 = d(X) = d(Y), K = d(X-Y) = d(Y-X)$ and $\bar{d}(X,Y) = 1$. From this $d(X, V-(X \cup Y)) = d(Y, V-(X \cup Y))$ follows, for otherwise if, say, the left-hand side were smaller, then we would have $d(Y-X) < d(X-Y) = K$, contradicting (2.1). Therefore $d(X \cup Y) = 2d(X, V-(X \cup Y))+1$, an odd number. Hence $X \cup Y \neq V$.

Suppose indirectly that $X \cup Y$ is not dangerous, that is, $d(X \cup Y) \geq K + 2$. Here we must have strict inequality since $d(X \cup Y)$ was shown to be odd. Applying (1.2) we get $(K+1) + (K+1) = d(X) + d(Y) \geq d(X \cap Y) + d(X \cup Y) \geq K + (K+3)$ and this contradiction proves the claim. ♠

From the Claim it follows that the union M of all dangerous $t\bar{s}$-sets is dangerous. Now there is an edge $e = su$ with $u \notin M$ since otherwise $d(V-M) = d(M+s) < d(M) - 2 \leq K - 1$ contradicting (2.1). By the construction of M and by (*) e, f are splittable. ♠♠♠

Proof of the Weak Orientation Theorem. Assume that G is $2k$-edge-connected. We use induction on the number of nodes. We may assume that G has at least two nodes and is minimal with respect to this property, that is, $G - e$ is not $2k$-edge-connected for every edge e of G. It is an easy exercise to see that such a graph contains a node s of degree $2k$. By applying k times Theorem 2.1, we obtain a $2k$-edge-connected graph G' that has one less node. By induction G' has a k-edge-connected orientation and this orientation determines a k-edge-connected orientation of G. ♠

The proof of Theorem 2.1 is a prototype for proofs of splitting off theorems. An extension was used [Frank, 1992b] to derive W. Mader's (undirected) splitting off theorem which is a generalization of L. Lovász'. In what follows $U = V + s$ will denote the node set of the graphs in question.

THEOREM 2.2 [Mader, 1978] *Let* $G = (V + s, E)$ *be a (connected) undirected graph in which* $0 < d_G(s) \neq 3$ *and there is no cut-edge incident with* s. *Then there exists a pair of edges* $e = su, f = st$ *so that* $\lambda(x, y; G) = \lambda(x, y; G^{ef})$ *holds for every* $x, y \in V$. ♠

Recently we proved [Bang-Jensen, Frank and Jackson, 1993] an extension of Mader's theorem to mixed graphs. Let $M = (V + s, A \cup E)$ be a mixed graph composed from a digraph $D = (V + s, A)$ and an undirected graph $G = (V + s, E)$ so that s is incident only with undirected edges.

For $x, y \in U$ and for integer $k \geq 2$ let us define

$$r_M(x, y) := \min(k, \lambda(x, y; M)) \tag{2.2}$$

Let

$$T(M) := \{x \in V : \varrho_M(x) \neq \delta_M(x)\}, \tag{2.3}$$

that is, $T(M)$ is the set of non-di-Eulerian nodes of M. (If M is an undirected graph, then every node is di-Eulerian). Assume that

$$\lambda(x, y; M) \geq k \text{ for every } x, y \in T(M). \tag{2.4}$$

With this notation we see that $\lambda(x, y; M) \geq r_M(x, y)$ for every pair of nodes x, y.

THEOREM 2.3 *Suppose that in* $M = (V + s, A \cup E)$ *node* s *is incident only with undirected edges,* $0 < d_M(s) \neq 3$, *and*

$$\text{there is no cut-edge incident with } s. \tag{2.5}$$

Let $k \geq 2$ *be an integer satisfying (2.4). Then there is a pair of edges* $e = su, f = st$ *so that*

$$\lambda(x, y; M^{ef}) \geq r_M(x, y) \text{ for every } x, y \in V. \tag{2.6}$$

An equivalent form of Theorem 2.3 is as follows.

THEOREM 2.3A *In a mixed graph* $M = (V + s, A \cup E)$ *a node* s *is incident only with undirected edges,* $d(s)$ *is even, and (2.5) holds. Let* $k \geq 2$ *be an integer satisfying (2.4). Then the set of edges incident with* s

can be matched into $d(s)/2$ disjoint pairs so that $\lambda(x,y;M^+) \geq r_M(x,y)$ for every $x,y \in V$ where M^+ denotes the mixed graph arising from M by splitting off all these pairs.

COROLLARY 2.4 Suppose that in a mixed graph $M = (V+s, A \cup E)$ node s is incident only with undirected edges, $0 < d(s) \neq 3$, and there is no cut-edge incident with s. Let $k \geq 2$ be an integer so that $\lambda(x,y;M) \geq k$ for every $x,y \in V$. Then there is a pair of edges $e = su$, $f = st$ so that $\lambda(x,y;M^{ef}) \geq k$ for every $x,y \in V$. ♠

COROLLARY 2.5 Suppose that in a mixed graph $M = (V+s, A \cup E)$ node s is incident only with undirected edges, $0 < d(s) \neq 3$, there is no cut-edge incident with s, and $\varrho_M(v) = \delta_M(v)$ for every node $v \in V$. Then there is a pair of edges $e = su$, $f = st$ so that $\lambda(x,y;M^{ef}) = \lambda(x,y;M)$ for every $x,y \in V$. ♠

Note that already this corollary is a generalization to Mader's Theorem 2.2. Now let us consider results concerning directed splittings. The following important result is also due to W. Mader [1982]:

THEOREM 2.6 Let $D = (V+s, A)$ be a directed graph for which $\lambda(x,y;D) \geq k$ for every $x,y \in V$ and $\varrho(s) = \delta(s)$. Then for every edge $f = st$ there is an edge $e = us$ so that $\lambda(x,y;D^{ef}) \geq k$ for every $x,y \in V$. ♠

By repeated applications we get:

THEOREM 2.6A Let $D = (V+s, A)$ be a directed graph for which $\lambda(x,y;D) \geq k$ for every $x,y \in V$ and $\varrho(s) = \delta(s)$. Then the edges entering and leaving s can be partitioned into $\varrho(s)$ pairs so that splitting off all these pairs leaves a k-edge-connected digraph. ♠

It would be tempting to extend theorem 2.6 so as to preserve local edge-connectivities as well, analogously to the situation with undirecting splittings. Such an extension, however, is possible only for di-Eulerian digraphs.

The next theorem was proved by A. Frank [1989] and B. Jackson [1988]:

THEOREM 2.7 *Let $D = (V + s, A)$ be a di-Eulerian digraph, that is, $\varrho(x) = \delta(x)$ for every node x of D. Then for every edge $f = st$ there is an edge $e = us$ so that $\lambda(x, y; D^{ef}) = \lambda(x, y; D)$ for every $x, y \in V$.* ♠

In [Bang-Jensen, Frank and Jackson, 1993] we give a common generalization of these two theorems.

THEOREM 2.8 *Let $M = (V + s, A \cup E)$ be a mixed graph and k an integer ≥ 1 satisfying (2.4). Assume that s is incident only with directed edges and $\varrho_M(s) = \delta_M(s) > 0$. Then for every edge $f = st$ there is an edge $e = us$ so that*

$$\lambda(x, y; M^{ef}) \geq r_M(x, y) \text{ for every } x, y \in V \tag{2.6}$$

If $M = D$ is directed graph and $\lambda(x, y; D) \geq k$ for every $x, y \in V$, then $r_M(x, y) = k$ and we are back at Theorem 2.6. If $M = D$ is a directed Euleraian graph and $k := \max(\lambda(x, y; D) : x, y \in V)$, then we are back at Theorem 2.7.

An equivalent form is:

THEOREM 2.8A *Let $M = (V + s, A \cup E)$ be a mixed graph graph satisfying (2.4). Assume that s is incident only with directed edges and $\varrho_M(s) = \delta_M(s)$. Then the edges entering and leaving s can be matched into $\varrho_M(s)$ disjoint pairs so that $\lambda(x, y; M^+) \geq r_M(x, y)$ for every $x, y \in V$ where M^+ denotes the mixed graph arising from M by splitting off all these pairs.*

In Sections 6 and 7 we will show how these splitting off theorems may be applied to obtain results concerning edge-connectivity augmentation problems and disjoint paths problems, respectively.

3. ORIENTATIONS

In the preceding section we saw a proof of the weak orientation theorem of Nash-Williams. Here we provide an overview on orientation results.

In Sections 5 and 7 some applications of the orientation techniques will be shown.

Let us start with orientations satisfying upper and lower bounds on the in-degrees. Let $G = (V, E)$ be an undirected graph and $f : V \to \mathbf{Z}_+$, $g : V \to \mathbf{Z}_+ \cup \{+\infty\}$ two functions so that $f \leq g$.

THEOREM 3.1 (a) *There exists an orientation of G whose in-degree function ϱ satisfies*

$$\varrho(v) \geq f(v) \text{ for every } v \in V \qquad (3.1a)$$

if and only if

$$|S(X)| \geq f(X) \text{ for every } X \subseteq V. \qquad (3.2a)$$

(b) *There exists an orientation of G for which*

$$\varrho(v) \leq g(v) \text{ for every } v \in V \qquad (3.1b)$$

if and only if

$$|E(X)| \leq g(X) \text{ for every } X \subseteq V. \qquad (3.2b)$$

(c) *There exists an orientation of G satisfying both (3.1a) and (3.1b) if and only if there is one satisfying (3.1a) and there is one satisfying (3.1b) (or equivalently, both (3.2a) and (3.2b) hold).*

Proof. First we prove (a). If the desired orientation exists, then for every set $X \subseteq V$ we have $f(X) \leq \sum(\varrho(v) : v \in V) = |S(X)| - \delta(X) \leq |S(X)|$ and the necessity of (3.2a) follows.

Suppose now that (3.2a) holds. Start with any orientation of G. If $\varrho(v) \geq f(v)$ for every $v \in V$, we are done. So let s be a "bad" node for which $\varrho(s) < f(s)$. Let X be the set of nodes reachable from s in the given orientation of G. There is a node $t \in X$ with $\varrho(t) > f(t)$ since otherwise $f(X) > \sum(\varrho(v) : v \in X) = |S(X)| - \delta(X) = |S(X)|$ contradicting (3.2a). Choose any directed path from s to t and reverse the orientation of its edges. By this operation the sum $\sum(f(v) - \varrho(v) : f(v) > \varrho(v))$ is reduced; therefore after at most $|E|$ such changes we get a desired orientation.

To see part (b), apply part (a) to $f(v) := d(v) - g(v)$ $(v \in V)$. Then (3.2a) transforms into (3.2b) and thereby there exists an orientation with $\varrho(v) \geq d(v) - g(v)$ or equivalently $\delta(v) \geq g(v)(v \in V)$. Reorienting every edge we get the required orientation.

To prove part (c) modify slightly the proof of Part (a) as follows. Because of (3.2b) there is an orientation satisfying $\varrho(v) \leq g(v)$ for every $v \in V$. In the proof of Part (a) start with such an orientation and observe that in the reorientation procedure $\varrho(v)$ can be increased only if $\varrho(v) < f(v)(\leq g(v))$. Therefore the final orientation satisfies both the lower and the upper bound requirements. ♠♠♠

We hasten to draw the attention to the phenomenon occuring in the third part of this theorem: we call it the **linking principle.** An earlier occurance of the linking principle is due to A.L. Dulmage and N.S. Mendelsohn [1959] who proved that there is a matching in a bipartite graph $(A, B; E)$ covering two specified subsets $X \subseteq A, Y \subseteq B$ if and only if there is a matching covering X and there is a matching covering Y. We are going to encounter many other examples of the linking principle and show that all these results follow from a theorem concerning polymatroids.

Theorem 3.1 is easy but has important consequences. For example, Hall's theorem on the existence of a perfect matching in a bipartite graph $G = (A, B; E)$ can easily be derived if we observe that a perfect matching of G corresponds to an orientation of G in which every in-degree in A is 1 and the in-degree of every node $v \in B$ is $d_G(v) - 1$. The following theorem, which first appeared in [Ford and Fulkerson, 1962], is also an easy consequence.

THEOREM 3.2 *Suppose that a mixed graph $M = (V, E \cup A)$ is composed from a directed graph $D = (V, A)$ and from an undirected graph $G = (V, E)$. The undirected edges of M can be oriented so that the resulting digraph is di-Eulerian if and only if every $d_G(v) + \varrho_D(v) + \delta(v)$ is even for every node v and*

$$d_G(X) \geq \varrho_D(X) - \delta_D(X) \text{ for every } X \subseteq V. \qquad (3.3)$$

Nash-Williams' weak orientation theorem answers the question of when a graph can be oriented so as to obtain a k-edge-connected digraph. What if we require less and want to have an orientation in which only from a specified node are there k edge-disjoint paths to every other node? The next result was proved in [Frank, 1978].

THEOREM 3.3 *Given an undirected graph $G = (V, E)$ with a specified node s, G has an orientation for which $\varrho(X) \geq k$ for every $X \subseteq V - s$ if and only if*

$$e_{\mathcal{F}} \geq k(t-1) \qquad\qquad (3.4)$$

holds for every partition $\mathcal{F} = \{X_1, \ldots, X_t\}$ of V where $e_{\mathcal{F}}$ denotes the number of edges connecting different members of $\mathcal{F}$.

In Section 5 we will point out an interesting relationship between Theorem 3.3, Tutte's theorem [1961] on disjoint trees and Edmonds' theorem on disjoint arborescences [1973].

Robbins' theorem tells us when a strongly connected orientation of a graph exists. From Theorem 3.1 we know a necessary and sufficient condition for the existence of an orientation satisfying degree constraints. It is tempting to try to combine these requirements. This task was accomplished in [Frank and Gyárfás, 1976].

Let $G = (V, E)$ be a 2-edge-connected undirected graph and $f : V \to \mathbf{Z}_+$, $g : V \to \mathbf{Z}_+ \cup \{+\infty\}$ two functions so that $f \leq g$. For $X \subseteq V$ let $c(X)$ denote the number of components of $G - X$.

THEOREM 3.4 (a) *There exists a strongly connected orientation of G whose in-degree function ϱ satisfies*

$$\varrho(v) \geq f(v) \text{ for every } v \in V \qquad\qquad (3.5a)$$

if and only if

$$|S(X)| - c(X) \geq f(X) \text{ for every } \emptyset \neq X \subseteq V. \qquad (3.6a)$$

(b) *There exists a strongly connected orientation of G for which*

$$\varrho(v) \leq g(v) \text{ for every } v \in V \qquad\qquad (3.5b)$$

if and only if

$$|E(X)| + c(X) \leq g(X) \text{ for every } \emptyset \neq X \subseteq V. \qquad (3.6b)$$

(c) *There exists a strongly-connected orientation of G satisfying both (3.5a) and (3.5b) if and only if there is one satisfying (3.5a) and there is one satisfying (3.5b).*

Naturally we may be interested in finding k-edge-connected orientations satisfying degree constraints. Such a theorem as well as all the foregoing orientation theorems of this section are consequences of the following general result [Frank, 1980]. Let $G = (V, E)$ be again an undirected graph. Let $h : 2^V \to \mathbf{Z}+$ be a non-negative integer valued set-function with $h(V) = h(\emptyset) = 0$. We say that h is **fully G-supermodular** if

$$h(X) + h(Y) \leq h(X \cup Y) + h(X \cap Y) + d(X, Y) \qquad (3.7)$$

for every pair $\{X, Y\}$ of subsets of V. If (3.7) is satisfied only by intersecting (respectively, crossing) pair of subsets X, Y, then h is called **intersecting (crossing) G-supermodular.** We call h **symmetric** if $h(X) = h(V - X)$ for every $X \subseteq V$.

THEOREM 3.5 *Let h be a non-negative crossing G-supermodular set-function. G has an orientation satisfying*

$$\varrho(X) \geq h(X) \text{ for every } X \subseteq V \qquad (3.8)$$

if and only if both

$$e_{\mathcal{F}} \geq \sum h(V_i) \qquad (3.9a)$$

and

$$e_{\mathcal{F}} \geq \sum h(\bar{V}_i) \qquad (3.9b)$$

hold for every partition $\mathcal{F} = \{V_1, \ldots, V_t\}$ of V where $\bar{V}_i = V - V_i$ and $e_{\mathcal{F}}$ denotes the number of edges connecting different sets V_i $(i = 0, 1, \ldots, t)$. If h is intersecting G-supermodular, then (3.9a) alone is necessary and sufficient. If h is fully G-supermodular or if h is symmetric crossing supermodular, then it suffices to require (3.9) only for partitions of two parts.

The following corollary is an extension of the weak orientation theorem of Nash-Williams. It is another occurance of the linking principle. Let $f : V \to \mathbf{Z}_+$, $g : V \to \mathbf{Z}_+ \cup \{+\infty\}$ be two functions so that $f \leq g$.

THEOREM 3.6 (a) *There exists a k-edge-connected orientation of G whose in-degree function ϱ satisfies*

$$\varrho(v) \geq f(v) \text{ for every } v \in V \qquad (3.10a)$$

if and only if

$$e_{\mathcal{F}} + |E(V_0)| \geq kt + f(V_0) \qquad (3.11a)$$

holds for every partition $\mathcal{F} = \{V_0, V_1, \ldots, V_t\}$ of V where only V_0 maybe empty.

(b) *There exists a k-edge-connected orientation of G whose in-degree function ϱ satisfies*

$$\varrho(v) \leq g(v) \text{ for every } v \in V \qquad (3.10b)$$

if and only if

$$e'_{\mathcal{F}} - |E(V_0)| \geq kt - f(V_0) \qquad (3.11b)$$

holds for every partition $\mathcal{F} = \{V_0, V_1, \ldots, V_t\}$ of V where only V_0 may be empty and $e'_{\mathcal{F}}$ denotes the number of edges connecting different members V_i of $\mathcal{F}$ $(i = 1, 2, \ldots, t)$.

(c) *There exists a k-edge-connected orientation of G satisfying both (3.10a) and (3.10b) if and only if there is one satisfying (3.10a) and there is one satisfying (3.10b).*

Theorem 3.6 may be derived from Theorem 3.5 if we define h as follows. For $X \subset V$ let $h(X) :\equiv k$ if $2 \leq |X|, |V - X|$. If $X = \{v\}$ for some $v \in V$, let $h(X) := f(v)$. If $X = \{V - v\}$ for some $v \in V$, let $h(X) := d(v) - g(v)$. The derivation consists of showing that for this choice of h (3.8) is equivalent to (3.10) and that (3.9) is equivalent to (3.11).

In Theorem 3.5, set-function h is required to be non-negative. This apparently natural and harmless requirement imposes however a real restriction in applications. For example, Boesch and Tindell [1980], extending Robbins' theorem, proved that the undirected edges of a mixed graph $M = (V, E \cup A)$ can be oriented so as to get a strongly connected digraph if and only if there are no cut-edges and directed cuts in M. This is a rather simple result but it does not seem to be the consequence of Theorem 3.5. (We note that a simple greedy-type procedure finds the desired orientation: consider the undirected edges in an arbitrary order and orient them in such a way that no directed cut arises. It can be shown that among the two possible orientations of the current edge at least one will always do).

More generally, we can pose the problem of finding a necessary and sufficient condition for the existence of an orientation of a mixed graph that is k-edge-connected. Suppose that M is composed from a digraph $D = (V, A)$ and from an undirected graph $G = (V, E)$. Let $h(X) := k - \varrho_D(X)$ for $\emptyset \neq X \subset V$ and $h(\emptyset) = h(V) = 0$. Clearly, the orientation problem is equivalent to finding an orientation of the undirected graph $G = (V, E)$ so that the in-degree function ϱ satisfies $\varrho(X) \geq h(X)$ for every $X \subseteq V$. This function h is crossing G-supermodular (actually, crossing supermodular) but Theorem 3.5 cannot be applied since h is not necessarily non-negative. However we have the following more complicated characterization.

Let $G = (V, E)$ be again an undirected graph. Let $h : 2^V \to \mathbf{Z} \cup \{-\infty\}$ be an integer-valued **crossing G-supermodular** set-function with $h(V) = h(\emptyset) = 0$.

THEOREM 3.7 G has an orientation satisfying

$$\varrho(X) \geq h(X) \text{ for every } X \subseteq V \tag{3.12}$$

if and only if

$$s_t \geq \sum_{i=1}^{t} \left(\sum_j h(V_i^j) - e_i \right) \tag{3.13}$$

for every sub-partition $\{V_1, V_2, \ldots, V_t\}$ of V where each V_i is the intersection of a family of pairwise co-disjoint sets $V_i^1, V_i^2, \ldots$, s_t denotes the number of edges entering a V_i, and for a given i e_i denotes the number of edges connecting different sets V_i^j $(j = 1, 2, \ldots)$.

This theorem, as well as its min-cost version, can be proved with the help of submodular flows. We will return to it in Section 8. With the help of this theorem one can derive a necessary and sufficient condition for the existence of an orientation in a mixed graph so that the resulting graph is k-edge-connected and satisfies degree-constraints.

When non-negativity is left out from the hypotheses of Theorem 3.5 the statement is not true anymore. What if we do not require h to be integer-valued? The following counter-example shows that neither can this be be done. Let G be the Petersen graph and let $h(X) := d_G(X)/4$. It can be shown that h is G-supermodular, symmetric and satisfies (3.9) but no orientation of G satisfying (3.9), exists. But why on earth would we be interested in non-integer-valued functions? Because the 4-color problem can be formulated this way. Namely, G. Minty proved that the nodes of a graph G can be coloured by k colours if and only if there is an orientation of edges so that every circuit C of G contains at least $|C|/k$ edges in both directions. Applying this to the dual of a planar graph we obtain that a planar graph G is four-colourable if and only if there is an orientation of the edges of G so that the in-degree $\varrho(X) \geq h(X) := d(X)/4$ holds for every subset X of nodes. For a 2-edge-connected graph such an h satisfies the hypotheses of Theorem 3.6.

4. THE PAIRING THEOREM OF NASH-WILLIAMS

Let us turn now to the proof of Nash-Williams' strong orientation theorem. The proof we describe below is not revolutionary new since it uses ideas from Nash-Williams' and Mader's proofs, as well as from my proof of Mader's splitting-off theorem. These steps, however, are combined in a slightly more economical way and hence the present proof is perhaps simpler than its predecessors. But the main reason why this proof has been used here is that I want to encourage readers to find a really simple proof and an ultimate answer to Nash-Williams' hopes cited in Section 1. And, also, one should not forget that the problem of min-cost and/or degree-constrained well-balanced orientation is still open.

Proof of Theorem 1.2. The starting idea of Nash-Williams' approach is an observation that the theorem is trivial for Eulerian graphs. Indeed, an Eulerian graph always has a orientation so that the resulting directed graph is di-Eulerian and this orientation satisfies (1.1). If the graph is not Eulerian, then first, as Nash-Williams argues, let us try to make it Eulerian by adding a suitable matching M on the subset of nodes of odd degrees, second, find an Eulerian orientation of the resulting Eulerian graph, and finally leave out the new edges. Naturally, the resulting orientation of the original graph can be expected to satisfy (1.1) only if the auxiliary matching M fulfills certain requirements. To formulate these we need the following notation.

For any integer or integer-valued function f let $\hat{f} := 2\lfloor f/2 \rfloor$. Let us define a set-function $R = R_G$ as follows. $R(\emptyset) := R(V) := 0$ and for $\emptyset \subset X \subset V$ let $R(X) := \max(\lambda(x,y;G) : X$ separates $x,y)$. We note that function R was successfully used in a simple proof of Mader's splitting of theorem [Frank, 1992b] as well as in solving augmentation problems

[Frank, 1992a]. By the directed edge-version of Menger's theorem, (1.1) is equivalent to the requirement that

$$\varrho(X) \geq \hat{R}_G(X)/2 \text{ holds for every } X \subseteq V. \tag{4.1}$$

Let M be a matching that pairs the nodes of G of odd-degree. We call M a **feasible odd-node pairing** if

$$d_M(X) \leq b_G(X) := d_G(X) - \hat{R}_G(X) \text{ holds for every } X \subseteq V \tag{4.2}$$

where $d_M(X)$ denotes the number of elements of M with precisely one end-node in X. Clearly, $b_G \geq 0$ and $d_M(X) \equiv b_G(X) \pmod{2}$ for every $X \subseteq V$. Hence (4.2) always holds if $|X|$ or $|V - X|$ is at most one. We call such an X **trivial**.

The pairing theorem of Nash-Williams is as follows.

PAIRING THEOREM 4.1 *Every undirected graph has a feasible odd-node pairing M.*

A key observation is that Theorem 4.1 easily implies the Strong Orientation Theorem. Indeed, consider an Eulerian orientation D' of the Eulerian graph $G + M$. Let ϱ' and ϱ denote the in-degree functions of D' and $D := D' - M$, respectively. Note that D is an orientation of G. We have $\varrho(X) \geq \varrho'(X) - d_M(X) = (d_G(X) + d_M(X))/2 - d_M(X) = (d_G(X) - d_M(X))/2 \geq \hat{R}_G(X)/2$, from which (4.1) and Theorem 1.2 follows.

Proof of Theorem 4.1. We may assume that G is connected. We further assume that G is 2-edge-connected as if this is not the case, then replace each cut-edge by three parallel edges. It can easily be checked that a feasible pairing of the resulting graph is a feasible pairing of G. We use induction on $|E| + |V|$. The following easy lemma may be proved directly using the definition of $\hat{R}$.

LEMMA 4.2 *For any $X, Y \subseteq V$ at least one of the following four statements holds.*

$$\hat{R}(X) \leq \hat{R}(X \cup Y) \text{ and } \hat{R}(Y) \leq \hat{R}(X \cap Y), \tag{4.3a}$$

$$\hat{R}(Y) \le \hat{R}(X \cup Y) \text{ and } \hat{R}(X) \le \hat{R}(X \cap Y), \qquad (4.3b)$$

$$\hat{R}(X) \le \hat{R}(X - Y) \text{ and } \hat{R}(Y) \le \hat{R}(Y - X), \qquad (4.3c)$$

$$\hat{R}(Y) \le \hat{R}(X - Y) \text{ and } \hat{R}(X) \le \hat{R}(Y - X). \qquad (4.3d)$$

This immediately implies:

LEMMA 4.3 For any $X, Y \subseteq V$ at least one of the following two ineqal-ities holds.

$$\hat{R}(X) + \hat{R}(Y) \le \hat{R}(X \cap Y) + \hat{R}(X \cup Y), \qquad (4.4\alpha)$$

$$\hat{R}(X) + \hat{R}(Y) \le \hat{R}(X - Y) + \hat{R}(Y - X). \qquad (4.4\beta)$$

Moreover, if

$$R(X) \le \min(R(Y), R(X \cap Y), R(X \cup Y)), \qquad (4.5)$$

then (4.4α) holds. ♠

Assume first that there is a non-trivial set $X \subset V$ for which $b_G(X) = 0$. Let G_1 and G_2 denote the graphs arising from G by contracting X and $(V - X)$, respectively, into one node. It is easy to see that, for $Z \subseteq V - X$, $R_{G_1}(Z) \ge R_G(Z)$ and hence $b_{G_1}(Z) \le b_G(Z)$. Similarly $b_{G_2}(Z) \le b_G(Z)$ for $Z \subseteq X$. By induction there is a feasible odd-node pairing M_i of G_i $(i = 1, 2)$. We claim that the odd-node pairing $M := M_1 + M_2$ of G is feasible, that is, $d_M(Y) \le b_G(Y)$ holds for an arbitrary subset $Y \subseteq V$. To see this we may assume that (4.4α) holds since otherwise we can replace Y by its complement and then (4.4β) transforms into (4.4α). Using (1.2) and that $\hat{R}$ is symmetric we get $d_M(Y) = d_M(X \cap Y) + d_M(X \cup Y) = d_{M_2}(X \cap Y) + d_{M_1}(V - (X \cup Y)) \le b_{G_2}(X \cap Y) + b_{G_1}(X \cup Y) \le b_G(X \cap Y) + b_G(X \cup Y) \le b_G(X) + b_G(Y) = b_G(Y)$, as required.

Therefore, we are at the following case:

$$b_G(X) > 0 \text{ for every non-trivial } X \subset V. \qquad (4.6)$$

Assume now that there is an edge $f = uv$ of G connecting two nodes of odd degree and let $G' := G - f$. By induction there is a feasible odd-node pairing of M'. Suppose that there are nodes x, y for which $\hat{\lambda}(x, y; G') < \hat{\lambda}(x, y; G)$. Then $\lambda(x, y; G') = \lambda(x, y; G) - 1$ and $\lambda(x, y; G)$ is even. By

Menger's theorem there is an $x\bar{y}$-set X for which $d_G(X) = \lambda(x, y; G)$ and X separates u and v. By taking the complement, if necessary, we may assume that $|X| \leq |V - X|$. Now $b_G(X) = 0$ and hence $|X| = 1$, that is, $X = \{u\}$ or $X = \{v\}$. But this contradicts the assumption that $d(u)$ and $d(v)$ are odd.

Hence $\hat{\lambda}(x, y; G') = \hat{\lambda}(x, y; G)$ holds for every $x, y \in V$. Then $b_{G'}(X) = b_G(X) - 1$ (resp., $b_{G'}(X) = b_G(X)$) for every $X \subseteq V$ separating (non-separating) u and v, and hence $M' + f$ is a feasible pairing.

Alternatively, if

$$\text{one end of every edge is of even degree,} \qquad (4.7)$$

let T denote the set of nodes with degree three and let $S = V - T$. Let us call a subset $X \subseteq V$ **non-essential** if X is trivial or there is a node $v \in T \cap X$ for which (*) $d(v, X - v) \leq 1$ or there is a node $v \in T - X$ for which (**) $d(v, X) \geq 2$. Otherwise X is called **essential**. We claim that a pairing M of odd nodes is feasible if (4.2) holds for essential sets. Indeed, (4.2) holds for trivial sets. For a non-trivial, non-essential set X let $\tilde{X} \subseteq T$ be the set of nodes satisfying (*) and (**). Then for $X' := X \oplus \tilde{X}$ (symmetric difference) we have $\hat{R}(X') = \hat{R}(X)$, $d(X') \geq d(X) + |\tilde{X}|$ and $d_M(X') \leq d_M(X) + |\tilde{X}|$. Since X' is essential, (4.2) follows for X'.

Let s be a node of S with minimum degree. If S has only one element, then s is connected to every node $t \in T$ by three parallel edges and G has no any other edge. In this case $d_G(s)$ is even and any pairing of the odd nodes is feasible. So let $|S| \geq 2$. Let $\lambda_S := \min \lambda(x, y; G) : x, y \in S)$. Clearly, $\lambda_S \leq d_G(s)$.

CASE 1

$$\lambda_S = d_G(s). \qquad (4.8)$$

By Mader's Theorem (2.2) there is a pair of edges $e = su, f = st$ whose splitting results in a graph G' for which $\lambda(x, y; G') = \lambda(x, y; G)$ for every $x, y \in V - s$.

CLAIM $\hat{R}_{G'}(X) = \hat{R}_G(X)$ for any essential set X.

Proof. By taking the complement if necessary, we may assume that $s \in X$. Since X is essential, there is an element $y \in S - X$. We cannot have $X \cap S = \{s\}$ for otherwise $d_G(s) \geq \lambda_S \geq \lambda(x, y; G) \geq d_G(X)$ and hence, by (4.7), $d_G(t, V - X) \geq 2$ for any $t \in T$, that is, X would not be essential. So there is an element $x \in S \cap X - s$. Now $\lambda(x, y; G') = \lambda(x, y; G) \geq d_G(s)$ and the claim follows. ♠

By induction there is a pairing M of odd nodes feasible with respect to G'. Since $d_{G'} \leq d_G$, the claim implies that $b_{G'} \leq b_G$ and hence M is feasible with respect to G as well.

CASE 2

$$\lambda_S < d_G(s). \tag{4.9}$$

Let $x, y \in S$ be two nodes for which $\lambda(x, y; G) = \lambda_S$ and let A be an $x\bar{y}$-set for which $d_G(A) = \lambda(x, y; G)$. By (4.9) A is non-trivial. Furthermore $R_G(A) \geq \lambda(x, y; G) = d_G(A) \geq R_G(A)$ and hence $d_G(A) = R_G(A)$. From (4.6) we obtain that $R_G(A)$ is odd and that $b_G(A) = 1$. From the choice of A it also follows that $R(A) \leq R(Z)$ holds for any set Z separating two elements of S.

Let G_1 and G_2 denote the graphs arising from G by contracting A and $(V - A)$ respectively into one node. For $Z \subseteq V - A$, $R_{G_1}(Z) \geq R_G(Z)$ and hence $b_{G_1}(Z) \leq b_G(Z)$. Similarly $b_{G_2}(Z) \leq b_G(Z)$ for $Z \subseteq A$. By induction there is a feasible odd-node pairing M_i of G_i ($i = 1, 2$).

Let e_i denote the edge of M_i that matches a contracted node and let v_i be the other end-node of e_i ($i = 1, 2$). Then $M := M_1 + M_2 - e_1 - e_2 + v_1 v_2$ is an odd-node pairing of G. We claim that M is feasible, that is, $d_M(X) \leq b_G(X)$ holds for an each subset $X \subseteq V$. As we have noted earlier, it suffices to prove this for essential sets.

By replacing X with its complement, if necessary, we may assume that $d_M(A, X) = 0$. Because X is essential and by (4.7) $S \cap A \cap X \neq \emptyset$ and $S \cap (V - (A \cup X)) \neq \emptyset$. Hence (4.5) follows and therefore (4.4α) holds for A, X. Using (1.2) and that $\hat{R}$ is symmetric we get $d_M(X) = d_M(A \cap X) + d_M(A \cup X) - d_M(A) + 2d_M(A, X) = d_M(A \cap X) + d_M(A \cup X) - 1 = d_{M_2}(A \cap X) + d_{M_1}(V - (A \cup X)) - 1 \leq b_{G_2}(A \cap X) + b_{G_1}(A \cup X) - 1 \leq b_G(A \cap X) + b_G(A \cup X) - 1 \leq b_G(A) + b_G(X) - 1 = b_G(X)$, as required. ♠♠♠

5. PACKING AND COVERING WITH TREES AND ARBORESCENCES

The first results in this area are due to C.St.J.A. Nash-Williams [1964] and W.T. Tutte [1961] who found necessary and sufficient conditions for the existence of k covering trees and k disjoint trees in an undirected graph. A very powerful theorem on packing directed trees is due to J. Edmonds [1973].

THEOREM 5.1 [Edmonds, 1973] *In a digraph $D = (V, A)$ with a special node s there are k disjoint spanning arborescences of root s if and only if*

$$\varrho(X) \geq k \text{ holds for every } X \subseteq V - s \qquad (5.1)$$

(or, equivalently, there are k edge-disjoint paths from s to every other node of D.) ♠

The following nice proof is an early appearance of the submodular proof-technique.

Proof of sufficiency [Lovász, 1976]. Induction on k. Starting from s we are going to build up a sub-arborescence F of D rooted at s so that

$$\varrho_{E-F}(X) \geq k - 1 \text{ holds for every } X \subseteq V - s. \qquad (*)$$

If we can find such a spanning arborescence, then, by applying the induction hypothesis to $G - F$ (with $k - 1$), we are done.

In the general step let F be an arborescence satisfying $(*)$ and suppose that $V \neq V(F)$. We are going to find a one edge larger arborescence F' satisfying $(*)$. Call a set $X \subseteq V - s$ **critical** if $\varrho_{E-F}(X) = k - 1$. By (5.1) any critical set intersects $V(F)$.

We claim that the intersection of two intersecting critical sets X and Y is critical. Indeed, one has $(k-1) + (k-1) = \varrho_{E-F}(X) + \varrho_{E-F}(Y) \geq \varrho_{E-F}(X \cap Y) + \varrho_{E-F}(X \cup Y) \geq (k-1) + (k-1)$ from which equality must hold everywhere, in particular, $\varrho_{E-F}(X \cap Y) = k - 1$.

Let T be a minimal critical set not included in $V(F)$. (If no such a set exists, let $T := V$). There is an edge $e = uv$ with $u \in V(F) \cap T, v \in T - V(F)$ for otherwise $\varrho(T - V(F)) = \varrho_{E-F}(T - V(F)) \leq k - 1$ contradicting (5.1).

Now e cannot enter any critical set X for otherwise $X \cap T$ would be critical contradicting the minimal choice of T. Therefore $F' := F + e$ is an arborescence satisfying (*) and F' is larger than F. ♠♠♠

First, we are going to show several applications of this fundamental result. In the second part of this section some generalizations of Edmonds' theorem will also be mentioned.

THEOREM 5.2 [Frank, 1978] *The edge-set of a directed graph $D = (V, E)$ can be covered by k branchings if and only if*

$$\varrho(v) \leq k \text{ for every } v \in V \text{ and} \tag{5.2a}$$

$$|E(X)| \leq k(|X| - 1) \text{ holds for every } X \subseteq V. \tag{5.2b}$$

Proof. The necessity is an easy exercise. The sufficiency is proved by an elementary construction. Adjoin a new node s to D and adjoin $k - \varrho(v)$ parallel edges from s to v for each $v \in V$. In the resulting digraph D' we have $\varrho'(X) = \varrho(X) + \sum(k - \varrho(v) : v \in X) = \varrho(X) - \varrho(X) - |E(X)| + k|X| \geq k$ for every $X \subseteq V$. By Theorem 5.1 there are k disjoint spanning arborescences of root s in D'. Restricteｊd to D, these determine k branchings covering the edges of D. ♠♠♠

The following theorem on covering forests is an easy consequence.

THEOREM 5.3 [Nash-Williams, 1964] *The edge-set of an undirected graph $G = (V, E)$ can be covered by k forests if and only if*

$$|E(X)| \leq k(|X| - 1) \text{ holds for every } X \subseteq V. \tag{5.3}$$

Proof. By Theorem 3.1 there is an orientation of G in which $\varrho(v) \leq k$ for every $v \in V$. By Theorem 5.2 Nash-Williams' theorem follows. ♠♠♠

Using another orientation theorem we can derive the following counterpart:

THEOREM 5.4 [Tutte, 1961] *An undirected graph $G = (V, E)$ includes k disjoint spanning trees if and only if*

$$e_{\mathcal{F}} \geq k(t - 1) \tag{5.4}$$

holds for every partition $\mathcal{F} = \{X_1, \ldots, X_t\}$ of V, where $e_{\mathcal{F}}$ denotes the number of edges connecting different members of $\mathcal{F}$.

Proof. Note that (3.4) and (5.4) are the same. Let s be an arbitrary node of G. By Theorem 3.3 there is an orientation of G so that in the resulting digraph D, $\varrho(X) \geq k$ holds for every $X \subseteq V - s$. By Theorem 5.1 D includes k spanning arborescences of root s and these form the k disjoint trees in the original G. ♠♠♠

Note that, conversely, Tutte's theorem, too, implies Theorem 3.3. Indeed, when we have k disjoint spanning trees we may orient each so as to form an arborescences of root s. Clearly such an orientation satisfies the requirement of Theorem 3.3.

One may be interested in finding edge-disjoint spanning arborescences when the roots may be different.

THEOREM 5.5 [Frank, 1978] *In a digraph* $D = (V, E)$ *there are* k *disjoint spanning arborescences if and only if*

$$\sum_{i=1}^{t} \varrho(V_i) \geq k(t - 1) \tag{5.5}$$

holds for every sub-partition $\mathcal{F} = \{V_1, \ldots, V_t\}$ *of* V.

What if lower and upper bounds are imposed on the nodes to limit the number of arborescences rooted at the nodes? Let $f, g : V \to \mathbf{Z}_+$ be two functions on V with $f(v) \leq g(v) \leq k$ for every $v \in V$. The next theorem is another appearance of the linking principle.

THEOREM 5.6 *In a digraph* $D = (V, E)$ *there is a family* $\mathcal{F}$ *of* k *disjoint spanning arborescences so that*
      (i) [Frank, 1978] *every node* v *is the root of at most* $g(v)$ *members of* $\mathcal{F}$ *if and only if*

$$\varrho(X) + g(X) \geq k \text{ for every } X \subseteq V \tag{5.6}$$

and (5.5) holds

(ii) [Cai Mao-Cheng, 1983] *every node v is the root of at least $f(v)$ members of $\mathcal{F}$ if and only if*

$$\sum_{i=1}^{t} \varrho(X_i) - k(t-1)f(\cup_{i=1}^{t} X_i) \tag{5.7}$$

holds for every for every sub-partition $\{X_0, X_1, \ldots, X_t\}$ $(t = 0, 1, \ldots)$ of V where only X_0 may be empty,

(iii) *every node v is the root of at least $f(v)$ and at most $g(v)$ members of $\mathcal{F}$ if and only if (5.5), (5.6) and (5.7) hold.*

Given a digraph $D = (V, E)$, we call a vector $m : V \to \mathbf{Z}_+$ with $m(V) = k$ a **root vector** if there are k disjoint spanning arborescences of D so that each node v is the root of $m(v)$ of them. We will see in Section 8 that the convex combination of root vectors form the basis facet of a polymatroid.

Let us turn back to the case when the arborescences are required to be rooted at a specified node s but we have upper and lower bounds on the edges. Let $f : E \to \mathbf{Z}_+$ and $g : E \to \mathbf{Z}_+ \cup \{\infty\}$ be two functions such that $f \leq g$. Since no edge entering s may occur in an s-arborescence we assume that in D there are no such edges. The following theorem will be derived from Edmonds' theorem by an elementary construction. Note that in this case the linking principle does not hold, in general.

THEOREM 5.7 [Frank and Tardos, 1989] *There is a family $\mathcal{A}$ of k spanning arborescences rooted at s such that each edge e of D is contained in at least $f(e)$ and at most $g(e)$ members of $\mathcal{A}$ if and only if $\varrho_f(v) \leq k$ for every $v \in V - s$ and*

$$k - \varrho_f(X) \leq \sum [k - \varrho_f(v) : v \in A] + \sum [g(e) - f(e) : e = uv \in E, v \in$$
$$I(X) - A, u \in V - X] \tag{5.8}$$

holds whenever $X \subseteq V - s$ and $A \subseteq I(X)$ where $I(X) := \{v \in X : \text{there is an edge } uv \in E \text{ with } u \in V - X\}$.

Proof. We can assume that $g(e) \leq k$ for every $e \in E$. Construct a digraph $D' = (V', E')$ as follows. Let $V' := V \cup \{v' : v \in V - s\}$. Let E' consist of the following type of edges. For each $v \in V - s$ there are k edges from v to v' and $k - \varrho_f(v)$ edges from v' to v, and for every edge $uv \in V$ there are $f(uv)$ parallel edges from u to v and $g(uv) - f(uv)$ parallel edges from u to v'.

If in D' there are k edge-disjoint spanning arborescences rooted at s, then the arborescences in D corresponding to these ones will satisfy the requirements. If no such a family exists, then, by Edmonds' Theorem 5.1, there is a set $Y' \subseteq V' - s$ for which $\varrho'(Y') < k$. Suppose that Y' is maximal. Obviously, v' is in Y', then so is v. Furthermore, if $v \in Y'$ and $v' \notin Y'$, then there is edge $uv \in E$ with $u \notin Y'$. For otherwise $\varrho'(Y' + v') \leq \varrho'(Y')$ contradicting the maximal choice of Y'.

Consequently, Y' has the following form: $Y' = \{v, v' : v \in X - A\} \cup \{v : v \in A\}$ for some $X \subseteq V - s$ and $A \subseteq I(X)$. Now we have $k > \varrho'(Y') = \sum_{v \in A} [k - \varrho_f(v)] + \sum [g(uv) : u \notin X, v \in I(X) - A] + \sum [f(uv) : u \notin X, v \in A] = \sum_{v \in A} [k - \varrho_f(v)] + \sum [g(uv) - f(uv) : u \notin X, v \in I(X) - A] + \varrho_f(X)$, contradicting (5.8). ♠♠♠

An interesting special case ($f \equiv 1, g \equiv \infty$) which can be considered as a covering counter-part of Edmonds' packing theorem, was proved earlier by K. Vidyasankar [1978]:

COROLLARY 5.8 *Let $D = (V, A)$ be a digraph with a special node s such that there is no edge entering s. The edge-set of D can be covered by k spanning arborescences rooted at s if and only if $\varrho(v) \leq k$ for every $v \in V$ and $k - \varrho(X) \leq \sum [k - \varrho(v) : v \in I(X)]$ for every $X \subseteq V - s$.*

The following possible generalization of Edmonds' theorem naturally emerges. Beside s, let us be given a subset $T \subseteq V - s$ so that $\varrho(X) \geq k$ for every subset $X \subseteq V - s, X \cap T \neq \emptyset$. It might be tempting to conjecture that in this case there are k disjoint arborescences so that each contains every element of T but Lovász [1973] found a counter-example. However, invoking the splitting Theorem 2.8, we were able to prove in [Bang-Jensen, Frank, Jackson, 1993] the following:

THEOREM 5.9 *Let $D = (V, A)$ be a digraph with a special node s, and subset $T \subseteq V - s$ so that $\varrho(x) \geq \delta(x)$ for every $x \in V - T - s$. Assume that $\lambda(s, x) \geq k (\geq 1)$ for every $x \in T$ (or equivalently, that $\varrho(X) \geq k$ for every $X \subseteq V - s$ with $X \cap T \neq \emptyset$). Then there is a family $\mathcal{F}$ of k disjoint (possibly not spanning) arborescences rooted at s so that every node $x \in V$ belongs to at least $r(x) := \min(k, \lambda(s, x))$ members of $\mathcal{F}$.*

Note that we are back at Theorem 5.1 when $T = V - s$. In network flows it is a basic observation that a flow from s to t decomposes into path-flows. The following corollary to Theorem 5.9 may be considered as an analogous result on pre-flow graphs. We call a digraph $D = (V, A)$ with root s a **pre-flow digraph** if $\varrho(x) \geq \delta(x)$ holds for every $x \in V - s$.

COROLLARY 5.10 *In a pre-flow digraph* $D = (V, A)$, *for any integer* $k(\geq 1)$ *there is a family* $\mathcal{F}$ *of* k *disjoint arborescences of root* s *so that every node* x *belongs to* $\min(k, \lambda(s, x))$ *members of* $\mathcal{F}$. *In particular, if* $k := \max(\lambda(s, x) : x \in V - s)$, *then every* x *belongs to* $\lambda(s, x)$ *members of* $\mathcal{F}$. ♠

Another corollary of Theorem 5.9 will be found in Section 7. We finish this section by mentioning another interesting extension of Edmonds' theorem, due to A. Schrijver, whose proof relies on submodular functions. Let $D = (V, E)$ be a digraph and $\{S, T\}$ a bipartition of V. We call a sub-forest B of D a **bi-branching** if $\varrho(X) \geq 1$ for every $X \subseteq T$ and $X \supseteq T$ ($X \neq \emptyset, V$). (Note that a bi-branching is a spanning arborescence if $|S| = 1$.)

THEOREM 5.11 [Schrijver, 1985] *In* D *there are* k *edge-disjoint bi-branchings if and only if* $\varrho(X) \geq k$ *for every* $X \subseteq T$ *and* $X \supseteq T$ ($X \neq \emptyset, V$).

6. AUGMENTATIONS

What is the minimum number γ (or more generally, the minimum cost) of edges to be added to a graph (directed or not) so that in the resulting graph the local edge-connectivity $\lambda(u, v)$ between every pair of nodes u, v is at least a prescribed value $r(u, v)$? A strongly related variation of this problem is when the newly added set of edges is not required to be minimum but must satisfy some degree constraints.

These types of problems arise in paractice and are interesting for their own sake. It is another area where submodular functions help a lot. Our purpose is to survey the results and their relationship to submodular functions.

Let us first consider directed augmentations. Suppose we are given a digraph D with a source s and a target t. Let $r(u,v) = k$ if $u = s$, $v = t$ and $r(u,v) = 0$ otherwise. In this case the augmentation problem requires us to add a minimum cost set of edges so that in the resulting digraph there are k edge-disjoint paths from s to t. This problem can easily be reduced to a minimum cost flow problem in the union graph of the new and the original edges where the costs of the original edges are defined to be zero.

A more difficult problem consists of augmenting a digraph by adding a minimum cost of new edges so as to have k edge-disjoint paths from a specified source-node s to each other node. (That is, $r(u,v) = k$ if $u = s$ and $r(u,v) = 0$ otherwise.) An equivalent problem is that of finding a minimum cost subset F of edges so that in the subgraph of F there are k edge-disjoint paths from s to every other node.

This problem, in turn, can be reduced to a weighted matroid intersection problem where the first matroid is k times the circuit-matroid of the underlying undirected graph (that is, a subset of edges is independent if it is the union of k forests) while the second matroid is a partition matroid where a subset of edges is independent if it contains no more than k edges entering the same node. A common basis of these matroids corresponds to a subgraph D' of D which is the union of k disjoint spanning trees (in the undirected sense) and in which every in-degree $\varrho(v)$ $(v \in V - s)$ is precisely k. By Theorem 5.2 D' is the union of k disjoint spanning arborescences of root s. Therefore our augmentation problem is equivalent to a weighted matroid intersection problem. Since there are good algorithms for this [Edmonds, 1979], the augmentation problem is also solvable in strongly polynomial time.

One may consider the openly disjoint counterpart of the preceding problem; that is, improve a digraph by adding a minimum cost set of new edges so as to have k openly disjoint paths from a specified source-node to each other node. The problem was solved in [Frank and Tardos 1989] with the help of submodular flows.

K.P. Eswaran and R.E. Tarjan [1976] described a method of making a digraph strongly connected by adding a minimum number of edges. They also noticed that the minimum cost version of this problem includes as a special case the directed Hamiltonian path problem and therefore is NP-complete. However, the problem is tractable if we are allowed to add a new directed edge (u, v) only if (v, u) is an original edge of the digraph. In this case a weighted version of the following fundamental theorem helps us.

THEOREM 6.1 [Lucchesi and Younger, 1978] *In a directed graph D the maximum number, ν, of disjoint directed cuts is equal to the minimum number, τ, of edges covering all the directed cuts.*

This theorem has an extension by J. Edmonds and R. Giles [1977] to describe minimum weight coverings of directed cuts and in [Frank, 1981] a strongly polynomial algorithm is constructed to compute the minimum. (A polynomial time algorithm is called **strongly polynomial** if it uses, beside ordinary data manipulation, only basic operations such as comparing, adding, subtracting, multiplying, and dividing numbers, and the number of these operations is independent of the numbers occuring in the input.) The link between the problem of minimum-cost directed cuts and the problem of making a digraph srongly connected is the easy observation that a subset of edges is a covering of dicuts if and only if the addition of its elements in a reverse way makes the digraph strongly connected.

The directed augmentation problem when $r(u, v) \equiv k$ was solved by D.R. Fulkerson and L.S. Shapley [1971] when the starting digraph is $D = (V, \emptyset)$, by Y. Kajitani and S. Ueno [1986] when the starting digraph is a directed tree and by Frank [1992a] for an arbitrary starting digraph.

THEOREM 6.2 *A directed graph $D = (V, E)$ can be made k-edge-connected by adding γ new edges if and only if*

$$\sum(k - \varrho(X_i)) \leq \gamma \text{ and } \sum(k - \delta(X_i)) \leq \gamma \qquad (6.1)$$

holds for every sub-partition $\{X_1, \ldots, X_t\}$ of V.

The proof in [Frank, 1992a] consists of showing how the minimization problem can be reduced to the following feasibility problem; let m_{in} and m_{out} be two non-negative integer-valued functions on V such that $\gamma = m_{in}(V) = m_{out}(V)$.

THEOREM 6.3 *A directed graph $D = (V, E)$ can be made k-edge-connected by adding a set F of new edges satisfying*

$$\varrho_F(v) = m_{in}(v) \text{ and } \delta_F(v) = m_{out}(v) \qquad (6.2)$$

for every node $v \in V$ if and only if both

$$\varrho(X) + m_{in}(X) \geq k \text{ and } \delta(X) + m_{out}(X) \geq k \quad (6.3)$$

hold for every $X \in V$.

Proof. This theorem is nothing but a reformulation of Mader's directed splitting-off Theorem 2.6A. Indeed, extend D by a new node s and for each $v \in V$ adjoin $m_{in}(v)$ (respectively, $m_{out}(v)$) parallel edges from s to v (from v to s). Now by (6.3) the hypotheses of Theorem 2.6A are satisfied and hence we can split off γ pairs of edges to obtain a k-edge-connected digraph. The resulting set of γ new edges (connecting original nodes) satisfies the requirement. ♠

Our next problem is to find an augmentation of minimum cardinality if upper and lower bound are imposed both on the in-degrees and on the out-degrees of the digraph of newly added edges. Let $f_{in} \leq g_{in}$ and $f_{out} \leq g_{out}$ be four non-negative integer-valued functions on V (infinite values are allowed for g_{in} and g_{out}).

THEOREM 6.4 *Given a directed graph $D = (V, E)$ and a positive integer k, D can be made k-edge-connected by adding a set F of precisely γ new edges so that both*

$$f_{in}(v) \leq \varrho_F(v) \leq g_{in}(v) \quad (6.4a)$$

and

$$f_{out}(v) \leq \delta_F(v) \leq g_{out}(v) \quad (6.4b)$$

hold for every node v of D if and only if both

$$k - \varrho(X) \leq g_{in}(X) \quad (6.5a)$$

and

$$k - \delta(X) \leq g_{out}(X) \quad (6.5b)$$

hold for every subset $\emptyset \subset X \subset V$ and both

$$\sum(k - \varrho(X_i) : i = 1, \ldots, t) + f_{in}(X_0) \leq \gamma \quad (6.6a)$$

and

$$\sum(k - \delta(X_i) : i = 1, \ldots, t) + f_{out}(X_0) \leq \gamma \quad (6.6b)$$

hold for every partition $\{X_0, X_1, X_2, \ldots, X_t\}$ *of* V *where only* X_0 *may be empty.*

(We remark that both in this theorem and in each later theorem concerning degree-constrained augmentations loops are allowed to be added to D).

One may be interested in degree-constrained augmentations when there is no requirement for the number of new edges.

THEOREM 6.5 *Given a directed graph* $D = (V, E)$ *and a positive integer* k, D *can be made* k*-edge-connected by adding a set* F *of new edges satisfying (6.4) if and only if (6.5) holds and and*

$$\sum (k - \varrho(X_i) : i = 1, \ldots, t) + f_{in}(X_0) \le \alpha \qquad (6.7a)$$

and

$$\sum (k - \delta(X_i) : i = 1, \ldots, t) + f_{out}(X_0) \le \alpha \qquad (6.7b)$$

hold for every partition $\{X_0, X_1, X_2, \ldots, X_t\}$ *of* V *where only* X_0 *may be empty and* $\alpha := \min(g_{out}(V), g_{in}(V))$.

What can one say if the demand for k-edge-connectivity is replaced by a more complicated requirement? Unfortunately, already very simple-sounding versions are NP-complete.

THEOREM 6.6 *The following two problems are NP-complete: Let* $D = (V, E)$ *be a directed graph,* s *a specified node of* D, $T \subset V$ *a specified subset of nodes and let* γ *be a positive integer. Decide if it is possible to add at most* γ *new edges to* D *so as to have a path (i) from* s *to every element of* T, *(ii) from any element of* T *to any other element of* T.

These negative results do not leave too much space for possible generalizations, but not everything is lost. The idea behind the generalization is that Theorem 2.8 is a splitting theorem stronger than Mader's Theorem 2.6.

Let $D = (V, E)$ be a digraph and let $T(D) := \{v \in V : \varrho_D(v) \ne \delta_D(v)\}$ be the set of non-di-Eulerian nodes. Let k be a positive integer and $r(x, y)$, $(x, y \in V)$ a non-negative integer-valued demand function satisfying

$$r(x, y) = r(y, x) \le k \text{ for every } x, y \in V \text{ and} \qquad (6.8a)$$

$$r(x, y) \equiv k \text{ for every } x, y \in T(D). \tag{6.8b}$$

Let $R(\emptyset) = R(V) = 0$ and for $X \subseteq V$ let $R(X) := \max(r(x, y) : X$ separates x and $y)$. Let us define $q_{in}(X) := R(X) - \varrho_D(X)$, $q_{out}(X) := R(X) - \delta_D(X)$.

THEOREM 6.7 [Bang-Jensen, Frank and Jackson, 1993] *Given a digraph $D = (V, E)$, positive integers k, γ, and a demand function $r(x, y)$ satisfying (6.8), D can be extended to D^+ by adding γ new directed edges so that*

$$\lambda(x, y; D^+) \geq r(x, y) \text{ for every } x, y \in V \tag{6.9}$$

if and only if both

$$\sum q_{in}(X_i) \leq \gamma \tag{6.10a}$$

and

$$\sum q_{out}(X_i) \leq \gamma \tag{6.10b}$$

hold for every sub-partition $\{X_1, \ldots, X_t\}$ of V. ♠

COROLLARY 6.8 *Given an Eulerian digraph $D = (V, E)$, positive integers k, γ, and a demand function $r(x, y)$ satisfying (6.8a), D can be extended to an Eulerian digraph D^+ by adding γ new directed edges so that (6.9) holds if and only if (6.10) is satisfied.*

COROLLARY 6.9 *Given a digraph $D = (V, A)$, positive integers k, γ, and a subset $T \subseteq V$ so that $\varrho_D(v) = \delta_D(v)$ holds for every $v \in V - T$, then D can be extended to a digraph D^+ by adding γ new directed edges so that $\lambda(x, y; D^+) \geq k$ for every $x, y \in T$ if and only if both*

$$\sum(k - \varrho_D(X_i)) \leq \gamma \text{ and } \sum(k - \delta_D(X_i)) \leq \gamma \tag{6.11}$$

hold for every sub-partition $\{X_1, \ldots, X_t\}$ of V for which $X_i \cap T, T - X_i \neq \emptyset$ $(i = 1, \ldots, t)$. ♠

In the second part of this section we are concerned with augmentation when only undirected edges are allowed to be added. Let N be a mixed graph composed from a directed graph $D = (V, A)$ and an undirected graph $G = (V, E)$, and $r(x, y)$ a demand function satisfying (6.8). We say that a component C of N is **marginal** (with respect to r) if $r(u, v) \leq$

$\lambda(u, v; N)$ for every $u, v \in C$ and $r(u, v) \leq \lambda(u, v; N) + 1$ for every u, v separated by C. Let $\beta_N(X) := \min(\varrho_D(X) + d_G(X), \delta_D(X) + d_G(X))$. The following result was proved in [Frank, 1992a] for the special case when the starting graph is undirected.

THEOREM 6.10 [Bang-Jensen, Frank and Jackson, 1993]
Given a mixed graph N, integers $k \geq 2$, $\gamma \geq 0$, and a demand function $r(x, y)$ satisfying (6.8) so that there is no marginal components, then N can be extended to a mixed graph N^+ by adding γ new undirected edges so that

$$\lambda(x, y; N^+) \geq r(x, y) \text{ for every } x, y \in V \qquad (6.12)$$

if and only if

$$\sum(R(X_i) - \beta_N(X_i)) \leq 2\gamma \qquad (6.13)$$

holds for every sub-partition $\{X_1, \ldots, X_t\}$ of V. ♠

If N has a marginal component, then the above min-max theorem is not true but there exists a very simple reduction method to get rid of marginal components.

When $r \equiv k$ for an integer $k \geq 2$ Theorem 6.10 specializes to:

COROLLARY 6.11 *Let $N = (V, A \cup E)$ be a mixed graph and $k \geq 2, \gamma \geq 1$ integers. N can be made k-edge connected by adding γ new undirected edges if and only if*

$$\sum(k - \beta_N(X_i)) \leq 2\gamma$$

holds for every sub-partition $\{X_1, .., X_t\}$ of V. ♠

Further specializing this result to the case when N is an undirected graph (i.e. $A = \emptyset$) we obtain:

COROLLARY 6.12 [T. Watanabe and A. Nakamura, 1987] *An undirected graph $G = (V, E)$ can be made k edge-connected ($k \geq 2$) by adding γ new edges if and only if $\sum(k - d(X_i)) \leq 2\gamma$ holds for every sub-partition $\{X_1, .., X_t\}$ of V.* ♠

Degree-constrained versions of these augmentation problems can also be handled. To close this section we consider minimum cost k-edge-connected augmentations. As we have mentioned, if costs are assigned to the edges, the problem is NP-complete even if $k = 1$. Suppose now that $c_{in} : V \to R_+$ and $c_{out} : V \to R_+$ are two non-negative cost functions on the node-set V of a digraph D. The problem is to find a k-edge-connected augmentation of D for which $\sum \varrho_F(v)c_{in}(v) + \sum \delta_F(v)c_{out}(v)$ is minimum where F denotes the set of newly added edges.

Analogous problems can be posed concerning undirected augmentations. In [Frank, 1992a] it was shown how these problems can be reduced to a polymatroid optimization problem. (See Section 8). Hence the minimum node-cost augmentation problem can be solved in strongly polynomial time.

7. EDGE-DISJOINT PATHS

Edge-disjoint paths problems form an interesting class of combinatorial optimization problems. In a survey paper [Frank, 1990] it was shown how submodular functions can be used to prove results in this area. Here we briefly mention some more recent applications of submodular functions.

The **edge-disjoint paths problem** is as follows. Given a graph or digraph and k pairs of nodes $(s_1, t_1), (s_2, t_2), \ldots, (s_k, t_k)$, find k pairwise edge-disjoint paths connecting the corresponding pairs (s_i, t_i). Sometimes it is convenient to mark the terminal pairs to be connected by an edge. The graph $H = (U, F)$ formed by the marking edges is called a **demand graph** while the original graph $G = (V, E)$ is the **supply graph**. If G and H are directed, then a demand edge $t_i s_i$ of H indicates that we want to have a path in G from s_i to t_i. In this terminology the edge-disjoint paths problem is equivalent to seeking in $G + H$ for $|F|$ edge-disjoint circuits each of which contains exactly one demand edge. We will call such a circuit **good**.

Actually, this kind of problem can be considered as a feasibility problem. The **maximization problem** is one where no demands are specified and one is interested in finding a maximum number of paths connecting the corresponding terminal pairs.

The undirected edge-disjoint paths problem is NP-complete even if $G + H$ is Eulerian or if $G + H$ is planar [Middendorf and Pfeiffer, 1993] but it is solvable in polynomial time if k is fixed [Robertson and Seymour, 1986]. For digraphs, Fortune, Hopcroft and Wyllie [1980] proved that the (edge-) disjoint paths problem is NP-complete for $k = 2$. They also showed that for acyclic digraphs the (edge-) disjoint paths problem can be solved in polynomial time if k is fixed.

Both in the undirected and the directed case there is a natural necessary condition for the solvability:

CUT CRITERION $d_G(X) \geq d_H(X)$ for every $X \subseteq V$.

DIRECTED CUT CRITERION $\varrho_G(X) \geq \delta_H(X)$ for every $X \subseteq V$.

These criteria are not sufficient, in general, but, by the undirected and the directed edge-versions of Menger's theorem, they are sufficient provided that $s_1 = \ldots = s_k$ and $t_1 = \ldots = t_k$. The next interesting case is when the demand graph consists of two sets of parallel edges. That is, H has k_i edges from t_i to s_i $(i = 1, 2)$.

B. Rothschild and A. Whinston [1966], extending an earlier, slightly weaker result of T.C. Hu [1963], proved the following:

THEOREM 7.1 *If the demand graph H consists of two pairs of parallel edges and $G + H$ is Eulerian, then the edge-disjoint paths problem has a solution if and only*

$$d_G(X) \geq d_H(X) \text{ for every } X \subseteq V. \tag{7.1}$$

There is an amazingly simple proof of this theorem by C.St.J.A. Nash-Williams. To my best knowledge, he never published this proof and the only place where it appeared in print is the problem and exercise book of Lovász' [1979]. Since Nash-Williams' proof is so elegant I outline it here.

Proof. Form a mixed graph M from G by adding k_i parallel directed edges from t_i to s_i. It can easily be checked that (7.1) implies (3.3) and hence, by Theorem 3.2, there is an Eulerian orientation $\vec{M}$ of M. Let $\vec{G}$ denote the orientation of G defined by $\vec{M}$. (7.1) implies that $\varrho_{\vec{G}}(X) \geq k_1$ for every set containing t_1 but not s_1. By Menger's theorem there are k_1 edge-disjoint directed paths in $\vec{G}$ from s_1 to t_1. Leave out all the edges of these paths from $\vec{G}$ and add k_2 parallel directed edges from t_2 to s_2. By construction, the resulting digraph G' is Eulerian and hence it decomposes into edge-disjoint circuits. Therefore G' includes k_2 edge-disjoint paths from s_2 to t_2, as required. ♠

Logically, this proof of Nash-Williams consists of two parts. First he points out that there is a very easy directed counterpart of Theorem 7.1 and in the second part he shows that the original graph can be oriented so that the directed theorem applies. Let us formulate this directed version:

THEOREM 7.2 *If G and H are digraphs, $G + H$ is di-Eulerian, and H consists of k_i parallel edges from t_i to s_i $(i = 1, 2)$, then the directed cut criterion is necessary and sufficient for the solvability of the directed edge-disjoint paths problem.*

Inspired by Nash-Williams' proof, recently we found an analogous approach to the following theorem:

THEOREM 7.3 [Okamura and Seymour, 1981] *Suppose that G is undirected and planar, $G + H$ is Eulerian, and each terminal is on one face of G. Then the cut criterion is necessary and sufficient for the solvability of the edge-disjoint paths problem.*

THEOREM 7.4 [A. Frank, D. Wagner and K. Weihe, 1993] *Suppose that G and H are digraphs, G is planar with no clock-wise directed circuit, $G + H$ is di-Eulerian, and the terminal nodes are positioned on the boundary B of the outer face of G so that, with respect to a reference point r of B, each triple (r, t_i, s_i) is clock-wise. Then the directed cut criterion is necessary and sufficient for the solvability of the directed edge-disjoint paths problem.*

The nice thing in the proof of this theorem is that the paths can be constructed in a greedy way and if the desired solution does not exist, then a violating cut may also be found greedily. Equally easily the edges of the undirected graphs G and H in Theorem 7.3 can be oriented so as to satisfy the hypotheses of Theorem 7.4. Therefore the theorem of Okamura and Seymour is an immediate consequence of Theorem 7.4.

There are a great number of other theorems asserting that the undirected cut criterion is sufficient in certain circumstances. The directed cut criterion turns out to be sufficient only in few cases. A possible reason for this is that for the directed edge-disjoint paths problem there is a natural neccessary condition that is stronger than the directed cut criterion:

COVERING CRITERION Every subset of edges of $G + H$ covering all good circuits must have at least as many elements as the number k of demand edges.

It can be shown that for undirected graphs the covering criterion is equivalent to the cut criterion. The covering criterion is not sufficient in general, however one has the following;

THEOREM 7.5 *If G is acyclic and $G + H$ is planar, then the covering criterion is necessary and sufficient for the solvability of the directed edge-disjoint paths problem.*

Proof. We prove only sufficiency. Let D denote the directed planar dual graph of $G + H$. Note that directed cuts of D correspond to directed circuits of $G + H$. Therefore the Theorem 6.1 of C. Lucchesi and D. Younger [1978] for D can be formulated in terms of $G + H$. That is, the maximum number ν of edge-disjoint directed circuits of $G + H$ is equal to the minimum number τ of edges of $G + H$ covering all directed circuits of $G + H$.

By the covering criterion, $\tau \geq k$. Therefore $\nu \geq k$ and there are k edge-disjoint circuits in $G + H$. Since G is acyclic, each of these circuits must contain a demand edge. Since the number of demand edges is k, each of the k circuits is good and we have the k desired edge-disjoint paths.
♠♠♠

We close our list of feasibility type results by a new sufficient condition. Y. Shiloach [1979], pointed out that Edmonds' Theorem 5.1 immediately implies the following pretty result. Given k pairs $(s_1, t_1), \ldots, (s_k, t_k)$ of nodes in a k edge-connected digraph D, there are edge-disjoint paths from s_i to t_i $(i = 1, \ldots, k)$.

Using Theorem 2.8 we can derive the following generalization.

THEOREM 7.6 [Bang-Jensen, Frank and Jackson, 1993]
Let $(s_1, t_1), \ldots, (s_k, t_k)$ be k pairs of nodes in a digraph $D = (U, A)$ so that for every node x with $\varrho(x) < \delta(x)$ or $x = t_i$ there are edge-disjoint paths from s_i to x $(i = 1, \ldots, k)$. Then there are edge-disjoint paths from s_i to t_i $(i = 1, \ldots, k)$. ♠

So far we have studied disjoint paths problems of feasibility type. One can also be interested in the maximization form. Let $G = (V, E)$

and $H = (T, F)$ be the supply and the demand graphs, respectively, with $E \cap F = \emptyset$ and $T \subseteq V$. A path in G is called **H-admissible** (a **T-path**) if it connects two end-nodes of a demand edge (two distinct nodes in T). Throughout we assume that the pair (G, T) is **inner Eulerian** , that is, $d_G(v)$ is even for every $v \in V - T$.

The **maximization form** of the edge-disjoint paths problem consists of finding a maximum number $\mu = \mu(G, H)$ of edge-disjoint H-admissible paths. We can easily get an upper bound on μ. Let us call a sub-partition $\{X_1, X_2, \ldots, X_k\}$ of V **admissible** if $T \subseteq \cup X_i$ and each $X_i \cap T$ is stable in $H(i = 1, \ldots, k)$. Clearly,

$$\mu(G, H) \le \sum d_G(X_i)/2. \tag{8.2}$$

Let us call $\sum d_G(X_i)/2$ the **value** of the sub-partition and let $\tau = \tau(G, H)$ denote the minimum value of an admissible sub-partition. We have $\mu \le \tau$ and although equality does not hold in general, in important special cases it does.

For a subset $A \subseteq T$ the notation $\lambda(A, T - A; G)$ will be abbreviated to $\lambda(A; G)$ or to $\lambda(A)$ when no confusion can arise. B.V. Cherkasskij [1977] and L. Lovász [1976] proved the following theorem:

THEOREM 7.7 *For an inner Eulerian pair (G, T) the maximum number of edge-disjoint T-paths is equal to $(\sum \lambda(t) : t \in T)/2$. Furthermore, there is a family of disjoint sets $\{X(t) : t \in T\}$ such that $t \in X(t) \subseteq V$ and $d_G(X_t) = \lambda(t)$ $(t \in T)$.*

An equivalent formulation of the first part is:

THEOREM 7.7A *There is a family $\mathcal{F}$ of edge-disjoint T-paths such that $\mathcal{F}$ contains $\lambda(t)$ paths ending at t for each $t \in T$.*

In other words there is one single family of edge-disjoint T-paths that simultaneously contains a maximal family of edge-disjoint $(t, T - t)$-paths for each $t \in T$.

A.V. Karzanov [1984] and M.V. Lomonosov [1985] extended this theorem. To formulate their result let us say that a family $\mathcal{F}$ of edge-disjoint T-paths **locks** a subset $A \subseteq T$ if $\mathcal{F}$ contains $\lambda(A)$ $(A, T - A)$-paths. Furthemore, we say that $\mathcal{F}$ **locks a family** $\mathcal{T}$ of subsets of T if $\mathcal{F}$ locks all members of $\mathcal{T}$. Call a family $\mathcal{T}$ of subsets of T **3-cross free** if it has no three pairwise crossing members. (Two subsets A and B of T are called **crossing** if none of $A - B, B - A, A \cap B, T - (A \cup B)$ is empty).

THEOREM 7.8 [Karzanov, 1984], [Lomonosov, 1985] *Let (G, T) be inner Eulerian and T a 3-cross free family of subsets of T. Then there is a family of edge-disjoint T-paths that locks T.*

A simple proof relying on the splitting off technique can be found in [Frank, Karzanov, Sebő, 1992]. Using this locking theorem and the polymatroid intersection theorem of J. Edmonds [1970] (see Section 8), in the same paper we described a somewhat unexpected proof of a difficult theorem of Karzanov and Lomonosov.

Let us call a graph $H = (T, F)$ **bi-stable** if the family of maximal stable sets of H can be partitioned into two parts, each consisting of disjoint sets. Clearly a clique, or more generally a complete k-partite graph, is bi-stable and $2K_2$ is also bi-stable.

THEOREM 7.9 [Karzanov, 1985], [Lomonosov, 1985]
Suppose that (G, T) is inner Eulerian and $H = (T, F)$ is bi-stable. Then $\mu(G, H) = \tau(G, H)$.

We remark that A.V. Karzanov and P.A. Pevzner [1979] showed that if $H = (T, F)$ is not bi-stable, then there is a supply graph $G = (V, E)$, with $T \subseteq V$ and (G, T) inner Eulerian, so that $\mu(G, H) < \tau(G, H)$.

8. SUBMODULAR FUNCTIONS AND POLYHEDRA

In this section some basic notions and results will be summarized along with their relationship to applications appearing in earlier sections. For general accounts, see, for example, Fujishige [1991], Lovász [1983], Frank and Tardos [1988].

Throughout this section we will assume that every set-function is integer-valued and zero on the empty set. Let S be a finite ground-set and $b : 2^S \to \mathbf{Z} \cup \{\infty\}$ be a set-function. We call b **fully (intersecting, crossing) submodular** if

$$b(X) + b(Y) \geq b(X \cap Y) + b(X \cup Y) \tag{8.1}$$

holds for every (intersecting, crossing) $X, Y \subseteq S$. A finite, monotone increasing, fully submodular function is called a **polymatroid function.** We remark that many of the results below remain valid for real-valued functions, but in this paper we are concerned with combinatorial applications and hence we are mainly interested in integer-valued set-functions and polyhedra which are spanned by their integer points. We call such polyhedra **integral polyhedra.**

A set function p is called **supermodular** if $-p$ is submodular. We say that p is **x-supermodular** if for every $X, Y \subseteq S$ at least one of the following inequalities holds:

$$p(X) + p(Y) \le p(X \cap Y) + p(X \cup Y),$$

$$p(X) + p(Y) \le p(X - Y) + p(Y - X).$$

Note that intersecting supermodular functions are x-supermodular.

We say that a pair (p, b) of set-functions is a **strong pair** if p (resp. b) is fully supermodular (submodular) and they are **compliant,** that is,

$$b(X) - p(Y) \ge b(X - Y) - p(Y - X) \qquad (8.2)$$

holds for every $X, Y \subseteq S$. If p and b are intersecting super- and submodular functions and (8.2) holds for intersecting X, Y, then (p, b) is called a **weak pair.**

A finite-valued set function m is called **modular** if $m(X) + m(Y) = m(X \cap Y) + m(X \cup Y)$ holds for every $X, Y \subseteq S$. A modular function m with $m(\emptyset) = 0$ is determined by its value on the singletons as $m(X) = \sum(m(s) : s \in X)$.

Let $D = (V, E)$ be a directed graph. Let $f : E \to \mathbf{Z} \cup \{\infty\}$ and $g : E \to \mathbf{Z} \cup \{-\infty\}$ be such that $f \le g$. Furthermore let $b : 2^V \to \mathbf{Z} \cup \infty$ be a fully submodular function. We call a vector $x : E \to \mathbf{R}$ a **submodular flow** if

$$\varrho_x(A) - \delta_x(A) \le b(A) \qquad (8.3)$$

holds for every $A \subseteq V$. (Here $\varrho_x(A) := \sum(x(e) : e$ enters $A)$ and $\delta_x(A) := \sum(x(e) : e$ leaves A.) A submodular flow x is **feasible** if $f \le x \le g$. The set of feasible submodular flows is called a **submodular flow polyhedron** and is denoted by $Q(f, g; b)$. Submodular flows were introduced by R. Giles and J. Edmonds [1977]. Actually, they used crossing submodular functions which give much more flexibility in applications. On the other hand, as we shall see, crossing submodular functions define the same class of polyhedra.

FEASIBILITY THEOREM 8.1 *There exists an integer-valued feasible submodular flow if and only if*

$$\varrho_f(A) - \delta_g(A) \le b(A) \tag{8.4}$$

holds for every $A \subseteq V$.

Proof. Let x be a feasible submodular flow. Then $\varrho_f(A) - \delta_g(A) \le \varrho_x(A) - \delta_x(A) \le b(A)$ and (8.4) follows. To see the sufficiency of (8.4) we need the following lemma:

LEMMA 8.2 *Let $p(A) := \varrho_f(A) - \delta_g(A)$. Then $p(A) + p(B) \le p(A \cap B) + p(A \cup B)$ for every subsets A, B of V and equality holds if and only if $f(e) = g(e)$ for every edge e with one end in $A - B$ and the other end in $B - A$.*

Proof. The lemma follows from the following identity: $p(A) + p(B) = p(A \cap B) + p(A \cup B) + \sum(f(e) - g(e) : e$ has one end in $A - B$ and the other end in $B - A)$. This can be proved by showing that each edge of E has the same contribution to the two sides of the identity. ♠

Call an edge e **tight** if $f(e) = g(e)$ and a subset A of V **tight** if $p(A) = b(A)$. Suppose that this is untrue and that D is a counter-example for which the number of tight edges plus the number of tight sets is maximum.

CASE 1. *Every edge is tight.* Define $x := f$. Now $\varrho_x(A) - \delta_x(A) = \varrho_f(A) - \delta_f(A) = \varrho_f(A) - \delta_g(A) \le b(A)$, that is x is a feasible submodular flow, a contradiction.

CASE 2. $f(e_0) < g(e_0)$ *for some edge* e_0. We claim that there exists a tight set A entered by e_0, for if no such set exists, we can increase $f(e_0)$ without violating (8.4) until either a new tight set A (entered by e_0) arises or e_0 becomes tight. For the revised lower bound f' the sum of tight edges and tight sets is bigger, and as this cannot be a counter-example, there is a submodular flow x feasible with respect to f' and g. Obviously x is feasible with respect to f and g, a contradiction.

It can similarly be seen that there is a tight set B left by e_0. Now we have $b(A) + b(B) = p(A) + p(B) \leq p(A \cap B) + p(A \cup B) \leq b(A \cap B) + b(A \cup B) \leq b(A) + b(B)$. Therefore we have equality everywhere and in particular, $p(A) + p(B) = p(A \cap B) + p(A \cup B)$; but this contradicts the lemma since $f(e_0) < g(e_0)$. This contradiction shows that no counter-example can exist. ♠♠♠

When we apply Theorem 8.1 to $b \equiv 0$ we obtain A. Hoffman's classical theorem on the existence of feasible circulations. It is easily seen that in this special case a vector x is a submodular flow if and only if $\varrho_x(v) = \delta_x(v)$ for every $v \in V$. Such an x is called a **circulation**. Another important corollary is as follows:

DISCRETE SEPARATION THEOREM 8.3 *Let S be a groundset, $p_1 : 2^S \to \mathbf{Z} \cup \{-\infty\}$ a supermodular function and $b_1 : 2^S \to \mathbf{Z} \cup \{\infty\}$ a submodular function for which $p_1(\emptyset) = b_1(\emptyset) = 0$ and $p_1 \leq b_1$. Then there exists an integer-valued modular function m for which $p_1 \leq m \leq b_1$.*

Proof. Let S' and S'' be two copies of S. Let $V := S' \cup S''$, $E := \{s's'' : s \in S\}$ and $D = (V, E)$. Define $-f := g := \infty$ and for $X' \subseteq S', X'' \subseteq S''$ let $b(X' \cup X'') := b_1(X) - p_1(Y)$. Obviously, b is submodular. Apply Theorem 8.1 to this submodular flow problem. We claim that (8.4) holds since for a set $A = X' \cup Y''$, where $X \neq Y$ $(X, Y \subseteq S), \varrho_f(A) - \delta_g(A) = -\infty$. For a set $A = X' \cup X''$ $(X \subseteq S), \varrho_f(A) - \delta_g(A) = 0$, and then (8.4) follows from the assumption that $p_1(X) \leq b_1(X)$. By Theorem 8.1 there is a feasible submodular flow x, integer-valued if p_1 and b_1 are integer-valued. Defining $m(s) := x(s's'')$ we obtain the required modular function. ♠♠♠

It was proved in [Cunningham and Frank, 1985] that the faces of a submodular flow polyhedron are also submodular flow polyhedra. This and the feasibility theorem imply:

THEOREM 8.4 [Edmonds and Giles, 1977] *Submodular flow polyhedra are integral polyhedra.*

Note that Edmonds and Giles proved the stronger result that an optimization problem over a submodular flow polyhedron is "totally dual integral", that is, the dual linear program always has an integer-valued optimum if it has an optimum. The weighted version of the Lucchesi-Younger theorem (Theorem 6.1) is an interesting special case.

There are simpler polyhedra associated with sub- and supermodular functions. For set-functions p and b let us define the following polyhedra:

$$P(b) := \{x \in R^S : x \geq 0, x(A) \leq b(A) \text{ for every } A \subseteq S\} \qquad (8.5)$$

$$S(b) := \{x \in R^S : x(A) \leq b(A) \text{ for every } A \subseteq S\} \qquad (8.6)$$

$$B(b) := \{x \in R^S : x(S) = b(S), x(A) \leq b(A) \text{ for every } A \subseteq S\} \qquad (8.7)$$

$$C(p) := \{x \in R^S : x \geq 0, x(A) \geq p(A) \text{ for every } A \subseteq S\} \qquad (8.8)$$

$$Q(p,b) := \{x \in R^S : p(A) \leq x(A) \leq b(A) \text{ for every } A \subseteq S\}. \qquad (8.9)$$

When b is a polymatroid function, $P(b)$ is called a **polymatroid.** If b is fully submodular, $S(b)$ is called a **submodular polyhedron** and $B(b)$ a **basis polyhedron.** If p is fully supermodular and monotone increasing, then $C(P)$ is a **contra-polymatroid.** Finally if (p,b) is a strong pair, then $Q(p,b)$ is called a **generalized polymatroid** (in short, **g-polymatroid.)**

The name is justified by the observation that polymatroids, contra-polymatroids, basis polyhedra, and submodular polyhedra are g-polymatroids. Properties of these polyhedra were extensively studied in [Frank and Tardos, 1988]. Here we briefly mention some of these.

Since $p \leq b$ follows from (8.2), the Discrete Separation Theorem immediately shows that $Q(p,b)$ is never empty. Because sometimes g-polymatroids are defined by weaker functions, it will be convenient to declare the empty set to be a g-polymatroid. It is not difficult to prove that non-empty g-polymatroid Q uniquely determines its defining strong pair, namely, $b(X) := \max(x(X) : x \in Q)$ and $p(X) := \min(x(X) : x \in Q)$.

This generalizes an earlier result of J. Edmonds [1970] for polymatroids. Even the more general statement that g-polymatroids are integral poly-hedra is true.

In applications we often encounter intersecting, crossing or x-sub-modular (or supermodular) functions. How can we work with them? The major tool is truncation and bi-truncation. The (lower) **truncation** of a set-function b' is defined by $b(X) := \min(\sum b'(X_i) : \{X_1, \ldots, X_t\}$ a parti-tion of X). (The upper truncation is defined by replacing min with max. We will exclusively use lower truncation for submodular functions and upper truncation for supermodular functions and therefore the adjectives "lower" and "upper" will be left out.)

TRUNCATION THEOREM 8.5 [Lovász, 1977] *The truncation b of an intersecting submodular function b' is fully submodular and $S(b') = S(b)$.*

Combining the Truncation and the Discrete Separation Theorems we get:

THEOREM 8.6 *Let b' be an intersecting submodular function and p' an intersecting supermodular function (both integer-valued). There exists a (finite) integer-valued modular function m for which $p' \leq m \leq b'$ if and only if $\sum p'(F_i) \leq \sum b'(G_j)$ holds for every two sub-partitions of S $\{F_i\}$, $\{G_j\}$ for which $\cup F_i = \cup G_j$.*

Suppose now that b'' is a crossing submodular function with $k = b''(S)$ so that the polyhedron $B(b'')$ is non-empty. Define p'' by $p''(X) := k - b''(S - X)$ and let p' denote the upper truncation of p''. Let b' be defined by $b'(X) = k - p'(S - X)$ and let b denote the lower truncation of b'. We call b the **bi-truncation** of b''.

By two applications of the Truncation Theorem we obtain:

BI-TRUNCATION THEOREM 8.7 *The bi-truncation b of a crossing submodular function b'' is fully submodular and $B(b) = B(b'')$.*

Using this concept one can derive:

THEOREM 8.8 *For any weak pair (p', b') the polyhedron $Q = Q(p', b')$ is an integral g-polymatroid. Q is non-empty if and only if*

$$\sum b'(Z_i) \geq p'(\cup Z_i) \text{ and} \qquad (8.10a)$$

$$\sum p'(Z_i) \le b'(Z_i) \qquad\qquad (8.10b)$$

holds for every sub-partition $\{Z_i\}$ of S.

An important corollary of this theorem is:

COROLLARY 8.9 *Let $f : S \to \mathbf{Z} \cup \{-\infty\}, g : S \to \mathbf{Z} \cup \{+\infty\}$ be two functions with $f \le g$. Let $B_1 := \{x \in \mathbf{R}^S, x \ge f\}, B_2 := \{x \in \mathbf{R}^S, x \le g\}$ and Q a g-polymatroid. Then $Q_{12} := Q \cap B_1 \cap B_2$ is a g-polymatroid. Q_{12} is non-empty if and only if both $Q \cap B_1$ and $Q \cap B_2$ are non-empty.*

This theorem is the basis of each occurance of the linking principle mentioned in the paper. Another consequence of Theorem 8.7 is:

COROLLARY 8.10 *Given a digraph $D = (V, E)$ and a crossing submodular function b'' with $b''(V) = 0$, the polyhedron $Q(f, g; b'')$ is an integral submodular flow polyhedron. [Edmonds, 1970] If b' is a non-negative finite-valued intersecting submodular function, then $P(b')$ is a polymatroid.*

By using truncation and bi-truncation it is possible to extend the Feasibility Theorem to submodular flow polyhedra given by intersecting or crossing submodular functions.

THEOREM 8.11 *When b' is an intersecting submodular function, the submodular flow polyhedron $Q(f, g; b')$ contains an integer point if and only if*

$$\varrho_f(\cup X_i) - \delta_g(\cup X_i) \le \sum b'(X_i) \qquad\qquad (8.11a)$$

holds for every sub-partition $\{X_i\}$ of V.

When b'' is a crossing submodular function, $Q(f, g; b'')$ contains an integer point if and only if

$$\varrho_f(\cup X_i) - \delta_g(\cup X_i) \le \sum b'(X_{ij}) \qquad\qquad (8.11b)$$

holds for every sub-partition $\{X_i\}$ of V such that every X_i is the intersection of co-disjoint sets $X_{i1}, X_{i2}, \dots, X_{il}$.

A statement analogous to the second part of Corollary 8.10 can be formulated for contra-polymatroids and intersecting supermodular functions. Rather than stating this result we formulate a more general one:

THEOREM 8.12 *Let p^* be an x-supermodular function. Then $C(p^*)$ is a contra-polymatroid whose unique (monotone, fully supermodular) defining function p is given by $p(X) := \max(\sum p^*(X_i))$ where the maximum is taken over all sub-partitions $\{X_i\}$ of X.*

This theorem (proved in [Frank, 1992a] made it possible that minimum node-cost and degree-constrained augmentation problems to become tractable.

An equivalent form of Theorem 8.8 was proved by S. Fujishige [1984]:

THEOREM 8.8A *Let b'' be a crossing submodular function. The basis polyhedron $B = B(b'')$ contains an integer point if and only if both*

$$\sum b''(Z_i) \geq b''(S) \text{ and} \tag{8.12a}$$

$$\sum b''(S - Z_i) \geq (t-1)b''(S) \tag{8.12b}$$

hold for every partition $\{Z_1, \ldots, Z_t\}$. If b'' is integer-valued and (8.12) holds, B contains an integer-valued element.

J. Edmonds [1970] proved that one can optimize a linear objective function over a polymatroid with the help of a greedy algorithm. The same can be proved for g-polymatroids provided that Q is defined by a strong pair. When Q is defined by a weak pair, than a more sophisticated procedure is required (see the bi-truncation algorithm in [Frank and Tardos, 1988]). The polymatroid intersection theorem proved by J. Edmonds [1970] can be extended to g-polymatroids as well. This is stated in the next theorem along with an interesting relationship between submodular flow polyhedra and g-polymatroid intersections.

THEOREM 8.13 *The intersection of two integral g-polymatroids is an integral polyhedron. A polyhedron Q is a submodular flow polyhedron if and only if Q is the projection of the intersection of two g-polymatroids. Where (p_i, b_i) $(i = 1, 2)$ are strong pairs, the intersection of two polymatroids $Q(p_i, b_i)$ $(i = 1, 2)$ is non-empty if and only if $p_1 \leq b_2$ and $p_2 \leq b_1$.*

9. APPLICATIONS

Let us turn to the relationship of this theory with the applications in preceding sections.

9.1 ORIENTATIONS Let $G = (V, E)$ be an undirected graph and $m : V \to \mathbf{Z}$ an integer-valued vector. We say that m is an **in-degree vector** if there is an orientation of G in which $\varrho(v) = m(v)$ for every $v \in V$. By Theorem 3.1 m is an in-degree vector if and only if

$$m(X) \geq |E(X)| \text{ for every } X \subseteq V \qquad (9.1a)$$

and

$$m(V) = |E| \qquad (9.1b)$$

Proof of Theorem 3.5. The second part easily follows from the first one and we prove only the sufficiency of (3.9). Define p'' by $p''(X) = h(X) + |E(X)|$. Then p'' is crossing supermodular. (3.9a) is equivalent to $\sum p''(X_i) = \sum h(V_i) + |E| - e_{\mathcal{F}} \geq |E| = p''(V)$. (3.9b) is equivalent to $\sum p''(\bar{V}_i p) = \sum h(\bar{V}_i) + (n-1)|E| = (n-1)p''(V)$ where $n := |V|$. By applying Fujishige's Theorem 8.8A to $b'' := -p''$, we obtain that there is an integer vector m. Since $h \geq 0$, (9.1) holds for this m and hence there is an orientation with in-degree vector m. This orientation satisfies (3.8). ♠♠♠

This derivation also shows that the in-degree vectors that belong to orientations satisfying (3.8) form the integer points of a g-polymatroid (in particular, a basis polyhedron). Therefore Corollary 8.9 may be applied and the linking principle formulated in 3.6 follows.

Proof of Theorem 3.7. First let us take an arbitrary orientation D of G. We use this as a reference orientation. Let A denote the set of oriented edges. An orientation of G will be defined by a vector $x : D \to \{0, 1\}$ such that $x(a) = 0$ means that we leave a alone while $x(a) = 1$ means that we reverse the orientation of a. The orientation defined this way will satisfy (3.12) if and only if $\varrho_D(X) - \varrho_x(X) + \delta_x(X) \geq h(X)$ for every $X \subseteq V$. Equivalently, $\varrho_x(X) - \delta_x(X) \leq b''(X) := \varrho_D(X) - h(X)$.

Now b'' is crossing submodular. An easy calculation shows that (3.13) is equivalent to (8.11b). By Theorem 8.11 There is an integer point in the submodular flow polyhedron and this corresponds to an orientation satisfying the requirements in Theorem 3.7. ♠

Because there are strongly polynomial time algorithms for solving the minimum cost submodular flow problem via the reduction described above, we can find algorithmically a minimum cost orientation of G satisfying (3.12) in the case when the two possible orientations of every edge have different costs. This algorithm needs a certain oracle (or subroutine) concerning h in (3.12). In the special case when we want to make a mixed graph k-edge-connected by orienting the undirected edges, a strongly polynomial algorithm is available for the desired oracle.

9.2 PACKING AND COVERING WITH ARBORESCENCES

Let $D = (V, E)$ be a digraph and k a positive integer. We call a vector $m :\to \mathbf{Z}_+$ a **root-vector** if there is a family $\mathcal{F}$ of k disjoint spanning arborescences in D so that each node $v \in V$ is the root of $m(v)$ members of $\mathcal{F}$. It directly follows from Edmonds' Theorem 5.1 that m is a root-vector if and only if

$$m(X) \geq p'(X) := k - \varrho(X) \qquad (9.2)$$

holds for every $\emptyset \subset X \subseteq V$ and $m(V) = k$. Define $p'(\emptyset) = 0$. Now p' is an intersecting supermodular function and hence the the root-vectors are precisely the integer points of the following g-polymatroid $Q := \{x \in \mathbf{R}_+ : x(X) \geq p'(X) \text{ for every } X \subseteq V\}$.

Let $B_1 := \{x \in \mathbf{R}^S, x \geq f\}$, $B_2 := \{x \in \mathbf{R}^S, x \leq g\}$. Then $Q \cap B_i$ is a g-polymatroid ($i = 1, 2$), applying Theorem 8.8 to B_1 and to B_2 we obtain parts (i) and (ii) of Theorem 5.8. Part (iii) follows from Corollary 8.9. ♠♠♠

From this approach we can see that minimum cost versions of the arborescence problem can also be solved.

9.3 AUGMENTATIONS Let G be a connected undirected graph and $r(x, y)$ a demand function. To indicate the link between augmentations and g-polymatroids we derive Theorem 6.10 in the special case when the starting graph is undirected:

It is possible to add a set F of γ new undirected edges to G so that in the resulting graph G', $\lambda(x,y;G') \geq r(x,y)$ for every $x,y \in V$ if and only if

$$\sum(R(X_i) - d_G(X_i)) \geq 2\gamma \qquad (9.3)$$

for every sub-partition $\{X_i\}$ of V, where $R(X) := \max(r(x,y) : X$ separates x and y).

Proof. We call a set F **feasible** if it satisfies the requirements of the theorem. Let us call a vector $m : V \to \mathbf{Z}_+$ an **augmentation vector** if there is a feasible subset F of edges such that $m(v) = d_F(V)$ for every $v \in V$. As a direct consequence of Mader's undirected splitting theorem we have that m is an augmentation vector if and only if $m(V)$ is even and

$$m(X) + d_G(X) \geq R(X). \qquad (9.4)$$

(Indeed, add to G $m(v)$ parallel edges between a new node s and v for every $v \in V$ and apply Theorem 2.2 of Mader $\gamma := m(V)/2$ times. The resulting set F of γ new edges is feasible and $m(v) = d_F(v)$ for every $v \in V$.)

Let $p^*(X) := R(X) - d_G(X)$ if $\emptyset \subset X \subset V$ and $p^*(\emptyset) = p^*(V) := 0$. Then p^* is an x-supermodular function. By Theorem 8.12 $C(p^*)$ is a contra-polymatroid.

Combining Theorem 8.12 with Theorem 8.8 we obtain that the basis polyhedron $C(p^*) \cap \{x \in \mathbf{R}^V : x(V) = 2\gamma\}$ contains an integer point m if and only if

$$\sum p^*(X_i) \leq 2\gamma \qquad (9.5)$$

holds for every sub-partition $\{X_i\}$ of V. But (9.5) follows from (9.3), and hence there is an augmentation vector m with $m(V) = 2\gamma$ and the theorem follows. ♠♠♠

An analogous reduction technique can be applied to directed augmentation problems. The relationship between augmentations and g-polymatroids makes it clear why minimum node-cost and degree-constrained augmentation problems are also tractable.

REFERENCES

[1993] J. Bang-Jensen, A. Frank and B. Jackson, Preserving and increasing local edge-connectivity in mixed graphs, SIAM J. Discrete Mathematics, submitted.

[1980] F. Boesch, R. Tindell, Robbins's theorem for mixed multigraphs, Am. Math. Monthly 87 (1980) 716-719.

[1983] Cai Mao-Cheng, Arc-disjoint arborescences of digraphs, J. Graph Theory, Vol.7, No.2, (1983), 235-24.

[1977] B.V. Cherkassky, A solution of a problem of multicommodity flows in a network, Ekon. Mat. Metody 13 (1), 143-151 (in Russian).

[1985] W. Cunningham and A. Frank, A primal-dual algorithm for submodular flows, Mathematics of Operations Research, Vol. 10, No. 2 (1985), 251-261.

[1959] A.L. Dulmage and N.S. Mendelsohn, A structure theory of bipartite graphs of finite exterior dimension, Transactions of the Royal Society of Canada, Third series, Section III, 53 (1959)1-13.

[1970] J. Edmonds, Submodular functions, matroids, and certain polyhedra in: Combinatorial Structures and their applications (R. Guy, H. Hanani, N. Sauer, and J. Schönheim, eds.) Gordon and Breach, New York, 69-87.

[1973] J. Edmonds, Edge-disjoint branchings in: Combinatorial Algoithms (B. Rustin, ed.), Acad. Press, New York, (1973), 91-96.

[1979] J. Edmonds, Matroid intersection, Annals of Discrete Math. 4, (1979) 39-49.

[1977] J. Edmonds and R. Giles, A min-max relation for submodular functions on graphs, Annals of Discrete Mathematics 1, (1977), 185-204.

[1976] K.P. Eswaran and R.E. Tarjan, Augmentation problems, SIAM J. Computing, Vol. 5, No. 4 December 1976, 653-665.

[1962] L.R. Ford, D.R. Fulkerson, Flows in Networks, Princeton Univ. Press, Princeton NJ. 1962.

[1980] S. Fortune, J. Hopcroft and J. Wyllie, The directed subgraph homeomorphism problem, Theoretical Computer Science, 10 (1980) 111-121.

[1978] A. Frank, On disjoint trees and arborescences, in: Algebraic Methods in Graph Theory, Colloquia Mathematica, Soc. J. Bolyai, 25 (1978) 159-169. North-Holland.

[1980] A. Frank, On the orientation of graphs, J. Combinatorial Theory, Ser B. , Vol. 28, No. 3 (1980) 251-261).

[1981a] A. Frank, How to make a digraph strongly connected, Combinatorica 1 No. 2 (1981) 145-153.

[1981b] A. Frank, Generalized polymatroids, in: Finite and infinite sets (Eger 1981), Colloquia Mathematica Soc. J. Bolyai, 37, 285-294 North-Holland.

[1989] A. Frank, On connectivity properties of Eulerian digraphs, Annals of Discrete Math. 41 (1989) 179-194 (North Holland).

[1990] A. Frank, Packing paths, circuits and cuts - a survey, in: "Paths, Flows and VLSI-Layouts" (B. Korte, L. Lovász, H-J. Prömel, A. Schrijver, eds) pp. 47-100. (1990) Springer Verlag.

[1992a] A. Frank, Augmenting paths to meet edge-connectivity requirements, SIAM J. Discrete Mathematics, (1992 February), Vol.5, No 1., pp.22-53.

[1992b] A. Frank, On a theorem of Mader, Proceedings of the Graph Theory Conference held in Denmark, 1990, Annals of Discrete Mathematics, Vol. 101,(1992) 49-57.

[1993] A. Frank, Submodular functions in graph theory, Proceedings of the conference held in Marseille, 1993, to appear.

[1976] A. Frank and A. Gyárfás, How to orient the edges of a graph, in: Combinatorics, (Keszthely 1976), Coll. Math. Soc. J. Bolyai 18, 353-364, North-Holland.

[1992] A. Frank, A. Karzanov and A. Sebő, On multiflow problems, in: Proceedings of the 2nd IPCO Conference, Pittsburgh, 1992.

[1988] A. Frank and É. Tardos, Generalized polymatroids and submodular flows, Mathematical Programming, Ser. B. 42 (1988) 489-563.

[1989] A. Frank and É. Tardos, An application of submodular flows, Linear Algebra and its Applications, 114/115, (1989) 329-348.

[1993] A. Frank, D. Wagner, K. Weihe, Edge-disjoint paths in planar digraphs, in preparation.

[1991] S. Fujishige, Submodular functions and optimization, (Monograph), Annals of Discrete Mathematics, 47, North-Holland.

[1984] S. Fujishige, Sructures of polyhedra determined by submodular functions on crossing families, Mathematical Programming, 29 (1984) 125-141.

[1971] D.R. Fulkerson and L.S. Shapley, Minimal k-arc-connceted graphs, Networks 1, (1971) 91-98.

[1963] T.C. Hu, Multicommodity network flows, Oper.Res. 11 (1963) 344-360.

[1986] Y. Kajitani and S. Ueno, The minimum augmentation of directed tree to a k-edge-connected directed graph, Networks, 16 (1986) 181-197.

[1985] A.V. Karzanov, On multicommodity flow problems with integer-valued optimal solutions, Dokl. Akad. Nauk SSSR Tom 280, No. 4 (English translation: Soviet Math. Dokl. Vol. 32 (1985) 151-154).

[1979] A.V. Karzanov, Combinatorial methods to solve cut-determined multiflow problems. in: Combinatorial methods for flow problems (Inst. for System Studies, Moscow, iss.3) 6-69 (in Russian).

[1984] A.V. Karzanov, A generalized MFMC-property and multicommodity cut problems, in: Finite and Infinite Sets, (Proc. 6th Hungarian Combinatorial Coll., Eger, North-Holland, Amsterdam, 443-486.

[1978] A.V. Karzanov and M.V. Lomonosov, Multiflows in undirected graphs, in: Mathematical Programming, The problems of the social and economic systems, Operations Research model 1 (in Russian) The Institute for System Studies, Moscow.

[1979] A.V. Karzanov and V.A. Pevzner, A characterization of the class of cut-non-determined maximum multicommodity flow problems, in: Combinatorial methods for flow problems, iss.3, Inst. for System Studies, Moscow 70-81, 1979 (in Russian).

[1985] M.V. Lomonosov, Combinatorial approaches to multiflow problems, Discrete Applied Mathematics, Vol 11, (special issue), No. 1, 1-93.

[1974] L. Lovász, Conference on Graph Theory, lecture, Prague, 1974.

[1976] L. Lovász, On two min-max theorems in graph theory, J. Combinatorial Theory, Ser. B, 21 (1976) 26-30.

[1976] L. Lovász, On some connectivity properties of Eulerian graphs, Acta Mat. Akad. Sci. Hungaricae. 28 (1976) 129-138.

[1977] L. Lovász, Flats in matroids and geometric graphs, in: P.J. Cameron (ed.) Combinatorial Surveys, Proceedings of the Sixth British Combinatorial Conference, academic Press, London, 1977, 45-86.

[1979] L. Lovász, Combinatorial Problems and Exercises, North-Holland, 1979.

[1983] L. Lovász, Submodular functions and convexity. In: Mathematical Programming - The State of the Art (eds.: A. Bachem, M. Grötschel. B. Korte) Springer 1983, 235-257.

[1978] C.L. Lucchesi and D.H. Younger, A minimax relation for directed graphs, J. London Math. Soc. (2) 17 (1978) 369-374.

[1978] W. Mader, A reduction method for edge-connectivity in graphs, Ann. Discrete Math. 3 (1978) 145-164.

[1982] W. Mader, Konstruktion aller n-fach kantenzusammenhängenden Digraphen, Europ. J. Combinatorics (1982) 3, 63-67.

[1993] M. Middendorf and F. Pfeiffer, On the complexity of the disjoint paths problem, Combinatorica, 13 (1993) 97-104.

[1960] C.St.J.A. Nash-Williams, On orientations, connectivity and odd vertex pairings in finite graphs, Canad. J. Math. 12 (1960) 555-567.

[1964] C.St.J.A. Nash-Williams, Decomposition of finite graphs into J. London Math. Soc. 39 (1964) 12.

[1969] C.St.J.A. Nash-Williams, Well-balanced orientations of finite graphs and unobtrusive odd-vertex-pairings, in: Recent Progress of Combinatorics, pp. 133-149.

[1939] H.E. Robbins, A theorem on graphs with an application to a problem of traffic control, American Math. Monthly 46 (1939) 281-283.

[1986] N. Robertson and P.D. Seymour, Graph minors XIII: The disjoint paths problem, J. Combinatorial Theory (B), to appear.

[1966] B. Rothschild and A. Whinston, Feasibility of two-commodity network flows, Operations Research 14, 1121-1129.

[1985] A. Schrijver, Supermodular colorings, in Matroid Theory, (A. Recski and L. Lovász, eds.) Colloq. Math. Soc. J. Bolyai, 40, North-Holland, 1985, 327-344.

[1979] Y. Shiloach, Edge-disjoint branchings in directed multigraphs, Inform. Proc. Letters 8, (1979) 24-27.

[1961] W.T. Tutte, On the problem of decomposing a graph into n connected factors, J. London Math. Soc. 36 (1961) 221-230.

[1978] K. Vidyasankar, Covering the edge-set of a directed graph with trees, Discrete Math. 24 (1978) 79-85.

WEIGHTED QUASIGROUPS

Anthony J.W. Hilton and Jerzy Wojciechowski

in honour of

Crispin St.J.A. Nash–Williams

Abstract.   We introduce the concept of a weighted quasigroup.   We show that, corresponding to any weighted quasigroup, there is a quasigroup from which it can be obtained in a certain natural way, which we term amalgamation.   Not all commutative weighted quasigroups can be obtained from commutative quasigroups by amalgamation; however, given a commutative weighted quasigroup whose weights are all even, we give a necessary and sufficient condition for the existence of a commutative quasigroup from which it can be obtained by amalgamation.   We also discuss conjugates of weighted quasigroups.

We also introduce the concept of a simplex zeroid, and relate this concept to that of a weighted quasigroup.

1.   <u>Weighted quasigroups.</u>

Suppose that we have a finite set S with a closed binary operation. If $a,b \in S$, we shall denote the result of this binary operation acting on a and b by ab. If the binary operation has the two properties

(i)   for each $a,b \in S$, the equation $ax = b$ is uniquely solvable for x, and

(ii)   for each $a,b \in S$ the equation $ya = b$ is uniquely solvable for y,

then S is a **quasigroup**. It is well—known, and easy to see, that S is a quasigroup if and only if its multiplication table is a latin square. The properties (i) and (ii) amount to the assertion that, in the multiplication table, each element of S occurs exactly once in each row and exactly once in each column.

Here we introduce a generalization of this concept. Let $S = \{d_1,...,d_s\}$. Suppose that each element a of S is assigned a weight $w(a)$, where $w(a)$ is a positive integer. The product ab is a multiset $\{d_1^{\alpha_1(a,b)}, d_2^{\alpha_2(a,b)},...,d_s^{\alpha_s(a,b)}\}$, where the notation $d_i^{\alpha_i(a,b)}$ means that d_i occurs $\alpha_i(a,b)$ times in the product ab, $\alpha_i(a,b)$ being a non—negative integer; we refer to $\alpha_i(a,b)$ as the **multiplicity** of d_i in the product ab. We require that

$$\alpha_1(a,b) + \alpha_2(a,b) + ... + \alpha_s(a,b) = w(a)w(b).$$

Thus the product ab contains $w(a)w(b)$ elements, counting multiplicities. For each pair $a,b \in S$ and for each $i \in \{1,...,s\}$, let $\gamma_i(a,b)$ be the number of times that b occurs in the product ad_i.

Then $\gamma_1(a,b) + \gamma_2(a,b) + ... + \gamma_s(a,b)$ is the total number of times that b occurs in the products $ad_1, ad_2, ..., ad_s$, i.e. in the row of the multiplication table of the weighted quasigroup S corresponding to the premultiplicand a. Similarly, for each pair $a,b \in S$ and for each $i \in \{1,...,s\}$ let $\delta_i(a,b)$ be the number of times that b occurs in the product $d_i a$. Then $\delta_1(a,b) + \delta_2(a,b) + ... + \delta_s(a,b)$ is the total number of times that b occurs in the products $d_1 a, d_2 a, ..., d_s a$, i.e. in the column of the multiplication table of the weighted quasigroup S corresponding to the postmultiplicand a. S is a **weighted quasigroup**, or a **W–quasigroup**, if

$$\text{(i)} \qquad \gamma_1(a,b) + \gamma_2(a,b) + ... + \gamma_s(a,b) \;=\; w(a)w(b) \,,$$

and

$$\text{(ii)} \qquad \delta_1(a,b) + \delta_2(a,b) + ... + \delta_s(a,b) \;=\; w(a)w(b) \,.$$

Then (i) indicates that b occurs a total of $w(a)w(b)$ times in "row a" of the multiplication table, and (ii) indicates that b occurs $w(a)w(b)$ times in "column a" of the multiplication table.

We remark that when $w(d_i) = 1$ $(1 \le i \le s)$ then conditions (i) and (ii) in the definition of a W–quasigroup reduce to conditions (i) and (ii) in the definition of a quasigroup.

A further minor comment is that if the elements of the W–quasigroup are $d_1,...,d_s$ then $\alpha_i(d_j,d_k) = \gamma_k(d_j,d_i) = \delta_j(d_k,d_i)$.

Let us give an example to illustrate the concept of a weighted quasigroup.   Let $S = \{a,b,c,d\}$ and let $w(a) = 3$, $w(b) = 2$, $w(c) = 1$, $w(d) = 1$.   Let the products be given by the chart in Figure 1.

	a	b	c	d
a	a a a a b b b c d	a a a b b c	a c d	a b d
b	a a a b c d	a b b d	a b	a c
c	b b c	a d	a	a
d	a a d	a c	b	b

Figure 1

Notice, for example, that cells $(1,1)$ and $(1,2)$ contain $9 = w(a)w(a)$ and $6 = w(a)w(b)$ elements, counting repetitions, respectively, and that b occurs $6 = w(a)w(b)$ times in row 1 and $4 = w(b)w(b)$ times in column 2.   Also, for example, if we call $a = d_1$, $b = d_2$, $c = d_3$ and $d = d_4$ we have that $\gamma_1(a,b) = 3$, $\gamma_2(a,b) = 2$, $\gamma_3(a,b) = 0$ and $\gamma_4(a,b) = 1$, so $\gamma_1 + \gamma_2 + \gamma_3 + \gamma_4 = 6 = 3 \times 2 = w(a)w(b)$.   Similarly $\delta_1(a,b) = 3$, $\delta_2(a,b) = 1$, $\delta_3(a,b) = 2$ and $\delta_4(a,b) = 0$, so $\delta_1 + \delta_2 + \delta_3 + \delta_4 = 6 = w(a)w(b)$.

In the same way that the multiplication table of a quasigroup is a latin square, the multiplication table of a weighted quasigroup is an outline latin square.   We now proceed to define these.

A **composition** of a positive integer n is a sequence $(a_1,...,a_r)$ of positive integers such that $a_1 + ... + a_r = n$.   Let $P = (p_1,p_2,...,p_s)$, $Q = (q_1,...,q_t)$ and $r = (r_1,...,r_u)$ be three compositions of the same

integer n.   A (P,Q,R)–outline latin rectangle is an s x t matrix on
symbols $\sigma_1,...,\sigma_u$ such that

(i) cell (i,j) contains $p_i q_j$ symbols, counting repetitions,

(ii) row i contains symbol σ_k $p_i r_k$ times, and

(iii) column j contains symbol σ_k $q_j r_k$ times.

The following proposition is easy.

Proposition 1.   Let   Q   be a weighted quasigroup with elements
$d_1,...,d_s$.   The multiplication table of Q is a (W,W,W)–outline latin
square, where $W = (w(d_1),$ $w(d_2),...,w(d_s))$.

Proof.   We need only observe that

(i) cell (i,j) of the multiplication table is the product $d_i d_j$, and so it
contains $w(d_i)w(d_j)$ elements, counting repetitions;

(ii) row i of the multiplication table contains element d_k $\gamma_1(d_i,d_k)$ +
$\gamma_2(d_i,d_k)$ + ... + $\gamma_s(d_i,d_k) = w(d_i)w(d_k)$ times; and

(iii) column j of the multiplication table contains element d_k $\delta_1(d_j,d_k)$
+ $\delta_2(d_j,d_k)$ + ... + $\delta_s(d_j,d_k) = w(d_j)w(d_k)$ times.   □

The weighted quasigroup depicted in Figure 1 can be obtained
in a natural way from the quasigroup depicted in Figure 2.

	a_1	a_2	a_3	b_1	b_2	c	d
a_1	a_1	a_2	a_3	b_1	b_2	c	d
a_2	b_1	b_2	a_1	c	a_2	d	a_3
a_3	c	d	b_2	a_1	a_3	a_2	b_1
b_1	a_2	a_3	b_1	b_2	d	a_1	c
b_2	d	c	a_2	a_3	b_1	b_2	a_1
c	b_2	b_1	c	d	a_1	a_3	a_2
d	a_3	a_1	d	a_2	c	b_1	b_2

Figure 2

All we have to do is to replace a_1, a_2 and a_3 with a, which has weight 3, and replace b_1 and b_2 with b, which has weight 2, and remove all the appropriate (i.e. the dotted) cell boundaries.

Our first result is that, for any weighted quasigroup Q, there is always a quasigroup from which Q can be obtained by this process. In order to state this fact formally we need one further definition and one further proposition.

Let $S = \{d_1,...,d_s\}$ and let $w(d_1),...,w(d_s)$ be positive integers. Let Q^* be a quasigroup on the elements

$$d_{11},...,d_{1w(d_1)}, d_{21},...,d_{2w(d_2)},...,d_{s1},...,d_{sw(d_s)}.$$

Let Q be the weighted quasigroup obtained from Q^* by replacing $d_{i1},...,d_{iw(d_i)}$ by d_i with weight $w(d_i)$ for each $i \in \{1,...,s\}$, and, for each $i,j \in \{1,...,s\}$, by defining the product $d_i d_j$ to be the multiset $\{d_1^{\alpha_1(d_i,d_j)},...,d_s^{\alpha_s(d_i,d_j)}\}$, where

$$\alpha_k(d_i, d_j) = \sum_{t=1}^{w(d_k)} \alpha_{(k,t)}(i,j)$$

and $\alpha_{(k,t)}(i,j)$ is the total number of times that d_{kt} occurs as a product in Q^* of one of $d_{i1},...,d_{iw(d_i)}$ with one of $d_{j1},...,d_{jw(d_j)}$ We call Q the **W–amalgamation** of Q^*, where W is the composition $(w(d_1),...,w(d_s))$ of $w(d_1) + ... + w(d_s)$.

<u>Proposition 2</u>. If Q is the W–amalgamation of Q^* (as described above), then Q is a weighted quasigroup on $d_1,...,d_s$ with weights $w(d_1),...,w(d_s)$.

<u>Proof</u>. Each of the $w(d_i)w(d_j)$ products of one of $d_{i1},...,d_{iw(d_i)}$ with one of $d_{j1},...,d_{jw(d_j)}$ in Q^* contributes one to the product d_id_j in Q. Thus the multiset d_id_j contains $w(d_i)w(d_j)$ elements.

Let a and b be an arbitrary pair of elements of Q. Suppose that the $w(a)$ elements of Q^* which were replaced by a in forming Q were $a_1,...,a_{w(a)}$ and that the $w(b)$ elements of Q^* which were replaced by b in forming Q were $b_1,...,b_{w(b)}$ In Q^* there is exactly one x satisfying $a_ix = b_j$ for each $i \in \{1,...,w(a)\}$ and $j \in \{1,...,w(b)\}$, and so in Q there are $w(a)w(b)$ solutions (counting repetitions) of the equation $ax = b$, i.e. if γ_i is the number of times that b occurs in the product ad_i then $\gamma_1 + ... + \gamma_s = w(a)w(b)$. Thus (i) of the definition of weighted quasigroups is satisfied. The proof of (ii) is similar.   □

We can now state our first noteworthy result; it is that the converse of Proposition 2 is true.

THEOREM 1.   Let Q be a weighted quasigroup on elements $d_1,...,d_s$
with weights $w(d_1),...,w(d_s)$.   Then there is a quasigroup Q^* with
$w(d_1) + ... + w(d_s)$ elements of which Q is the W—amalgamation.

In order to explain the proof we need one further definition, one
further   proposition   (Proposition   5),   and   a   lemma.   Let
$P = (p_1,...,p_s)$, $Q = (q_1,...,q_t)$ and $R = (r_1,...,r_u)$ again be three
compositions of   n.   Let   A   be a latin square of side   n   on symbols
$\sigma_{11},...,\sigma_{1r_1},\sigma_{21},...,\sigma_{2r_2},...,\sigma_{u1},...,\sigma_{ur_u}$.   The   **(P,Q,R)—amalgamation** of
the latin square   A   is the   s x t   matrix on symbols $\sigma_1,...,\sigma_u$ obtained
from   A   by placing each symbol σ_k in cell   (i,j)   each time one of
$\sigma_{k1},...,\sigma_{kr_k}$ occurs in the set of cells (ρ,τ) of   A,   where
$$\rho \in \{p_1 + ... + p_{i-1} + 1, ..., p_1 + ... + p_i\},$$
and
$$\tau \in \{q_1 + ... + q_{j-1} + 1, ..., q_1 + ... + q_j\}.$$
The following proposition is obvious.   The proof is given in   [9].

<u>Proposition 3</u>.   Let   B   be a (P,Q,R)—amalgamation of a latin square   A.
Then

 (i) cell (i,j) of   B   contains $p_i q_j$ symbols counting repetitions;

 (ii) row i of   B   contains symbol σ_k $p_i r_k$ times;

 (iii) column j of   B   contains symbol σ_k $q_j r_k$ times.

Proposition 3 can be reformulated as follows.

<u>Proposition 4</u>. Let B be a (P,Q,R)–amalgamation of a latin square A.
Then B is a (P,Q,R)–outline latin rectangle.

The next proposition draws together the notions of amalgamation
of quasigroups, and amalgamation of latin squares.

<u>Proposition 5</u>. Let Q be a weighted quasigroup with elements $d_1,...,d_s$
and weights $w(d_1),...,w(d_s)$.   Let $W = (w(d_1),...,w(d_s))$ and let   Q
have multiplication table   B.   Suppose that   B   is   the
(W,W,W)–amalgamation of a latin square A. Let Q^* be a quasigroup
with multiplication table A.   Then Q is the W–amalgamation of Q^*.

<u>Proof</u>.   Let   the   elements   of Q^* be
$d_{11},...,d_{1w(d_1)},d_2,...,d_{2w(d_2)},...,d_{s1},...,d_{sw(d_s)}$.   Let $\alpha_{(k,t)}(i,j)$ be the
number of times that d_{kt} occurs as a product in Q^* of one of
$d_{i1},...,d_{i(d_i)}$ with one of $d_{j1},...,d_{jw(d_j)}$. Since B is the multiplication
table of Q,   the symbols in   B   are $d_1,...,d_s$.   Since   B   is   the
(W,W,W)–amalgamation of A, we may suppose that B is obtained
from A by placing symbol d_k in cell (i,j) whenever one of
$d_{k1},...,d_{kw(d_k)}$ occur in the set of cells (ρ,τ) of A, where

$$\rho \in \{w(d_1) + ... + w(d_{i-1}) + 1, ..., w(d_1) + ... + w(d_i)\},$$

and

$$\tau \in \{w(d_1) + ... + w(d_{j-1}) + 1, ..., w(d_1) + ... + w(d_j)\}.$$

Let $\alpha_k(d_i,d_j) = \sum\limits_{t=1}^{w(d_k)} \alpha_{(k,t)}(i,j)$. Then d_k occurs $\alpha_k(d_i,d_j)$ times in cell (i,j). Since Q is the weighted quasigroup with multiplication table B, the product $d_i d_j$ is the multiset $\{d_1^{\alpha_1(d_i,d_j)},...,d_s^{\alpha_s(d_i,d_j)}\}$. Therefore Q is the W–amalgamation of Q^*.   □

The following result is proved in [6] (and in a slightly different form in [9]). The proof depends on a generalization due to de Werra [21, 22] of the well–known theorem of König [14] that the chromatic index of a bipartite multigraph equals its maximum degree.

Lemma 1. Let P, Q and R be compositions of n. Let B be a (P,Q,R)–outline latin rectangle. Then there is a latin square A whose (P,Q,R)–amalgamation is B.

We can now proceed to the proof of Theorem 1.

Proof of Theorem 1. Let Q be a weighted quasigroup on the elements $d_1,...,d_s$ with weights $w(d_1),...,w(d_s)$. Let $W = (w(d_1),...,w(d_s))$. Then, by Proposition 1, the multiplication table B of Q is a (W,W,W)–outline latin square. By Lemma 1, B is the (W,W,W)–amalgamation of a latin square A on $w(d_1) + ... + w(d_s)$ symbols. Let Q^* be a quasigroup which has A as its multiplication table. Then, by Proposition 5, Q is the W–amalgamation of Q^*.

                                                                        □

2.   Further algebraic analogues.

The truth of Theorem 1 raises the general question of what further algebraic properties can be added to the original quasigroup properties (i) and (ii) whilst retaining a theorem like Theorem 1. We show in Section 3 that, under certain circumstances, commutativity can be added, and in Section 5, via the use of conjugates, we find that, under certain circumstances, two other algebraic properties can be added. It would be interesting if yet further algebraic properties can be added; in particular it would be interesting if associativity can be added.

3.   Commutativity.

A weighted quasigroup with elements $d_1,...,d_s$ is **commutative** if $d_id_j = d_jd_i$ for $1 \le i < j \le s$. In this section we give an analogue of Theorem 1 for commutative weighted quasigroups. It is not true that every commutative weighted quasigroup is the W–amalgamation of some commutative quasigroup. However, under some not very restrictive conditions, such a statement is true.

In   a   commutative   W–quasigroup   we   have
$\alpha_i(d_j,d_k) = \alpha_i(d_k,d_j) = \gamma_k(d_j,d_i) = \gamma_j(d_k,d_i) = \delta_j(d_k,d_i) = \delta_k(d_j,d_i)$.

First a preliminary result.

<u>Proposition 6</u>. If Q^* is a commutative quasigroup and Q is the W–amalgamation of Q^* (as described just before Proposition 2) then Q is a commutative weighted quasigroup. Moreover, for $1 \le i \le s$, the number of elements with odd multiplicity in the product d_i^2 is at most $w(d_i)$.

<u>Proof</u>.   It is obvious that Q is commutative and it follows from Proposition 2 that Q is a weighted quasigroup.   Referring to the description just before Proposition 2, the product d_i^2 in Q is formed from the set of products of elements of $\{d_{i1},...,d_{iw(d_i)}\}$ with $\{d_{i1},...,d_{iw(d_i)}\}$. Since Q^* is commutative, if $j_1 \neq j_2$ the products $d_{ij_1} d_{ij_2}$ and $d_{ij_2} d_{ij_1}$ are the same. Thus if $d_{ij_1} d_{ij_2} = d_{kt}$ for some k and t, then the products $d_{ij_1} d_{ij_2}$ each contribute a separate d_k to d_i^2 , so together they contribute 2 to the multiplicity of d_k in d_i^2. The only products $d_{ij_1} d_{ij_2}$ which do not contribute 2 in this way are those with $j_1 = j_2$. There are $w(d_i)$ such products. Consequently the number of elements with odd multiplicity in the product d_i^2 is at most $w(d_i)$.   □

The interesting question is to what extent the converse of Proposition 6 is true.   If $w(d_1),...,w(d_s)$ are all even, the converse holds.

THEOREM 2.   Let Q be a commutative weighted quasigroup on elements $d_1,...,d_s$ with weights $w(d_1),...,w(d_s)$. Let $w(d_1),...,w(d_s)$ be even. Then there is a symmetric quasigroup Q^* with $w(d_1) + ... + w(d_s)$ elements of which Q is the W–amalgamation if and only if the number of elements with odd multiplicity in the product d_i^2 is at most $w(d_i)$.

Theorem 2 need not be true when the weights are not all even. For example, we give in Figure 3 a commutative weighted quasigroup Q

in which all the weights are 3 which is not the amalgamation of any commutative quasigroup.   To see that Q cannot be derived from a commutative quasigroup consider the multigraph G, shown in Figure 4, which corresponds to the element a, as follows.   If in   Q   the multiplicity of a in the product $d_i d_j$ is x, then d_i and d_j are joined by x edges in G.

	a	b	c	d	e	f
a	bbc cdd eef	aaa aab bcc	aaa abb bff	bbd ddd ddf	ccc cce fff	dee eee eff
b	aaa aab bcc	bbc cdd eef	aaa abc cff	ccd eef fff	bbb bdd dee	cdd dee eff
c	aaa abb bff	aaa abc cff	ccd dee eee	acc dde eee	ccc ddd fff	bbb bbd dff
d	bbd ddd ddf	ccd eef fff	acc dde eee	bbb bcc eee	aaa abb bcf	aaa acc fff
e	ccc cce fff	bbb bdd dee	ccc ddd fff	aaa abb bcf	dee eee eff	aaa aab bdd
f	dee eee eff	cdd dee eff	bbb bbd dff	aaa acc fff	aaa aab bdd	bbc ccc ccd

Figure 3.   The commutative weighted quasigroup Q

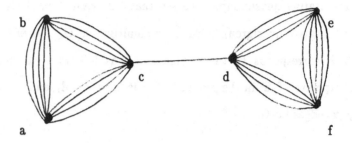

Figure 4   The multigraph G corresponding to a.

In any putative commutative quasigroup Q^* of which Q is the amalgamation, each vertex would be split into three vertices (say $a_1, a_2, a_3, b_1, b_2, b_3$, etc.), and the edges of G would be shared between each set of three vertices so that the graph would be simple (i.e. no multiple edges) and each vertex would have degree three. Moreover the edges would be coloured with a_1, a_2 and a_3 so that each vertex had one edge of each colour incident with it. Combining the vertices of each set together again would then yield an edge–colouring of G in which each vertex had three edges of each colour incident with it. Thus it would follow that G is the union of three edge–disjoint 3–regular spanning subgraphs, H_1, H_2 and H_3, say. But it is easy to see that this is not the case. For if the edge cd were in H_1, then H_1 must contain one bc edge, one ca edge and two ab edges. But then the nine remaining edges between the vertices a,b and c must be shared equally between H_2 and H_3, which is impossible.

Before proving Theorem 2 we need some more preliminary results.

The next proposition is the analogue for symmetric latin squares of Proposition 6. It corresponds to Proposition 3 for the non–symmetric case. The proof is similar to the proof of Proposition 6; it is discussed in [4]. As before, $P = (p_1,...,p_s)$ and $R = (r_1,...,r_u)$ are compositions of a positive integer n.

Proposition 7.   Let B be a (P,P,R)–amalgamation with symbols $\sigma_1,...,\sigma_u$ of a symmetric latin square A. Then

(i) cell (i,j) of B contains $p_i p_j$ symbols, counting repetitions;

(ii) row i and column i of B each contain symbol σ_k $p_i r_k$ times;

(iii) not more than p_i symbols occur in cell (i,i) an odd number of times.

In [4] the term **(P,R)–pseudo outline symmetric latin square** was coined for an s x s matrix on u symbols $\sigma_1,...,\sigma_u$ satisfying (i), (ii) and (iii) of Proposition 7. The term pseudo was included to emphasize the fact that such a square may not be the (P,P,R)–amalgamation of a symmetric latin square. Further examples to demonstrate this are given in [4]. With this terminology Proposition 7 can be reformulated thus:

Proposition 8.   Let B be a (P,P,R)–amalgamation of a symmetric outline latin square. Then B is a (P,R)–pseudo outline symmetric latin square.

Now we have defined pseudo outline symmetric latin squares, we can give the analogue we need of Proposition 1.

<u>Proposition 9</u>.   Let   Q   be a commutative weighted quasigroup with elements $d_1,...,d_s$ and weights $w(d_1),...,w(d_s)$.   Suppose also that, for $1 \leq i \leq s$,   the number of symbols with odd multiplicity in the product $d_i{}^2$ is at most $w(d_i)$.   Then the multiplication table of   Q   is a (W,W)–pseudo outline symmetric latin square, where $W = (w(d_1),...,w(d_s))$.

<u>Proof</u>.   Let   B   be the multiplication table of   Q.   It is obvious that   B is symmetric, and by Proposition 1,   B   is a   (W,W,W)–outline latin square.   The condition that the number of elements with odd multiplicity in the product d_i^2 translates into the condition that not more than $w(d_i)$ symbols occur an odd number of times in cell (i,i) of B.   It follows that   B   is a   (W,W)–pseudo outline symmetric latin square.   □

The   connection   between   amalgamations   of   commutative quasigroups and amalgamations of symmetric latin squares is stated explicitly in Proposition 10.

<u>Proposition 10</u>.   Let   Q   be a commutative weighted quasigroup with elements $d_1,...,d_s$ and   weights $w(d_1),...,w(d_s)$.   Let   W = $(w(d_1),...,w(d_s))$ and let   Q   have multiplication table B.   Suppose that B   is the   (W,W,W)–amalgamation of a symmetric latin square A.   Let

Q^* be a quasigroup with multiplication table A. Then Q is the W–amalgamation of Q^*.

Proof. This is the same as that of Proposition 5.

The next result lies at the core of the proof of Theorem 2. The proof of Lemma 2 is really quite difficult. It is given in [4] by Chetwynd and Hilton, and some of the main ideas stem from earlier work of Andersen and Hilton [1,2,3]. The proof depends on a generalization of Petersen's theorem [20] about 2–factorizing regular multigraphs of even degree, as well as on de Werra's theorem [21,22] on balanced edge–colourings of bipartite multigraphs.   □

Lemma 2.   Let B be a (P,R)–pseudo outline symmetric latin square. If $p_1,...,p_s$ are all even, then there is a latin square A whose (P,P,R)–amalgamation is B.

We can now prove Theorem 2.

Proof of Theorem 2.

Necessity.   Suppose Q^* is the W–amalgamation of Q. Then, by Proposition 6, the number of symbols with odd multiplicity in the product d_i^2 is at most $w(d_i)$.

Sufficiency.   Now suppose that Q is a commutative weighted quasigroup with the property that, for $1 \leq i \leq s$, the number of elements with odd multiplicity in the product d_i^2 is at most $w(d_i)$. Then, by Proposition 9, the multiplication table B of Q is a (W,W)–pseudo outline symmetric latin square.   By Lemma 2, since $w(d_1),...,w(d_s)$ are all even,   B   is the (W,W,W)–amalgamation of a symmetric latin square   A   on $w(d_1) + ... + w(d_s)$ symbols.   Let Q^* be a commutative quasigroup which has   A   as its multiplication table. Then,   by Proposition 10,   Q   is the W–amalgamation of Q^*.

□

4.   <u>Conjugates and orthogonal arrays</u>.

In this section we define conjugates of weighted quasigroups.   In the case when all the weights are one, our definition reduces to the usual definition of conjugates.

Let π be   a   permutation   of   (1,2,3).   The $(\pi(1),\pi(2),\pi(3))$–conjugate of   Q,   denoted by $Q_{\pi(1,2,3)}$ is formed as follows.   If in   Q d_k occurs in $d_i d_j$ $\alpha_k(d_i,d_j)$ times then in $Q_{\pi(1,2,3)}$ $d_{\pi(k)}$ occurs in $d_{\pi(i)} d_{\pi(j)}$ $\alpha_k(d_i,d_j)$ times, where $\pi(i)$, $\pi(j)$, $\pi(k)$ denote the permutation of the triple   (i,j,k)   such that   i   occurs in position $\pi(1)$,   j   occurs in position $\pi(2)$,   and   k   occurs in position $\pi(3)$.

To illustrate this, in Figure 5 we give the conjugates $Q_{(3,2,1)}$ and $Q_{(1,3,2)}$ of the weighted quasigroup   Q   of   Figure 1.   The multiplication tables of the remaining conjugates, $Q_{(2,1,3)}$, $Q_{(2,3,1)}$ and $Q_{(3,1,2)}$ are merely transposes of the multiplication tables of the quasigroups $Q = Q_{(1,2,3)}$, $Q_{(3,2,1)}$ and $Q_{(1,3,2)}$ respectively.

	a	b	c	d
a	a a a a b b b d d	a a a b c d	a b c	a b c
b	a a a b c c	a a b b	b d	a d
c	a b c	a d	a	b
d	a b d	b c	a	a

$$Q_{(3,2,1)}$$

	a	b	c	d
a	a a a a b b b c d	a a a b b d	a b c	a c d
b	a a a b c d	a b b c	a d	a b
c	b c d	a a	a	b
d	a a b	c d	b	a

$$Q_{(1,3,2)}$$

Figure 5

Just as a quasigroup can be represented by an orthogonal array, so can a weighted quasigroup be represented by a weighted orthogonal array. Let $\{d_1,...,d_s\}$ be a finite set, let $w(d_1),...,w(d_s)$ be positive integers, and let $W = (w(d_1),...,w(d_s))$ be a composition of $w(d_1) + ... + w(d_s)$. Let $w(d_1) + ... + w(d_s)$ be denoted by $|W|$. A **W-orthogonal array** is a $|W|^2 \times 3$ matrix in which in any pair of columns each ordered pair (d_i,d_j) occurs $w(d_i)\,w(d_j)$ times $(1 \leq i \leq s,$ $1 \leq j \leq s)$.

A W-quasigroup can be used to construct a W-orthogonal array as follows. The first pair of columns consists of each ordered pair (d_i,d_j) written $w(d_i)\,w(d_j)$ times. The third column of the $(w(d_i)$ $w(d_j)) \times 3$ submatrix, whose first and second columns consist entirely of the d_i's and d_j's respectively, contains each element d_k $\alpha_k(d_i,d_j)$

times.   Since

$$\alpha_1(d_i,d_j) + \dots + \alpha_s(d_i,d_j) = w(d_i)\,w(d_j)$$

it is clear that the third column is completely filled by this process.

<u>Proposition 11</u>. The $|W|^2 \times 3$ matrix constructed above is a W—orthogonal array.

<u>Proof</u>.   By definition, the first two columns contain each ordered pair (d_i,d_j) $w(d_i)\,w(d_j)$ times.   Now consider the first and third columns. The number of times that (d_i,d_j) occurs with d_i in the first column and d_j in the third column is the number of times that d_j occurs in the products $d_i d_1, d_i d_2, \dots, d_i d_s$, i.e. $\alpha_j(d_i,d_1) + \dots + \alpha_j(d_i,d_s)$.   But

$$\sum_{k=1}^{s} \alpha_j(d_i,d_k) = \sum_{k=1}^{s} \gamma_k(d_i,d_j) = w(d_i)\,w(d_j) \,.$$

The   argument   for   the   second   and   third   columns   is   similar.

□

As   an   illustration,   in   Figure 6   we   give   the   transpose   of   the W—orthogonal   array   corresponding   to   the   weighted   quasigroup   of Figure 1.

a a a a a a a a a a a a a a a a a a a a a b b b

a a a a a a a a a b b b b b b c c c d d d a a a . . .

a a a a b b b c d a a a b b c a c d a b d a a a

b b b b b b b b b b b c c c c c c c d d d d d d

. . . a a a b b b b c c d d a a a b b c d a a a b b c d

b c d a b b d a b a c b b c a d a a a a d a c b b

Figure 6

We remark that a W–orthogonal array can be viewed as a decomposition into edge–disjoint triangles of a tripartite multigraph with vertex sets $\{r_1,...,r_s\}$, $\{c_1,...,c_s\}$, $\{\sigma_1,...,\sigma_s\}$ (corresponding to the first, second and third columns respectively of the W–orthogonal array, or, equivalently, to the rows, columns and symbols of the multiplication table of the W–quasigroup), where each of the vertex pairs $\{r_i,c_j\}$, $\{r_i,\sigma_j\}$ and $\{c_i,\sigma_j\}$ are joined by $w(d_i)w(d_j)$ edges.

Finding the $\pi(1,2,3)$–conjugate of a W–quasigroup Q corresponds to performing a $\pi(1,2,3)$ permutation on the columns of the corresponding W–orthogonal array, and then treating the first two columns as the row and column indicators for the products. Thus, for example, if we interchange the first and last columns of the W–orthogonal array whose transpose is given in Figure 6, we obtain the W–orthogonal array corresponding to the $Q_{(3,2,1)}$ in Figure 5. If A is a W–orthogonal array, let $A_{\pi(1,2,3)}$ denote the W–orthogonal array obtained by applying the permutation $\pi(1,2,3)$ to the columns of A.

The next proposition is fairly obvious; we omit the proof, but we could argue directly or via orthogonal arrays.

<u>Proposition 12.</u>   Let Q be a W–quasigroup, and let Q^* be a quasigroup of which Q is the W–amalgamation. Let π be a permutation of (1,2,3). Then Q_π is the W–amalgamation of Q^*_π.

5.   Conjugates of commutative W–quasigroups.

It is well–known that the (3,1,2)–conjugate of a commutative quasigroup satisfies the equation $x(xy) = y$,   and that the (2,3,1)–conjugate satisfies the equation $(yx)x = y$.   The equation $x(xy) = y$ can be expressed by the pair of equations $xy = z$ and $xz = y$, and the equation $(yx)x = y$ can be expressed by the pair $yx = z$ and $zx = y$. For a commutative quasigroup, the (3,1,2)–conjugate and the (3,2,1)–conjugate are the same; similarly the (2,3,1)–conjugate and the (1,3,2)–conjugate are the same.

The commutativity condition for a W–quasigroup Q can be expressed by the equation $\alpha_i(d_j,d_k) = \alpha_i(d_k,d_j)$.   In the (3,1,2)–conjugate of Q, this translates to $\alpha_i(d_j,d_k) = \alpha_k(d_j,d_i)$. If Q is in fact a quasigroup, then this condition can be re–expressed as the pair $xy = z$ and $xz = y$ of equations. The condition that the number of elements with odd multiplicity in the product d_i^2 is at most $w(d_i)$ translates to the condition that the number of $k \in \{1,...,s\}$ such that d_i occurs with odd multiplicity in the product $d_k d_i$ is at most $w(d_i)$. From Proposition 12 and Theorem 2 we obtain the following result.

THEOREM 3.   Let Q be a W–quasigroup satisfying $\alpha_i(d_j,d_k) = \alpha_k(d_j,d_i)$ for all $i,j,k \in \{1,...,s\}$.   Let $w(d_1),...,w(d_s)$ be even.   Then there is a quasigroup Q^* satisfying the equation $x(xy) = y$ of which Q is the W–amalgamation if and only if, for each $i \in \{1,...,s\}$, the number of $k \in \{1,...,s\}$ such that d_i occurs with odd multiplicity in the product $d_k d_i$ is at most $w(d_i)$.

Similarly in the $(2,3,1)$–conjugate of the commutative W–quasigroup Q, the condition $\alpha_i(d_j, d_k) = \alpha_i(d_k, d_j)$ translates to the condition $\alpha_i(d_j, d_k) = \alpha_j(d_i, d_k)$. Consequently the condition that the number of elements with odd multiplicity in d_i^2 is at most $w(d_i)$ translates to the condition that the number of $k \in \{1,...,s\}$ such that d_i occurs with odd multiplicity in $d_i d_k$ is at most $w(d_i)$. So in the same way we have the following theorem.

THEOREM 4.   Let Q be a   W–quasigroup   satisfying $\alpha_i(d_j, d_k) = \alpha_j(d_i, d_k)$ for all $i,j,k \in \{1,...,s\}$.   Let $w(d_1),...,w(d_s)$ be even.   Then there is a quasigroup Q^* satisfying the equation $(yx)x = y$ of which Q is the W–amalgamation if and only if, for each $i \in \{1,...,s\}$, the number of $k \in \{1,...,s\}$ such that d_i occurs with odd multiplicity in the product $d_i d_k$ is at most $w(d_i)$.

6. Simplex zeroids.

In this section we look at weighted quasigroups from a completely different point of view. We define a simplex zeroid, and then show that a simplex zeroid is equivalent to a weighted quasigroup.

Let $S = \{d_1,...,d_s\}$ and let $\bar{S}$ be the set of all formal sums $$\sum_{i=1}^{s} \beta_i d_i,$$ where, for $1 \leq i \leq s$, β_i is a non–negative rational number and

$\sum\limits_{i=1}^{s} \beta_i = 1$. We let $\circ$ denote a closed binary operation on $\bar{S}$: thus

$\circ: \bar{S} \times \bar{S} \to \bar{S}$. The operation $\circ$ will moreover be **bilinear**, that is to

say,

$$\left[\sum_{i=1}^{s} \beta_i d_i \right] \circ \left[\sum_{j=1}^{s} \psi_j d_j \right] = \sum_{\substack{1 \le i \le s \\ 1 \le j \le s}} \beta_i \psi_j (d_i \circ d_j)$$

whenever $\beta_1,..., \beta_s, \psi_1,...,\psi_s$ are non–negative rationals satisfying

$\sum\limits_{i=1}^{s} \beta_i = 1$ and $\sum\limits_{j=1}^{s} \psi_j = 1$. An element $x \in \bar{S}$ is an **interior point** of $\bar{S}$

if $x = \sum\limits_{i=1}^{s} \beta_i d_i$ with all the β_i's being positive. An element $x_0 \in \bar{S}$ is

a zero element if

$$x_0 \circ d = d \circ x_0 = x_0$$

for all $d \in \bar{S}$. Note that in fact there is at most one zero, because if x_0'
and x_0'' are both zeros then $x_0' = x_0' \circ x_0'' = x_0''$.

We say that the triple $(S,\circ,x_0)$ is a **simplex zeroid** if S is a
finite set, $\circ$ is a closed binary bilinear operation on $\bar{S}$, and there is an
interior point x_0 which is the zero.

We show first that if we are given a weighted quasigroup we can
use it to define a simplex zeroid explicitly.

Given a weighted quasigroup $S = \{d_1,...,d_s\}$ with weight function w, let o be the bilinear operation on $\bar{S} \times \bar{S}$ defined by

$$a \circ b = \sum_{i=1}^{s} \frac{\alpha_i(a,b)}{w(a)\ w(b)} d_i \quad (\forall a,b \in S),$$

where $\alpha_i(a,b)$ is as in the definition of a weighted quasigroup. Let

$$\bar{w} = \sum_{i=1}^{s} w(d_i),$$

and let

$$x_0 = \frac{1}{\bar{w}} \sum_{i=1}^{s} w(d_i)d_i \ .$$

Lemma 3. The triple (S, o, x_0) defined above is a simplex zeroid.

Proof. First we show that the operation o on $\bar{S} \times \bar{S}$ is closed.

$$\left[\sum_{i=1}^{s} \beta_i d_i\right] \circ \left[\sum_{j=1}^{s} \psi_j d_j\right] = \sum_{\substack{1 \leq i \leq s \\ 1 \leq j \leq s}} \beta_i \psi_j d_i \circ d_j$$

$$= \sum_{\substack{1 \leq i \leq s \\ 1 \leq j \leq s}} \beta_i \psi_j \left[\sum_{k=1}^{s} \frac{\alpha_k(d_i, d_j)}{w(d_i)\ w(d_j)}\right] d_k = \sum_{k=1}^{s}\left[\sum_{\substack{1 \leq i \leq s \\ 1 \leq j \leq s}} \beta_i \psi_j \phi_{ijk}\right] d_k,$$

$$\tag{1}$$

writing $\phi_{ijk} = \dfrac{\alpha_k(d_i, d_j)}{w(d_i)\, w(d_j)}$. Recall that $\displaystyle\sum_{k=1}^{S} \alpha_k(d_i,d_j) = w(d_i)w(d_j)$,

so that $\displaystyle\sum_{k=1}^{s} \phi_{ijk} = 1$. The sum of the coefficients in (1) is

$$\sum_{\substack{1\leq i\leq s \\ 1\leq j\leq s \\ 1\leq k\leq s}} \beta_i\psi_j\phi_{ijk} = \sum_{\substack{1\leq i\leq s \\ 1\leq j\leq s}} \beta_i\psi_j \sum_{1\leq k\leq s} \phi_{ijk} = \sum_{\substack{1\leq i\leq s \\ 1\leq j\leq s}} \beta_i\psi_j = \left[\sum_{i=1}^{S}\beta_i\right]\left[\sum_{j=1}^{S}\psi_j\right] = 1.$$

Consequently o is closed.

Next we show that x_0 is the zero element. Firstly if $a \in S$ then (with $\gamma_i(a,b)$ as in the definition of a weighted quasigroup)

$$a \circ x_0 = a \circ \frac{1}{\bar{w}} \sum_{i=1}^{s} w(d_i)d_i = \frac{1}{\bar{w}} \sum_{i=1}^{s} w(d_i)\, a \circ d_i$$

$$= \frac{1}{\bar{w}} \sum_{i=1}^{s} w(d_i) \sum_{j=1}^{s} \frac{\alpha_j(a,d_i)}{w(a)w(d_i)} d_j = \frac{1}{\bar{w}} \sum_{\substack{1\leq i\leq s \\ 1\leq j\leq s}} \frac{\gamma_i(a,d_j)}{w(a)} d_j$$

$$= \frac{1}{\bar{w}} \sum_{j=1}^{s} \left[\sum_{i=1}^{s} \gamma_i(a,d_j)\right] \frac{1}{w(a)} d_j = \frac{1}{\bar{w}} \sum_{j=1}^{s} w(d_j)d_j = x_0.$$

Consequently, if $\sum\limits_{i=1}^{s} \psi_i d_i$ is an arbitrary element of $\bar{S}$, then

$$\left[\sum_{i=1}^{s} \psi_i d_i\right] \circ x_0 = \sum_{i=1}^{s} \psi_i(d_i \circ x_0) = \sum_{i=1}^{s} \psi_i x_0 = x_0.$$

Using the equality $\alpha_i(d_j,d_k) = \delta_i(d_k,d_i)$ it follows in an analogous way that $x_0 \circ d = x_0$ for any $d \in \bar{S}$.

Since the weights are positive, x_0 is an interior point of $\bar{S}$. Thus $(S,\circ,x_0)$ is a simplex zeroid, as required. $\square$

We shall say that the simplex zeroid $(S,\circ,x_0)$ is **generated** by the weighted quasigroup on S.

It is easy to see that if in a W–quasigroup Q there is an integer m such that $m^2 \mid \alpha_i(d_j,d_k)$ for all i,j,k, and $m \mid w(d_i)$ for all i, then the simplex zeroid generated by Q is the same as the simplex zeroid generated by Q^0, where in Q^0 we have multiplicities $\alpha_i{}^0(d_j,d_k) = \dfrac{1}{m^2}\alpha_i(d_j,d_k)$ and weights $w^0(d_i) = \dfrac{1}{m} w(d_i)$.

If there is no integer m such that $m^2 \mid \alpha_i(d_j,d_k)$ $(\forall i,j,k)$ and $m \mid w(d_i)$ for all i, then the W–quasigroup is called a **primitive generator** of $(S,\circ,x_0)$.

The converse of Lemma 1 is also true.

<u>Lemma 4.</u>   Every simplex zeroid is generated by a weighted quasigroup.

<u>Proof.</u>   Let $S = \{d_1,\ldots,d_s\}$, let $x_0 = \sum\limits_{i=1}^{s} w_i d_i$, where the w_i are

positive rational numbers with $\sum\limits_{i=1}^{s} w_i = 1$, and let (S, o, x_0) be a simplex zeroid.

Define numbers ϕ_{ijk} $(i,j,k \in \{1,...,s\})$ by the equations

$$d_i \circ d_j = \sum_{k=1}^{s} \phi_{ijk} d_k \qquad (i,j \in \{1,...,s\}).$$

Then

$$\sum_{k=1}^{s} \phi_{ijk} = 1 \qquad (i,j \in \{1,...,s\}).$$

For $i,j,k \in \{1,...,s\}$ let p_{ijk} be a non–negative integer, and let q be a positive integer, such that

$$\phi_{ijk} = \frac{p_{ijk}}{q^2}.$$

Similarly, for $1 \leq i \leq s$, let p_i be a non–negative integer and r a positive integer, such that

$$w_i = \frac{p_i}{r}.$$

Define a weight function $w: S \longrightarrow \mathbb{Z}^+$ by

$$w(d_i) = p_i q$$

and the functions $\alpha_k: S \times S \longrightarrow \mathbb{Z}$ $(1 \leq k \leq s)$ by

$$\alpha_k(d_i, d_j) = p_i p_j p_{ijk}.$$

For $1 \leq i \leq s$ the functions $\gamma_i: S \times S \longrightarrow \mathbb{Z}$ and $\delta_i: S \times S \longrightarrow \mathbb{Z}$ are defined by

$$\alpha_i(d_j, d_k) = \gamma_k(d_j, d_i) = \delta_j(d_k, d_i) .$$

To show that S with weight function w and product $ab = \{d_1{}^{\alpha_1(a,b)}, ..., d_s{}^{\alpha_s(a,b)}\}$ is a weighted quasigroup we need to show that, for $i, j, k \in \{1, ..., s\}$,

$$\sum_{k=1}^{s} \alpha_k(d_i, d_j) = w(d_i)w(d_j); \quad \sum_{j=1}^{s} \gamma_j(d_i, d_k) = w(d_i)w(d_k);$$

$$\sum_{i=1}^{s} \delta_i(d_j, d_k) = w(d_j)w(d_k).$$

First, for $i, j \in \{1, ..., s\}$,

$$\sum_{k=1}^{s} \alpha_k(d_i, d_j) = p_i p_j \sum_{k=1}^{s} p_{ijk} = p_i p_j q^2 \sum_{k=1}^{s} \phi_{ijk}$$

$$= p_i p_j q^2 = w(d_i)w(d_j) .$$

Next we notice that

$$\sum_{k=1}^{s} w_k d_k = x_0 = d_i \circ x_0 = \sum_{j=1}^{s} w_j(d_i \circ d_j) = \sum_{k=1}^{s} \left(\sum_{j=1}^{s} w_j \, \phi_{ijk} \right) d_k .$$

It follows that, for $i, k \in \{1, ..., s\}$,

$$w_k = \sum_{j=1}^{s} w_j \, \phi_{ijk} ;$$

therefore

$$\frac{p_k}{r} = \sum_{j=1}^{s} \frac{p_j}{r} \frac{p_{ijk}}{q^2},$$

and so

$$q^2 p_k = \sum_{j=1}^{s} p_j p_{ijk}.$$

Consequently, for $i,k \in \{1,...,s\}$,

$$\sum_{j=1}^{s} \gamma_j(d_i,d_k) = \sum_{j=1}^{s} \alpha_k(d_i,d_j) = p_i \sum_{j=1}^{s} p_j p_{ijk} = p_i p_k q^2 = w(d_i)w(d_k) .$$

The demonstration that $\sum_{i=1}^{s} \delta_i(d_j,d_k) = w(d_j)w(d_k)$ is similar.

Thus S with weight function w and product $ab = \{d_1^{\alpha_1(a,b)},...,d_s^{\alpha_s(a,b)}\}$ is a weighted quasigroup. It is clear that this weighted quasigroup generates the simplex zeroid (S, o, x_0).

$$\square$$

We remark that if the weighted quasigroup in Lemma 4 is not a primitive generator of (S, o, x_0) then we only need to divide by the largest m such that $m \mid w(d_i)$ for all i and $m^2 \mid \alpha_i(d_j,d_k)$ for all i,j,k, and we obtain a weighted quasigroup that is a primitive generator of (S, o, x_0). Thus to every simplex zeroid there is a weighted quasigroup that generates it primitively.

These are just preliminary remarks about simplex zeroids. Simplex zeroids seem to provide an alternative framework for discussing the themes broached in this paper, and of course Theorems 1–4 can be reexpressed in terms of simplex zeroids. A further discussion would take up too much space, so we hope to present it elsewhere.

One of the objects of this paper is to raise the question of whether there are theorems like the ones in this paper for quasigroups satisfying other algebraic identities. As we have seen, algebraic identities tend to take on a rather awkward form when interpreted for weighted quasigroups. On the other hand, in terms of simplex zeroids some algebraic identities at least tend to retain their simplicity. For example, the associative law for weighted quasigroups is awkward to express, but for a simplex zeroid the associative law is simply $xo(yoz) = (xoy) \ o \ z$.

7.   Some final remarks.

Zaverdinos [23,24] has undertaken some not dissimilar studies, but apart from that, so far as we are aware, the kind of algebraic results we have presented here is new. Whether or not further results along these lines can be obtained for quasigroups satisfying other identities is anyone's guess.

The proofs of Lemmas 1 and 2, which underlie the theorems in this paper, are graph theoretic. A certain number of similar graph theoretic and combinatorial results have been obtained. Hilton [8], and Hilton and Rodger [10] have obtained some analogous results about Hamiltonian decompositions of complete graphs and of complete s—partite graphs. Nash—Williams' work on detachments of graphs [15,16,17,18,19] is very closely related, and in fact he has generalized some of the Hamiltonian decomposition results. Hilton and Rodger [12,13] have used similar ideas in connection with Triple Systems of even index, and have used them to prove that a Partial Triple System of index λ, where $4|\lambda$, on n elements can be embedded in a Triple

System of index λ and order t whenever $t \geq 2n + 1$, $t \equiv 0$ or 1 (mod 3).   There seems to be further scope for applying these ideas in a combinatorial context.   Moreover there have been some "real life" applications to various kinds of timetables [5,7,11].

References.

1. L.D. Andersen and A.J.W. Hilton, Generalized latin rectangles,
    In: Graph Theory and Combinatorics, Research Notes in
    Mathematics (Pitman, London, 1979), 1–17.

2. L.D. Andersen and A.J.W. Hilton, Generalized latin rectangles I:
    construction and decomposition, Discrete Math., 31 (1980),
    125–152.

3. L.D. Andersen and A.J.W. Hilton, Generalized latin rectangles II:
    embedding, Discrete Math., 31 (1980), 235–260.

4. A.G. Chetwynd and A.J.W. Hilton, Outline symmetric latin
    squares, Discrete Mathematics, 97 (1991), 101–117.

5. Deng, Chai Ling and Lim, Cheong Keang, A result on generalized
    latin squares, Discrete Math., 72 (1988), 71–80.

6. A.J.W. Hilton, The reconstruction of latin squares, with
    applications to school timetabling and to experimental design,
    Math. Programming Study, 13 (1980), 68–77.

7. A.J.W. Hilton, School timetables, Studies on graphs and discrete
    programming, (P. Hansen, Ed.), North–Holland Pub. Co., (1981),
    177–188.

8. A.J.W. Hilton, Hamiltonian decompositions of complete graphs,
    J. Combinatorial Theory B, 36 (1984), 125–134.

9. A.J.W. Hilton, Outlines of latin squares, Ann. Discrete Math., 34
    (1987), 225–242.

10. A.J.W. Hilton and C.A. Rodger, Hamiltonian decompositions of
    complete regular s–partite graphs, Discrete Math., 58 (1986),
    63–78.

11. A.J.W. Hilton and C.A. Rodger, Matchtables, Annals of Discrete
    Math., 15 (1982), 239–251.

12.   A.J.W. Hilton and C.A. Rodger, Edge–colouring graphs and embedding partial Steiner triple systems of even index, in: (G. Hahn et al., eds.) "Cycles and Rays" NATO AS1 Ser. C., Kluwer Academic Publishers, Dordrecht, 1990, pp. 101–112.

13.   A.J.W. Hilton and C.A. Rodger, The embedding of partial triple systems of index λ when four divides λ, J. Combinatorial Theory A, 56 (1991), 109–137.

14.   D. König, Theorie der Endlichen und Unendlichen Graphen, Akademische Verlagsgesellschaft M.B.H., Leipzig, 1936.

15.   C.St.J.A. Nash–Williams, Acyclic detachments of graphs, Graph Theory and Combinatorics (Proc. Conf. Open University, Milton Keynes, 1978). Research Notes in Mathematics, 34 (Pitman, San Francisco, 1979), pp. 87–97.

16.   C.St.J.A. Nash–Williams, Connected detachments of graphs and generalized Euler trails, J. London Math. Soc. (2), 31 (1985), 17–29.

17.   C.St.J.A. Nash–Williams, Detachments of graphs and generalized Euler trails, in: Proc. 10th British Combinatorics Conference, Surveys in Combinatorics, (1986), 137–151.

18.   C.St.J.A. Nash–Williams, Amalgamations of almost regular edge–colourings of simple graphs, J. Combinatorial Theory B, 43 (1987), 322–342.

19.   C.St.J.A. Nash–Williams, Another proof of a theorem concerning detachments of graphs, Europ. J. Combinatorics, 12 (1991), 245–247.

20.   J. Petersen, Die Theorie der regulären Graphen, Acta Math., 15 (1891), 193–220.

21.   D. de Werra, Balanced schedules, INFOR 9 (1971), 230–237.

22.   D. de Werra, A few remarks on chromatic scheduling, in: B. Roy, ed., Combinatorial Programming: Methods and Applications (Reidel, Dordrecht, 1975), 337–342.

23.   C. Zaverdinos, The Three–Family Problem and Related
       Structuries, Ph.D. Thesis, University of Natal, Pietermaritzburg,
       1984.

24.   C. Zaverdinos, When is a Complex Matrix a Character Table? A
       Reduction to Vertex Independence, Ars Combinatoria, to appear.

Addresses

First author: Second author:

Department of Mathematics   Department of Mathematics
University of Reading   West Virginia University
Whiteknights   Morgantown
Reading RG6 2AX   West Virginia 26506
United Kingdom,   USA.

and

Department of Mathematics
West Virginia University
Morgantown
West Virginia 26506
USA

Graphs with Projective Subconstituents which contain Short Cycles

A.A.Ivanov

Abstract

We consider graphs whose automorphism groups act transitively on the set of vertices and the stabilizer of a vertex preserves on the neighbourhood of the vertex a projective spaces structure and induces on it a flag-transitive action. Such a graph is said to possess a projective subconstituent. As an additional condition we assume that there is a short cycle in the graph (of length at most 8). The canonical examples of graphs possessing projective subconstituents and containing short cycles are related to finite Lie groups of type A_n, D_n and F_4. There is also a remarkable "sporadic" series coming from the so-called P-geometries (geometries related to the Petersen graph). The automorphism groups of the graphs from this series are sporadic simple groups M_{22}, M_{23}, Co_2, J_4, F_2 and nonsplit extensions $3 \cdot M_{22}$, $3^{23} \cdot Co_2$ and $3^{4371} \cdot F_2$. The classification of flag-transitive P-geometries was recently completed by S.V.Shpectorov and the author. The consequences of this classification for the graphs possessing projective subconstituents and containing short cycles are reported in the present survey in the context of the general situation in the subject.

1. Introduction.

The subject of the survey can be viewed in a general framework of local characterization of graphs. That is characterization by structure of the neighbourhoods of vertices. Usually this structure is described by the subgraph induced by the neighbourhood. If the vertex stabilizer in the automorphism group of the considered graph induces on the neighbourhood a doubly transitive permutation group the induced subgraph is trivial (either the complete graph or the null graph). In this case the local structure still can be defined as a "virtual" one preserved by the action of the vertex stabilizer. We deal with the situation when this structure preserved is a projective space. To be more precise we are interested in pairs (Γ, G) satisfying the following

Hypothesis A. Γ *is an undirected, connected, locally finite graph and* $G \leq Aut(\Gamma)$ *acts on* Γ *vertex- and edge-transitively. Let x be a vertex of*

Γ, $G(x)$ be the stabilizer of x in G, $\Gamma(x)$ be the set of vertices adjacent to x in Γ and $G(x)^{\Gamma(x)}$ be the action induced by $G(x)$ on $\Gamma(x)$. Then $G(x)^{\Gamma(x)}$ is permutationally isomorphic to the action induced by a group H such that $SL_n(q) \leq H \leq \Gamma L_n(q)$, on the set of 1-dimensional subspaces of an n-dimensional $GF(q)$-space.

If Hypothesis A holds then we say that Γ possesses (with respect to the action of G) a projective subconstituent of rank $n-1$ (over $GF(q)$). In this case $PSL_n(q) \leq G(x)^{\Gamma(x)} \leq P\Gamma L_n(q)$.

We start with examples of graphs possessing projective subconstituents of rank ≥ 2 related to parabolic geometries of groups of Lie type.

Let $\mathcal{G}$ be a geometry over the set of types $\Delta = \{1, 2, ..., n\}$ and for $i \in \Delta$ let $\mathcal{G}^i$ be the set of elements of type i in $\mathcal{G}$. For $i, j \in \Delta$ define $\Gamma_{i,j}(\mathcal{G})$ to be the bipartite graph on the set $\mathcal{G}^i \cup \mathcal{G}^j$ of vertices, in which two vertices are adjacent if they are of different type and are incident in the geometry.

Consider the parabolic geometries of Lie type groups with the following diagrams:

$A_n(q)$: $\overset{1}{\circ}\!\!-\!\!\overset{2}{\circ}\ \cdots\ \overset{n-2}{\circ}\!\!-\!\!\overset{n-1}{\circ}\!\!-\!\!\overset{n}{\circ}$

$D_n(q)$: $\overset{1}{\circ}\!\!-\!\!\overset{2}{\circ}\ \cdots\ \overset{n-3}{\circ}\!\!-\!\!\overset{n-2}{\circ}$

$F_4(q)$: $\overset{1}{\circ}\!\!-\!\!\overset{2}{\circ}\!\!=\!\!\overset{3}{\circ}\!\!-\!\!\overset{4}{\circ}$

Then the graphs $\Gamma_{1,n}(A_n(q))$, $\Gamma_{n-1,n}(A_{2n-2}(q))$, $\Gamma_{n-1,n}(D_n(q))$ and $\Gamma_{2,3}(F_4(q))$ possess projective subconstituents with respect to extensions of the corresponding Lie type groups by diagram automorphisms permuting the types i and j involved in the construction of the graph (in the case of $F_4(q)$ we assume that q is even).

Three of the above series of graphs can be naturally described in classical terms. Namely, $\Gamma_{1,n}(A_n(q))$ is the incidence graph of points and hyperplanes of an n-dimensional projective $GF(q)$-space; the vertices of $\Gamma_{n-1,n}(A_{2n-2}(q))$ are $(n-1)$-dimensional and n-dimensional subspaces in a $(2n-1)$-dimensional $GF(q)$-vector space with two subspaces being adjacent if one of them is a proper subspace in the other one; finally the vertices of $\Gamma_{n-1,n}(D_n(q))$ are the maximal isotropic subspaces with respect to a nonsingular quadratic form of plus type in a $(2n)$-dimensional $GF(q)$-vector space with two subspaces being adjacent if their intersection is of codimension one in both.

The length of the shortest cycle (i.e. the girth) is equal to 4 in $\Gamma_{1,n}(A_n(q))$ and $\Gamma_{n-1,n}(D_n(q))$; to 6 in $\Gamma_{n-1,n}(A_{2n-2}(q))$ and to 8 in $\Gamma_{2,3}(F_4(q))$. In the present survey we deal, in particular, with the following question. Up to which extend the above examples are characterized by their girths in the class of graphs possessing projective subconstituents? To be more precise, we consider the graphs satisfying the following

Hypothesis B. Γ *possesses a projective subconstituent of rank* $(n-1) \geq 2$ *(with respect to a group* $G \leq Aut(\Gamma))$ *and the girth* g *of* Γ *is at most 8.*

We will mention all known examples of graphs satisfying Hypothesis B and discuss the available results on their characterization. We start with some further motivations for the interest in such graphs.

2. Motivation.

The vertex set of a graph Γ will be denoted by the same letter Γ. For $x \in \Gamma$ and an integer i put $\Gamma_i(x) = \{y | y \in \Gamma, d(x,y) = i\}$ where d is the natural metric in Γ. For the set $\Gamma_1(x)$ we will write $\Gamma(x)$. An s-path in Γ which joins y with z is a sequence $(y = x_0, x_1, ..., x_s = z)$ of vertices such that $x_i \in \Gamma(x_{i-1})$ for $1 \leq i \leq s$ and $x_{i+1} \neq x_{x-1}$ for $1 \leq i \leq s-1$. By $\Gamma_{\leq i}(x)$ we denote the union of $\Gamma_j(x)$'s for $j \leq i$. Let $G \leq Aut(\Gamma)$ and $\Delta \subseteq \Gamma$. Then $G(\Delta)$ and $G\{\Delta\}$ denote the elementwise and the setwise stabilizers of Δ in G. For $H \leq G\{\Delta\}$ by H^Δ we denote the action induced by H on Δ. For $x \in \Gamma$ and $y \in \Gamma(x)$ put $G_i(x) = G(\Gamma_{\leq i}(x))$ and $G_i(x,y) = G_i(x) \cap G_i(y)$. The subgroup $G_1(x,y)$ plays a special role in our discussion. Its nontrivial elements are called elations.

The action of $G \leq Aut(\Gamma)$ on Γ is called s-transitive if G acts transitively on the set of s-paths in Γ. In particular the action is 2-transitive if and only if it is vertex-transitive and the subconstituent $G(x)^{\Gamma(x)}$ is a doubly transitive permutation group. The action of G on Γ is distance-transitive if G acts transitively of the set $\{(x,y)|x,y \in \Gamma, d(x,y) = i\}$ for every i (in this case Γ is called a distance-transitive graph).

Suppose that the action of G on Γ is distance-transitive and also s-transitive but not $(s+1)$-transitive. Then for an $(s+1)$-path $(x_0, x_1, ..., x_{s+1})$ there is another t-path joining x_0 with x_{s+1} such that $t \leq s+1$. This means that the girth g of Γ satisfies $g \leq 2(s+1)$. It was shown in [Weil] that if Γ possesses a projective subconstituent of rank ≥ 2 with respect to a group G, then the action of G is not 4-transitive. So a distance-transitive graph with a projective subconstituent with respect to the automorphism group has girth at most 8. Notice the graphs $\Gamma_{1,n}(A_n(q))$, $\Gamma_{n-1,n}(A_{2n-2}(q))$, $\Gamma_{n-1,n}(D_n(q))$ (but not $\Gamma_{2,3}(F_4(q))$) are distance-transitive with respect to the extended Lie type groups. Thus any distance-transitive graph possessing a projective subconstituent of rank ≥ 2 satisfies Hypothesis B.

There is another motivation. Let Γ possesses with respect to G a projective subconstituent of rank $n-1$ over $GF(q)$. Consider the determination problem of the amalgam $\mathcal{A} = \{G(x), G\{x,y\}\}$ for $x \in \Gamma$, $y \in \Gamma(x)$. (We understand an amalgam just as a collection of groups with common identity element and with multiplications and the operations of inverse coinciding on intersections. In particular any family of subgroups in a group form an amalgam.) The determination problem of $\mathcal{A}$ is equivalent to description of all locally finite actions on infinite trees which give rise to projective subconstituents. A crucial step for solving the problem would be bounding the order of $G(x)$ by a function of the valency (equivalently, by a function of n and q). It is easy to show that such a bound exists if and only if there is a constant c such that $G_c(x) = 1$ for every locally finite action corresponding to a projective subconstituent. For about 10 years this was the only open case in the general class of actions with doubly transitive subconstituents [Wei4]. Some partial results were obtained in [Wei1], [Wei5].

Recently it was shown that $G_6(x) = 1$ for any locally finite action with a projective subconstituent. The result was announced in [Tro1] and the published part [Tro2] covers together with the above mentioned earlier results the case $char(GF(q)) \geq 5$. For various reasons the following conjecture seems to be realistic.

Conjecture C. *Every amalgam $\mathcal{A} = \{G(x), G\{x,y\}\}$ corresponding to a locally finite 2-transitive action with a projective subconstituent can be realized by means of such an action on a graph whose girth is at most 8.*

In other words it is conjectured that the universal completion $\tilde{G}$ of the amalgam $\mathcal{A}$, when acting on the corresponding tree $\tilde{\Gamma}$, preserves an equivalence relation I, such that the factor graph $\Gamma = \tilde{\Gamma}/I$ has girth at most 8 and the induced action of $\tilde{G}$ on Γ corresponds to the same amalgam $\mathcal{A}$.

A proof of the conjecture (if it is true at all) would follow from an explicit description of the possible amalgams $\mathcal{A}$. But if one would be able to prove the conjecture using just the equality $G_6(x) = 1$, the description of the amalgams would be reduced to study of graphs satisfying Hypothesis B. The ideal goal is to prove the conjecture using exclusively the projectivity of the subconstituent. This would provide a possibility to simplify considerably Trofimov's result $G_6(x) = 1$.

The above speculations are inspired by the brilliant classification of the amalgams $\mathcal{A}$ corresponding to projective subconstituents of rank 1 obtained in [Wei2], [DGS]. In that case, in the early stages of the classification it was shown that for $s \geq 4$ any s-transitive action of the considered type can be realized on a graph whose girth reaches the absolute minimum, namely $2(s - 1)$. Such a graph must be a generalized $(s - 1)$-gon. All s-transitive generalized $(s - 1)$-gons were earlier and independently proved to be isomorphic to the parabolic geometries of the rank 2 Lie type groups $A_2(q)$, $B_2(2^m)$ and $G_2(3^m)$ (here $s =$4, 5 and 7, respectively).

A possibility to realize a somewhat similar approach in the case of projective subconstituents of rank ≥ 2 is very attractive (cf. [Wei6]). From this point of view the graphs satisfying Hypothesis B would play a role similar to that of the s-transitive generalized $(s-1)$-gons.

3. Preliminaries.

In many cases (including the original examples coming from the Lie type groups) the projective subconstituent arises with respect to the full automorphism group of the graph. But this is not always the case. Sometimes a projective subconstituent can be obtained only by taking in $Aut(\Gamma)$ a certain proper subgroup. We present a number of such examples, where $m = (q^n - 1)/(q - 1)$.

Example 3.1. Let $\Gamma \cong K_{m+1}$ be the complete graph. Then G must be a doubly transitive permutation group on the vertex set of Γ such that $G(x)$ induces on $\Gamma - \{x\}$ a projective action. This means that G is a transitive (one point) extension of the projective group $G(x)$. All such extensions are classified [Hug].

Lemma 3.2. *Suppose that Hypothesis B holds and $g = 3$. Then $\Gamma \cong K_{m+1}$ and either $G \cong AGL_n(2)$ and $m = 2^n - 1$ or $M_{22} \leq G \leq Aut(M_{22})$ and $m = 21$.*

Example 3.3. Let $\Gamma \cong K_{m,m}$ be the complete bipartite graph. Then a 2-transitive action with a projective subconstitute can be realized by taking in $P\Gamma L_n(q)\ wr\ Z_2$ a transitive subgroup containing $PSL_n(q) \times PSL_n(q)$.

Example 3.4. Let $\Gamma = Q_m$ be the m-dimensional unit cube. The automorphism group of the graph is a semidirect product $A\lambda B$ where A is an elementary abelian normal subgroup which can be identified with the space of all subsets of $\Gamma(x)$ for $x \in \Gamma$ and $B \cong S_m$ stabilizes x and permutes $\Gamma(x)$. If $PSL_n(q) \leq H \leq P\Gamma L_n(q)$ is a subgroup of B then Q_m possesses with respect to $A\lambda H$ a projective subconstituent.

There are two obvious submodules in A: $A_0 = \{\emptyset, \Gamma(x)\}$ and $A_1 = \{X | X \subseteq \Gamma(x), |X| = 0 \mod 2\}$. If H_1 is an index 2 subgroup in H then the diagonal subgroup F lying between $A_1\lambda H_1$ and $A\lambda H$ acts 2-transitively on Q_m and also corresponds to a projective subconstituent. Now let M be an H-submodule in A such that the pairwise distances between vertices in an M-orbit are greater than 2. Then the factor graph Q_m/M possesses a projective subconstituent with respect to $(A/M)\lambda H$. The factorization over A_0 leads to the folded cube $\square_m$. If q is odd then there are no H-submodules besides A_0 and A_1 [Mor]. If q is even then $A = A_0 \oplus A_1$ and all H-submodules in A_1 are known [IP].

Lemma 3.5. *If q is even then the H-module A_1 is uniserial: $1 \leq R_{n-2} \leq$ ... $\leq R_2 \leq R_1$ where R_i is the submodule generated by the complements of all i-dimensional subspaces with respects to the projective space structure induced by H on $\Gamma(x)$.*

Notice that for $q = 2$ the modules R_i are the (contracted) Reed-Muller codes.

To the end of this section, let us recall a well-known construction. Let G act s-transitively of Γ, $s \geq 1$. Define the standard doubling $2.\Gamma$ as follows: $2.\Gamma = \{(x, \alpha) | x \in \Gamma, \alpha \in \{0, 1\}\}$; (x, α) and (y, β) are adjacent if x and y are adjacent in Γ and $\alpha \neq \beta$. Then $2.\Gamma$ admits an action of $G' = G \times <\tau>$ where the former factor is the natural extension on $2.\Gamma$ of the action of G and τ acts by the rule: $(x, \alpha)^\tau = (x, 1 - \alpha)$. The subconstituent of the action of G' on $2.\Gamma$ is isomorphic to that of G on Γ. If Γ is bipartite then $2.\Gamma$ is a disjoint union of two isomorphic copies of Γ otherwise $2.\Gamma$ is connected.

Lemma 3.6. *Let g be the girth of Γ and g' be the girth of $2.\Gamma$. If g is even then $g' = g$. If g is odd then $g' \geq g + 1$ and the equality holds precisely when Γ contains a cycle of length $g + 1$.*

Suppose we are classifying graphs satisfying Hypothesis B (or a similar one) and having a fixed even girth. Then we can switch from the original graphs to connected components of their standard doublings and assume that all graphs are bipartite. Having the complete list of the bipartite examples, usually it is easy to decide which are standard doublings of nonbipartite ones and to construct the corresponding foldings.

4. The main types and some technique.

In this section we consider the basic properties of graphs having projective subconstituents (for the details cf. [Wei1], [Tro2], [Ivn3]).

Let G act s-transitively on Γ and suppose that Γ possesses with respect to this action a projective subconstituent of rank $(n-1) \geq 2$ over $GF(q)$. For $x \in \Gamma$ the action of $G(x)$ induces on $\Gamma(x)$ a structure of a projective space. In what follows Π_x stands for this space while L_x and H_x denote its set of lines and hyperplanes, respectively. For $y \in \Gamma(x)$, by $L_x(y)$ and $H_x(y)$ we denote the set of lines and the set of hyperplanes passing through y. The structure of $(n-2)$-dimensional projective space induced on $L_x(y)$ will be denoted by $\Pi_x(y)$. The subgroup $G(x, y)$ induces on $L_y(x)$ a doubly transitive permutation group which contains $PSL_{n-1}(q)$. Since $G_1(x)$ is normal in $G(x, y)$, its action on $L_y(x)$ is either transitive or trivial. This determines the main two types.

Type 1. $G_1(x)$ is transitive on $L_y(x)$. Then $s = 3$ and it can be shown that the action contains $PSL_{n-1}(q)$. Let $z \in \Gamma(x) - \{y\}$. It is known

[Wei3] that $G_1(x, y)$ is a (possibly trivial) p-group and in the considered case $p = char(GF(q))$. In particular $G_1(x, y)$ acts trivially on $L_z(x)$ and there is a canonical collineation $\psi_x(y, z)$ of $\Pi_y(x)$ onto $\Pi_z(x)$ which commutes with the action of $G(x, y, z)$. So there is an $(n - 2)$-dimensional projective space Δ_x naturally associated with x and $G(x)$ induces a flag-transitive action on the direct product $\Pi_x \times \Delta_x$. The graphs $\Gamma_{n-1,n}(A_{2n-2}(q))$ and $\Gamma_{2,3}(F_4(q))$ are of this type.

Type 2. $G_1(x)$ is trivial on $L_y(x)$. Then $s = 2$ and there is a canonical isomorphism $\phi_{x,y}$ between $\Pi_x(y)$ and $\Pi_y(x)$ which commutes with the action of $G(x, y)$. The particular form of $\phi_{x,y}$ determines two subtypes.

Type 2.1. $\phi_{x,y}$ is a collineation. The situation is realized in the graph $\Gamma_{n-1,n}(D_n(q))$.

Type 2.2. $\phi_{x,y}$ is a correlation. Here $G_1(x) \neq 1$ and we can assume that $n \geq 4$. The situation is realized in the graph $\Gamma_{1,n}(A_n(q))$.

Now let us discuss briefly the technique used for the classification. We explore a geometrical approach to the classification which can be described as follows. The projectivity of the subconstituent enables us to relate with each vertex $x \in \Gamma$ a *local* geometry $\mathcal{G}(x)$ on which the stabilizer $G(x)$ acts flag-transitively. For $\mathcal{G}(x)$ we have the projective space Π_x in graphs of type 2 and the direct product $\Pi_x \times \Delta_x$ of projective spaces in graphs of type 1. Our strategy is to associate with Γ a *global* geometry $\mathcal{G} = \mathcal{G}(\Gamma)$ on which G acts flag-transitively with $\mathcal{G}(x)$ being canonically isomorphic to a suitable residue in $\mathcal{G}$. The elements of $\mathcal{G}$ are certain subgraphs in Γ. Which subgraphs to consider as elements of the geometry and how to define the incidence relation is dictated, up to a certain extent, by the examples related to the Lie type groups. Let us discuss the procedure for graphs of type 2.1 in more detail.

So we assume that Γ is of type 2.1. For $x \in \Gamma$ there is a structure Π_x of a projective space defined on $\Gamma(x)$. In what follows a subspace $\Lambda \subseteq \Pi_x$ will be identified with the set of its points (i.e. with vertices from $\Gamma(x)$). We are trying to construct the global geometry as a family $\mathcal{G}$ of connected subgraphs in Γ such that

(i) for $\Xi \in \mathcal{G}$ and $x \in \Xi$ the set $\Xi(x)$ is a subspace in Π_x whose dimension is independent of x and determines the type of Ξ in $\mathcal{G}$;

(ii) for each subspace Λ from Π_x there is a unique subgraph $\Xi \in \mathcal{G}$ containing x such that $\Xi(x) = \Lambda$ and $G(x) \cap G\{\Lambda\}$ stabilizes Ξ.

We will see in a moment that if such a system of subgraphs exists at all, then it is unique. Notice that by of such a system the geometry $D_n(q)$ can be reconstructed from the graph $\Gamma_{n-1,n}(D_n(q))$.

We start with a naive approach to reconstruction of the required system of subgraphs. Let $x \in \Gamma$ and Λ be proper subspace in Π_x. If Λ is a point y then $\Xi = \{x, y\}$ and we are done. So suppose that Λ has dimension

$i \geq 1$. We construct the subgraph Ξ_i by means adjoining vertices. We start with the empty set and adjoin the vertex x together with the vertices from Λ. If $y \in \Lambda$ then Ξ_i must contain an i-dimensional subspace M from Π_y stabilized by $G(x,y) \cap G\{\Lambda\}$. Since Γ is of type 2.1, there is a unique such subspace, namely $M = \phi_{x,y}(\Lambda)$. So we adjoin M to Ξ_i and continue in the same manner. Eventually we will either construct the subgraph possessing all the properties we need or adjoin to Ξ_i a subspace from Π_x distinct from Λ. In the latter case the construction fails.

In order to get a control over the situation when the naive construction fails, we modify the procedure. Define a graph Σ_i by the following rule. The vertices of Σ_i are all pairs (x, Λ) where $x \in \Gamma$ and Λ is an i-dimensional subspace in Π_x; the vertices (x, Λ) and (y, M) are adjacent if (a) x and y are adjacent in Γ; (b) $y \in \Lambda$ and $x \in M$; (c) $M = \phi_{x,y}(\Lambda)$. Then it is easy to check that the following three conditions are equivalent:

(1) Σ_i is disconnected;

(2) if $\Lambda \neq \Lambda'$ then (x, Λ) and (x, Λ') are in distinct connected components of Σ_i;

(3) if Φ is a maximal flag in Π_x containing $\{y\}$ and Λ then for any $g \in G\{x,y\}$ such that $\Phi^g = \phi_{x,y}(\Phi)$ the subgroup $\langle g, G(x) \cap G\{\Lambda\}\rangle$ in G is proper.

Suppose first that the conditions (1) - (3) are satisfied. Let ξ be the mapping of Σ_i onto Γ which maps (x, Λ) onto x. Then the restriction of ξ to the connected component of Σ_i containing (x, Λ) is an isomorphic embedding and the image of this restriction will be the required subgraph Ξ_i. Suppose now that i is minimal for which the conditions (1) - (3) fail. Then we are not in a position to define the full rank geometry in the original graph Γ. But instead we can start studying $\Sigma = \Sigma_i$. This graph is connected, G acts on it and with respect to the action Σ possesses a projective subconstituent of rank i. In a sense Σ carries the complete information about Γ since the latter can be obtained by factorizing Σ over a certain imprimitivity system of G on the set of vertices. Since i is chosen to be the minimal index for which the conditions (1) - (3) fails, it is easy to show (compare condition (3)) that the subgraph Σ_j defined with respect to Σ is disconnected for $1 \leq j < i$ and we can reconstruct in Σ the full rank geometry.

So switching from Γ to Σ, if necessary, we can assume that the conditions (1) - (3) are satisfied for all i. If this case we can define the elements of $\mathcal{G}(\Gamma)$ as the subgraphs Ξ_i for $i \geq 1$ together with all vertices and edges of Γ. The incidence be defined by means of inclusion. Then $\mathcal{G}$ belongs to a string diagram and the residue of x is isomorphic to Π_x. The rank 2 residue of type {vertices, edges} is the geometry of vertices and edges of the subgraph Ξ_1. Now we should try to use the condition on the girth to specify this residue. In some cases it is possible to show that the shortest cycle should be contained in Ξ_1.

In the case of $\Gamma_{n-1,n}(D_n(q))$ the subgraph Ξ_1 is the complete bipartite graph $K_{q+1,q+1}$. In order to obtain the standard D_n diagram we should slightly modify the definition of geometry. Namely we should consider the standard doubling (which does not change the girth since it is even) and split the vertices into two different types with two vertices of different type (from different parts of the bipartition) being incident if they are adjacent in the graph. It is well known (cf. [Tit]) that the geometry $D_n(q)$ is characterized by its standard diagram.

5. $g = 4$

The first and extremely important step in classification of graphs satisfying Hypothesis B was done in [CP] where the case $g = 4$ was considered.

Theorem 5.1. [CP] *Let Γ satisfy Hypothesis B and $g = 4$. Then one of the following holds:*

(i) $G_1(x) \neq 1$ and Γ is isomorphic to one of the following graphs: $\Gamma_{1,n}(A_n(q))$, $\Gamma_{n-1,n}(D_n(q))$ or $K_{m,m}$ (cf. Example 3.3);

(ii) Γ is the standard doubling of the complete graph K_{m+1} (cf. Lemma 3.2);

(iii) $G_1(x) = 1$ and if $d(x,y) = 2$ then $k = |\Gamma(x) \cap \Gamma(y)| = 2, 3, 4$ or 6. Moreover, if $k \geq 2$ then either $k = q = 3$ or 4 and $\Gamma(x) \cap \Gamma(y)$ is a line of Π_x with one point removed, or $k = q + 1 = 3, 4$ or 6, $n \geq 4$ and $\Gamma(x) \cap \Gamma(y)$ is a line of Π_x. $\square$

Notice that the graphs from Example 3.4 correspond to the case (iii) of the theorem for $k = 2$.

A considerable part of the work [CP] deals with the characterization of the graphs $\Gamma_{n-1,n}(D_n(q))$. The characterization begins with reconstruction of the abstract structure of the vertex stabilizer $G(x)$. After that for increasing values of i the action of $G(x)$ on $\Gamma_i(x)$ is being specified and simultaneously the structure of the subgraph induced by $\Gamma_{\leq i}(x)$ is being determined. The success of the approach is mainly due to the fact that the resulting graph turns out to be distance-transitive, even it was not assumed at the beginning.

An alternative characterization of $\Gamma_{n-1,n}(D_n(q))$ along the geometric approach discussed at the end of the previous section, is given in [Chi]. Also in that paper the classification of graphs corresponding to the case (iii) in Theorem 5.1 was essentially completed. In particular two new and very nice examples were constructed.

Example 5.2. The group $G_2(2) \cong U_3(3).2$ acting on the cosets of a subgroup $L_3(2)$ of index 72 has a subconstituent of length 7. The graph corresponding to the subconstituent is undirected with girth 4 and $k = 2$.

Example 5.3. The group $U_4(3).2_2$ (in the notation of [Atlas]) acting on the cosets of a subgroup $L_3(4)$ of index 324 has a subconstituent of length 21. The corresponding graph is undirected of girth 4 and $k = 4$. The full automorphism group of the graph is of the shape $U_4(3).(2^2)_{122}$.

Let us introduce additional notation. For the case $G_1(x) = 1$ let $\bar{G}(x)$ denote the normal subgroup of $G(x)$ isomorphic to $PSL_n(q)$. For a pair of adjacent vertices x, y in Γ let $\bar{G}$ denote the normal subgroup of G generated by $\bar{G}(x)$ and $\bar{G}(y)$.

Theorem 5.4. [Chi] *Let Γ correspond to the case (iii) in Theorem 5.1. Then one of the following holds:*

(i) Γ is isomorphic to the graph in Example 5.2 or to the graph in Example 5.3;

(ii) $\bar{G}$ is isomorphic to a factorgroup of the group $A_1 \lambda PSL_n(q)$ (cf. Example 3.4), in particular $O_2(G).G(x) \leq G \leq O_2(G).G(x).2$. $\quad\square$

In view of Lemma 3.5 it is possible to obtain the complete list of possibilities for the case (ii) of the theorem.

6. $g = 5$

After the paper [CP] was published there was a belief that there are no graphs satisfying Hypothesis B with $g = 5$ and $G_1(x) \neq 1$. But later it was noticed that such examples are provided by two distance-transitive graphs of valency 7. The first of the graphs, denoted by $\Gamma(M_{22})$ was constructed in [Big]. Its vertices are the 330 blocks of the Steiner system $S(5, 8, 24)$ which miss two given points of the system; two blocks are adjacent if their intersection is empty. The automorphism group of $\Gamma(M_{22})$, isomorphic to $Aut(M_{22})$, induces a 2-transitive action with projective subconstituent of rank 2 over $GF(2)$ and $G_1(x) \neq 1$. The second graph was constructed in [FII] within the classification of the distance-transitive graphs of valency 7. This graph is denoted by $\Gamma(3.M_{22})$ and it is a 3-fold cover of $\Gamma(M_{22})$ which corresponds to a nonsplit extension of M_{22} by a center of order 3.

The graphs $\Gamma(M_{22})$ and $\Gamma(3.M_{22})$ are of type 2.1 and in both cases the subgraph Ξ_1 is isomorphic to the Petersen graph. So within the geometric approach discussed in Section 4 we obtain geometries with the following diagram:

$$\underset{2}{\circ}\!\!-\!\!-\!\!-\!\!\underset{2}{\circ}\overset{P}{-\!\!-\!\!-}\underset{1}{\circ}$$

An arbitrary geometry which belongs to a string diagram, all whose nonempty edges except one are projective planes over $GF(2)$ and one terminal edge is the geometry of vertices and edges of the Petersen graph are called P-geometries. So the graphs $\Gamma(M_{22})$ and $\Gamma(3.M_{22})$ correspond to flag-transitive rank 3 P-geometries (denoted by $\mathcal{G}(M_{22})$ and $\mathcal{G}(3.M_{22})$). It is shown in [Shp1] that these two examples exhaust the flag-transitive rank 3 P-geometries.

In [Ivn1] other examples of girth 5 graphs with projective subconstituents were constructed (see also [IS1]: a graph $\Gamma(M_{23})$ on the 506 blocks of the Steiner system $S(5,8,24)$ missing a given point; a graph $\Gamma(Co_2)$ related to the Conway group, which possesses a natural description in terms of the Leech lattice; a graph $\Gamma(J_4)$ of valency 31 related to Janko's group and a graph $\Gamma(F_2)$ of the same valency related to the Baby Monster group. Each of these graphs possesses a projective subconstituent over $GF(2)$ with respect to the action of the corresponding sporadic group. The graphs are of type 2.1. For the graphs $\Gamma(M_{23})$, $\Gamma(Co_2)$ and $\Gamma(F_2)$ the procedure described in Section 4 leads to the full rank P-geometry. In the case of $\Gamma(J_4)$ the graph Σ_3 is connected and corresponds to a rank 4 P-geometry. As a result we obtain four P-geometries $\mathcal{G}(M_{23})$, $\mathcal{G}(Co_2)$, $\mathcal{G}(J_4)$ and $\mathcal{G}(F_2)$.

It can be shown that starting with an arbitrary flag-transitive P-geometry $\mathcal{G}$ of rank r one can construct a girth 5 graph $\Gamma(\mathcal{G})$ which possesses a projective subconstituent of rank $(r-1)$ over $GF(2)$ with respect to the action of the automorphism group of the geometry. This graph is called the derived graph of the P-geometry. Its vertices are the elements of the geometry of the type corresponding to the terminal vertex of the diagram which is contained in the Petersen-type edge; two elements are adjacent in the derived graph if they are incident to a common element of the adjacent type. It turns out that under certain additional assumption the reverse is also true.

Theorem 6.1. [Ivn2], [Ivn3] *Let Γ satisfy Hypothesis B, that is possess a projective subconstituent of rank $(n-1)$ over $GF(q)$ with respect to a group G. Suppose that $g = 5$ and $G_1(x) \neq 1$. Then $q = 2$, Γ is of type 2.1 and one of the following holds:*

(i) Γ is isomorphic to the derived graph of a flag-transitive P-geometry of rank n;

(ii) the graph Σ_{n-2} is connected and is isomorphic to a derived graph of a flag-transitive P-geometry of rank $(n-1)$.   □

The case (ii) is realized in the graph $\Gamma(J_4)$. Notice that the vertex stabilizer of J_4 acting on this graph is isomorphic to the vertex stabilizer of $D_5(2).2$ acting on $\Gamma_{n-1,n}(D_n(q))$ (cf. [SW]) but the amalgams $\mathcal{A}$ in these two cases are different.

The proof of Theorem 6.1 realizes the geometric approach from Section 4. First it was shown that the subgraph Ξ_1 is always reconstructable and contains a shortest cycle (that is a cycle of length 5). The subgroup $G\{\Xi_1\}$ acts on Ξ_1 3-transitively which implies that Ξ_1 is isomorphic to the Petersen graph. Finally it was shown that the conditions (1) - (3) might fail only if Λ is a hyperplane in Π_x.

So the classification in the case $g = 5$ and $G_1(x) \neq 1$ was reduced to the classification of the flag-transitive P-geometries. Recently the classification was completed (cf. [IS3]). During the classification two additional examples of P-geometries were constructed. These are universal 2-covers $\mathcal{G}(3^{23}.Co_2)$ [Shp2] and $\mathcal{G}(3^{4371}.F_2)$ [IS2] of the geometries $\mathcal{G}(Co_2)$ and $\mathcal{G}(F_2)$, respectively. The automorphism groups of the covers are nonsplit extensions of the corresponding sporadic groups by elementary abelian 3-groups. As a corollary of the classification of the flag-transitive P-geometries and Theorem 6.1 we have the following

Theorem 6.2. *Let Γ satisfy Hypothesis B with $g = 5$ and $G_1(x) \neq 1$. Then one of the following holds:*

(i) Γ is the derived graph of one of the following P-geometries: $\mathcal{G}(M_{22})$, $\mathcal{G}(3.M_{22})$, $\mathcal{G}(Co_2)$, $\mathcal{G}(3^{23}.Co_2)$, $\mathcal{G}(J_4)$, $\mathcal{G}(F_2)$ and $\mathcal{G}(3^{4371}.F_2)$;

(ii) Γ is the graph $\Gamma(J_4)$ of valency 31 related to Janko's group. $\square$

The only known example of girth 5 graph with faithful projective subconstituent (that is with $G_1(x) = 1$) is the derived graph $\Gamma(M_{23})$ of the P-geometry $\mathcal{G}(M_{23})$. We conclude the section by the following

Conjecture D. *A graph of girth 5 admitting a 2-transitive action with faithful projective subconstituent of rank ≥ 2 is isomorphic to the graph $\Gamma(M_{23})$.*

7. $g = 6$

Under Hypothesis B the girth 6 graphs of type 1 are completely classified. They turned out to be the graphs $\Gamma_{n-1,n}(A_{2n-2}(q))$.

In the case $G_1(x, y) \neq 1$ the classification is obtained in [Ivn3]. The main goal of that paper was to classify the distance-transitive graphs admitting elations and possessing a projective subconstituent of rank ≥ 2. But within the characterization of $\Gamma_{n-1,n}(A_{2n-2}(q))$ the the distance-transitivity condition was used exclusively in order to show that girth must be equal to 6. Let us present the general line of the arguments.

We assume that Γ is of type 1 and $G_1(x, y) \neq 1$. Let $W = (z, y, x, w)$ be a 3-path in Γ. Then $G(W)$ has exactly two orbits on the set $\Gamma(w) - \{x\}$. One of these orbits has length q and coincide with the image under $\psi_x(y, w)$ of the line from $L_y(x)$ which contains z. So $G(W)$ has two orbits, say $\mathcal{P}_1(W)$ and $\mathcal{P}_2(W)$, on the set of 4-paths which are continuations of W (we assume that $|\mathcal{P}_1(W)| = q$). The first step in reconstruction of the geometry is to prove existence of a connected subgraph $\Xi = \Xi(W)$ such that

(Ξ_1) Ξ contains W;

(Ξ_2) if W' is a 3-path contained in Ξ, then Ξ contains all paths from $\mathcal{P}_1(W')$ and no paths from $\mathcal{P}_2(W')$.

It is easy to see that if $\Xi(W)$ exists then it is unique and of valency $q+1$. Notice that an analogous construction can be applied for graphs of type 2.1 where one should start with a 2-path W. This construction leads to an alternative definition of the subgraph Ξ_1.

Lemma 7.1. [Ivn3] *Let Γ be a graph of type 1 and $G_1(x,y) \neq 1$. Then for a 3-path W there is a unique subgraph $\Xi(W)$ satisfying (Ξ_1) and (Ξ_2). Moreover, $G\{\Xi\}$ acts t-transitively on Ξ, $t \geq 4$ and $G(\Xi)$ acts transitively on $\Gamma(x) - \Xi$ for $x \in \Xi$.* □

Using the above lemma it was shown that distance-transitivity is possible only if $g = 6$. After that it was shown that if $g = 6$ then Ξ contains a cycle of length 6 and this immediately implies that Ξ is the incidence graph of the Desarguesian projective plane of order q.

In view of the arguments given at the end of Section 3 we can assume that Γ is bipartite with parts Γ^1 and Γ^2. We associate with Γ a geometry $\mathcal{G}$ which is a collection of subgraphs in Γ. If $\Theta \in \mathcal{G}$ and $i = i(\Theta) \in \{1,2\}$ then for $x \in \Xi$ and $x \in \Gamma^i$ we have $\Gamma(x) \subseteq \Xi$ and the image of $\Gamma_2(x) \cap \Xi$ in Δ_x is a (possibly empty) subspace Λ whose dimension is independent on x and together with $i(\Theta)$ determines the type of Θ in $\mathcal{G}$. The structure of Ξ implies that the required subgraph Θ can be defined as the subgraph induced by the vertices fixed by the elementwise stabilizer in $G_1(x)$ of the full preimage in $\Gamma_2(x)$ of the subspace $\Lambda \subseteq \Delta_x$. Let Θ_1 and Θ_2 be elements of $\mathcal{G}$. If $i(\Theta_1) = i(\Theta_2)$ then for Θ_1 and Θ_2 to be incident, one of the subgraphs must contain another one; otherwise it is sufficient for them to have a nonempty intersection.

Now the structure of Ξ implies that $\mathcal{G}$ and $A_{2n-2}(q)$ have the same diagram and hence (cf. [Tit]) these geometries are isomorphic.

Theorem 7.2. [Ivn3] *Let Γ satisfy Hypothesis B, be of type 1, girth 6 and suppose that $G_1(x,y) \neq 1$. Then Γ is isomorphic to $\Gamma_{n-1,n}(A_{2n-2}(q))$* □

Theorem 7.3. [Chi] *There are no graphs satisfying Hypothesis B which are of girth 6, type 1 and with $G_1(x,y) = 1$.* □

Discussion of girth 6 graphs of type 2 we start with examples.

Example 7.4. Let Γ be the derived graph of a flag-transitive P-geometry of rank $r \geq 3$. Then the standard doubling of Γ has girth 6 and possesses a projective subconstituent of rank $r - 1$.

Another very nice example is related to Fischer's group Fi_{22} and in the context of projective subconstituents it was mentioned in [Tro1].

Example 7.5. Consider the minimal 3-local parabolic geometry of the group Fi_{22} from [RS]. This is a rank 3 geometry which has one type of

points with stabilizers of the shape $3^{1+6}.2^2.(SL_2(3) \times SL_2(3)).2$ and two
types of planes with stabilizers of the shape $3^{3+3}.L_3(3)$. The two types of
planes are permuted by the outer automorphism of Fi_{22}. Define a graph
$\Gamma(Fi_{22})$ on the set of all planes, where two planes are adjacent if they are
of different type and are incident to a common point. Then the action of
$Aut(Fi_{22})$ on $\Gamma(Fi_{22})$ is 2-transitive and with respect to this action there is
a projective subconstituent of rank 2 over $GF(3)$. The graph $\Gamma(Fi_{22})$ is of
type 2.1 and its subgraph Ξ_1 is isomorphic to a 4-fold cover $4.K_{4,4}$ of the
complete bipartite graph. This is a subgraph of the incidence graph of the
projective plane over $GF(4)$ induced by the points which are not on a fixed
line and by the lines which miss a fixed point on that line. The girth of
$4.K_{4,4}$ as well as of $\Gamma(Fi_{22})$ is 6. Notice that $G_2(x)$ is of order 3^3.

Before considering the possibilities to characterize the above examples,
let us see what is known about finiteness of the graphs satisfying Hypothesis
B.

It is well known that a 2-transitive graph of valency k and girth 4 has at
most 2^k vertices and this bound is attained for the k-dimensional cube Q_k.
The number of vertices in a 3-transitive graph of girth 6 is also bounded
by a function of the valency. Both results are consequences of a general
proposition proved in [Ter].

Definition 7.6. For any integers c, s, a and k with $a + 2, s, c, k \geq 2$ an
(s, c, a, k)-graph Γ satisfies:

(1) the maximal valency is k;

(2) there are exactly c minimal paths between any two vertices at distance s apart;

(3) there are exactly a paths of length s connecting any two vertices at
distance $s - 1$ apart;

(4) the girth of Γ is $2s - 1$ if $a > 0$ and $2s$ if $a = 0$.

Theorem 7.7. [Ter] *Let Γ be an (s, c, a, k)-graph with $a \leq c$. Then Γ
is finite with diameter d bounded by an explicitly given function of s, c and
k.* $\square$

For a 2-transitive graph of girth 4 we have $s = 2$, $a = 0$, $c \geq 2$; for a
3-transitive graph of girth 6 we have $s = 3$, $a = 0$, $c \geq 2$. So in these two
case the finiteness is guaranteed by a general result. On the other hand
for the derived graph of a P-geometry we have $s = 2$, $a = 1$ and $c = 1$,
so Theorem 7.7 can not be applied. Actually there exist infinite, locally
finite 2-transitive graphs of girth 5 [Ivn1] and the fact that all P-geometries
turned out to be finite is quite remarkable.

Now let us turn to graphs in Examples 7.4 and 7.5.

Recall that for a geometry $\mathcal{G}$ over a set Δ of types the minimal circuit
diagram is a graph having Δ as the set of vertices; the multiplicity of the
edge joining i and j is m_{ij} where $2m_{ij}$ is the length of a minimal circuit in

the incidence graph of a rank 2 residue of type $\{i, j\}$ (the minimum should be taken over all such residues). A diagram is said to be spherical if it is the diagram of a finite Coxeter group and non-spherical otherwise. The standard convention is that if $m_{ij} = 2$ there is no edge between i and j and if $m_{ij} = 3$ we omit the label 3. Any diagram geometry [Bue] gives rise to an indexed complex by considering the elements of the geometry as vertices and flags of size i as $(i-1)$-simplexes. This enables one to define the Euler characteristic of a diagram geometry.

Theorem 7.8. [Ron] *If $\mathcal{G}$ is a rank 3 diagram geometry whose minimal circuit diagram is non-spherical and the Euler characteristic of $\mathcal{G}$ is not 1 then the universal cover of $\mathcal{G}$ is infinite.* $\square$

Let $\mathcal{G}$ be geometry whose elements are vertices edges and the subgraphs Ξ_1 in a graph Γ from Examples 7.4 or 7.5. Then the minimal diagram is the following:

This diagram is non-spherical and by Theorem 7.8 (since the Euler characteristic is not 1) we see that the universal cover of the geometry is infinite. On the other hand the cover corresponds to the cover $\tilde{\Gamma}$ of Γ with respect to the subgroup of the fundamental group which is generated by all cycles of length 6. Hence $\tilde{\Gamma}$ is also of girth 6 and possesses a projective subconstituent.

In view of the above discussions we see that in case of girth 6 graphs of type 2 it would be realistic to expect just a local characterization, that is the classification of amalgams of parabolic subgroups of the corresponding geometries.

Notice that for P-geometries of rank 3 the minimal circuit diagram is H_3, that is spherical.

8. $g = 7, 8$

As far as I know, in the cases left nothing serious is proved yet. I believe that there are no examples at all with $g = 7$ and that the graphs $\Gamma_{2,3}(F_4(2))$ can be characterized along the geometric approach. Probably one would need an additional condition like $G_2(x) \neq 1$.

Acknowledgement

The author wishes to thank the University of Michigan for its hospitality and financial support while a part of this work was being done and P.J.Cameron for his valuable comments on the paper.

References

[Big]   N.L.Biggs, Designs, factors and codes in graphs, Quart. J. Math. Oxford (2), **26** (1975), 113-119.

[Bue]   F.Buekenhout, Diagrams for geometries and groups, J. Combin. Theory A, **27** (1979), 121-151.

[CP]   P.J.Cameron and C.E.Praeger, Graphs and permutation groups with projective subconstituents, J. London Math. Soc. **25** (1982), 62-74.

[Chi]   K.Ching, Graphs of small girth which are locally projective spaces, Ph D. Thesis, Tufts Univ., Medford, MA, 1992.

[Atlas]   J.H.Conway,   R.T.Curtis,   S.P.Norton, R.A.Parker and R.A.Wilson, Atlas of Finite Groups, Clarendon Press, Oxford, 1985.

[DGS]   A.Delgado, D.M.Goldschmidt and B.Stellmacher, Groups and Graphs: New Results and Methods. Birkhauser Verlag, Basel, 1985.

[FII]   I.A.Faradjev, A.A.Ivanov and A.V.Ivanov, Distance-transitive graphs of valency 5, 6 and 7, Europ. J. Comb. **7** (1986), 303-319.

[Hug]   D.R.Hughes, Extensions of designs and groups: projective, symplectic and certain affine groups, Math. Z., **89** (1965), 199-205.

[Ivn1]   A.A.Ivanov, On 2-transitive graphs of girth 5, Europ. J. Combin. **8** (1987), 393-420.

[Ivn2]   A.A.Ivanov, Graphs of girth 5 and diagram geometries related to the Petersen graphs, Soviet Math. Dokl. **36** (1988), 83-87.

[Ivn3]   A.A.Ivanov, The distance-transitive graphs admitting elations, Math. of the USSR Izvestiya, **35** (1990), 307-335.

[IP]   A.A.Ivanov and C.E.Praeger, On finite affine 2-arc transitive graphs, Europ. J. Comb. (to appear)

[IS1]   A.A.Ivanov and S.V.Shpectorov, Geometries for sporadic groups related to the Petersen graph. II, Europ. J. Combin. **10** (1989), 347-362.

[IS2]   A.A.Ivanov and S.V.Shpectorov, The last flag-transitive P-geometry. Israel J. Math. (to appear)

[IS3]   A.A.Ivanov and S.V.Shpectorov, The flag-transitive tilde and Petersen type geometries are all known, Preprint, 1993.

[Mor]   B.Mortimer, The modular permutation representations of the known doubly transitive groups, Proc. London Math. Soc (3), **41** (1980), 1-20.

[Ron]   M.A.Ronan, On the second homotopy group of certain simplicial complexes and some combinatorial applications. Quart. J. Math. (2) **32** (1981), 225-233.

[RS]   M.A.Ronan, G.Stroth, Minimal parabolic geometries for the sporadic groups, Europ. J. Combin., **5** (1984), 59-91.

[Shp1]   S.V.Shpectorov, A geometric characterization of the group M_{22}. In: Investigations in Algebraic Theory of Combinatorial Objects, pp. 112-123, Moscow, 1985 [In Russian]

[Shp2]   S.V.Shpectorov, The universal 2-cover of the P-geometry $\mathcal{G}(Co_2)$, Europ. J. Combin. **13** (1992), 291-312.

[SW]   G.Stroth and R.Weiss, Modified Steinberg relations for the group J_4, Geom. Dedic. **25** (1988), 513-525.

[Ter]   P.Terwilliger, Distance-regular graphs and (s, c, a, k)-graphs, J. Combin. Theory B, **34** (1983), 151-164.

[Tit]   J.Tits, A local approach to buildings, In: The Geometric Vein (Coxeter - Festschrift), pp. 519-547, Springer Verlag, Berlin, 1982.

[Tro1]   V.I.Trofimov, Stabilizers of the vertices of graphs with projective suborbits, Soviet Math. Dokl. **42** (1991), 825-827.

[Tro2]   V.I.Trofimov, Graphs with projective suborbits, Math. of the USSR Izvestya, **39** (1992), 869-894.

[Wei1]   R.Weiss, Symmetric graphs with projective subconstituents, Proc. Amer. Math. Soc., **72** (1978), 213-217.

[Wei2]   R.Weiss, Groups with a (B,N)-pair and locally transitive graphs, Nagoya Math. J., **74** (1979), 1-21.

[Wei3]   R.Weiss, Elations of graphs, Acta Math. Acad Sci. Hungar., **34** (1979), 101-103.

[Wei4]   R.Weiss, s-transitive graphs, In: Algebraic Methods in Graph Theory, pp. 827-847, Amsterdam, North-Holland, 1981.

[Wei5]   R.Weiss, Graphs with subconstituent containing $L_3(p)$, Proc. Amar. Math. Soc., **85** (1982), 666-672.

[Wei6]   R.Weiss, Graphs with projective suborbits, AMS Abstracts **13** (1992), 433.

Author's address:
A.A.Ivanov
Institute for System Analysis,
117312, Moscow, Russia

On Circuit Covers, Circuit Decompositions and Euler Tours of Graphs

Bill Jackson

Department of Mathematical Studies, Goldsmiths' College,
London SE14 6NW, England

1 Introduction

I will review certain results and problems concerning the covering of the edge set of a graph with circuits or Euler tours. The areas I shall consider are:

- Faithful circuit covers of weighted graphs.

- Shortest circuit covers of bridgeless graphs.

- Compatible circuit decompositions and Euler tours of Eulerian graphs.

- Even circuit decompositions of Eulerian graphs.

I will describe interconnections between these areas and also indicate connections with other areas of combinatorics: matroids, isotropic systems, edge colouring of graphs, latin squares, nowhere-zero flows in graphs, and the Chinese postman problem.

Throughout this survey my graphs will be finite and without loops, although they may contain multiple edges. By a **closed walk** in a graph G, I will mean an alternating sequence of vertices and edges which starts and ends at the same vertex and is such that consecutive vertices and edges are incident. The **length** of the walk is the number of edges in the sequence. A **tour** is a closed walk which does not repeat edges. An **Euler tour** of G is a tour which traverses every edge of G. We shall say that a graph is **Eulerian** if it has an Euler tour. A **tour decomposition** of G is a partition of $E(G)$ into (edge sets of) tours. We will consider an Euler tour of G to be a tour decomposition into exactly one tour. A **circuit** is a tour which does not repeat vertices, except when the first vertex is repeated as the last. A **circuit cover (circuit decomposition)** of G is a list of circuits of G such that every edge of G belongs to at least (exactly) one circuit in the list. Note that:

- G is Eulerian if and only if G is connected and each vertex of G has even degree.

- G has a circuit decomposition if and only if each component of G is Eulerian.

- G has a circuit cover if and only if G has no bridges.

An **edge cut** of G is the set of edges between some proper subset U of $V(G)$ and its complement $\bar{U}$. The edge cut is **cyclic** if both the induced subgraphs $G[U]$ and $G[\bar{U}]$ contain circuits. We shall say that G is **k-edge connected (cyclically k-edge connected)** if each (cyclic) edge cut of G has order at least k. If U consists of a single vertex u then we shall denote the edge cut of G from U to $\bar{U}$ by $E_u(G)$, or simply E_u, when it is clear which graph we are referring to. I refer the reader to [BM] for graph theoretic terminology which is not defined here and also to [F4,5] for a comprehensive theory of Eulerian graphs.

2 Faithful circuit covers

Problem 2.1 Given a graph G and a weight function $w : E(G) \to Z^+$, determine whether G has a circuit cover X such that each edge e of G lies in exactly $w(e)$ circuits of X.

I shall call such a circuit cover X a **faithful** circuit cover for (G, w). The above problem was first considered in [S1] where Seymour gave two obvious necessary conditions for the existence of X:

(1) for each vertex v of G we have $\sum_{e \in E_v} w(e)$ is even, and

(2) for each edge cut S of G and each $e \in S$ we have $w(e) \leq \sum_{f \in S-e} w(f)$.

I shall say that w is an **admissible** weight function for G if it satisfies (1) and (2). There exist graphs with admissible weight functions for which there are no faithful circuit covers, for example the Petersen graph P with a weight function which takes the value two on a 1-factor of P and one elsewhere. If one avoids the Petersen graph, however, then the necessary condition of admissibility is also sufficient. The first result along these lines occurs in [S1] where Seymour used the 4-colour theorem[AH] to show that every admissible weighting for a planar graph has a faithful circuit cover. This was extended by Alspach, Goddyn and Zhang in [AGZ] where they were able to obtain the following result (without resorting to the 4-colour theorem).

Theorem 2.2 Let G be a graph with no subgraph homeomorphic to the Petersen graph and w be an admissible weighting for G. Then (G, w) has a faithful circuit cover.

An earlier version of this result was obtained by Alspach and Zhang[AZ] for the special case of cubic graphs with a weight function which takes the values one and two only (called a **(1,2)-weight function**). In proving his planar result, Seymour[S1] first reduced to the case when w is a (1,2)-weight function and then used a nowhere-zero 4-flow in the planar graph G (which exists by the 4-colour theorem) to construct a faithful circuit cover. Zhang[Z1] observed that this second idea can be used in general to give:

Theorem 2.3 Let G be a graph with a nowhere-zero 4-flow and w an admissible (1,2)-weight function for G. Then (G, w) has a faithful circuit cover.

Zhang also gives an example in [Z3] to show that Theorem 2.3 does not hold if one allows the weight function to take arbitrary values. The following immediate corollary of Theorem 2.3 highlights a relationship between a special case of Problem 2.1 and the 3-edge colourability of cubic graphs.

Corollary 2.4 Every admissible (1,2)-weight function for a cubic 3-edge colourable graph has a faithful circuit cover.

For the remainder of this section I will be mostly concerned with admissible (1,2)-weight functions for cubic graphs. (Note that a (1,2)-weight function w for a cubic graph G is admissible if and only if it satisfies (1) and G is bridgeless.) This special case of Problem 2.1 seems to be important since the proofs in both [S1] and [AGZ] reduce to the case of a (1,2)-weight function in a graph of maximum degree four. In addition, if G is cubic then Corollary 2.4 shows that the theory of edge colouring may be successfully employed (as, for example, the result of Ellingham [El] was used in [AZ]).

Suppose we are given an admissible (1,2)-weight function for a cubic graph and wish to decide if there exists a faithful circuit cover (it is not yet known whether this decision problem is NP- complete). If the graph has a cyclic edge cut of order at most three then the decision problem can be reduced to two smaller graphs as follows:

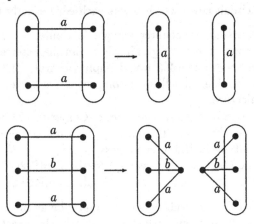

Thus we may suppose that the graph is cyclically 4-edge connected. We may construct an infinite family of cyclically 4-connected cubic graphs with admissible (1,2)-weight functions for which there is no faithful circuit cover by using an operation similar to the **dot product** for snarks, see [I], starting with two copies of the Petersen graph.

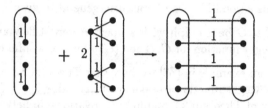

We do not, however, know of any cyclically 5-edge connected examples other than the Petersen graph.

Problem 2.5 Is the Petersen graph the only example of a cyclically 5-edge connected cubic graph which has an admissible (1,2)-weight function for which there is no faithful circuit cover?

Note that using Corollary 2.4 the search for other examples can be restricted to the family of cyclically 5-edge connected snarks. (Recall that a **snark** is a cyclically 4-edge connected cubic graph of girth at least five which is not 3-edge colourable.)

We next mention two outstanding conjectures from [S1].

Conjecture 2.6 Let w be an admissible weight function for a graph G such that $w(e)$ is even for all edges e of G. Then (G, w) has a faithful circuit cover.

A special case of this conjecture is the infamous

Conjecture 2.7 (The Circuit Double Cover Conjecture) Every bridgeless graph has a circuit cover which covers each edge exactly twice.

It is known that a smallest counterexample to Conjecture 2.7 would be a cyclically 5-edge connected cubic graph other than the Petersen graph. Thus a positive answer to Problem 2.5 would imply the truth of Conjecture 2.7. I refer the reader to Jaeger[Ja2] for a comprehensive survey on the circuit double cover conjecture.

We could also consider the special cases of Conjecture 2.6 when w takes a constant value other than two. The case $w(e) \equiv 4$ was resolved by Bermond, Jaeger and the author in [BJJ] as an easy consequence of Jaeger's 8-flow theorem[Ja1]. The case $w(e) \equiv 6$ was settled by Fan[Fa2] using Seymour's 6-flow theorem[S2]. Combining these two results Fan[Fa2] obtained

Theorem 2.8 Let G be a bridgeless graph and k be an integer greater than one. Then G has a circuit cover which covers each edge exactly $2k$ times.

I close this section by noting that two variations of Problem 2.1, when one is allowed to use sums of circuits with either, positive rational coefficients, or, positive and negative integer coefficients, to represent w are solved completely in [S1]. Extensions of these problems to matroids are surveyed by Goddyn in [G]. One conjecture from [G,4.10] which is related to Conjectures 2.6 and 2.7 is:

Conjecture 2.9 Let w be an admissible weight function for a graph G such that $w(e) \geq 2$ for all edges e of G. Then (G, w) has a faithful circuit cover.

3   Shortest circuit covers

Problem 3.1 Given a bridgeless graph G find a circuit cover for G such that the sum of the lengths of the circuits is as small as possible.

Define the **length** of a circuit cover to be the sum of the lengths of its circuits and let $scc(G)$ be the minimum length of a circuit cover of G. Problem 3.1 was first considered by Itai and Rodeh in [IR], where, amongst other results, they point out a relationship between Problem 3.1 and

Problem 3.2 (The Chinese Postman Problem) Given a connected graph G find a shortest closed walk which covers every edge of G.

Let $cp(G)$ denote the length of such a shortest walk. It can be seen that $cp(G)$ is given by the minimum value of $w(G)$ taken over all weight functions w for G which satisfy condition (1) of admissability (where $w(G)$ denotes the sum of the weights of the edges of G). On the other hand, $scc(G)$ is given by the minimum value of $w(G)$ taken over all admissible weight functions w for G such that (G, w) has a faithful circuit cover. This implies the following result from [IR].

Lemma 3.3 For any 2-edge connected graph G we have $scc(G) \geq cp(G)$.

The Petersen graph P is an example for which equality in Lemma 3.3 does not occur. Every solution to the Chinese postman problem for P corresponds to a weight function w which takes the value two on a 1-factor of P and the value one elsewhere. Thus $cp(P) = 20$. As pointed out in Section 2, there is no faithful circuit cover for (P, w) and so $scc(P) > 20$. (In fact $scc(P) = 21$.)

Since there exists a polynomial algorithm for solving the Chinese postman problem (due to Edmonds and Johnson[EJ]) but, as yet, no polynomial algorithm for solving Problem 3.1, it is useful to identify families of graphs for which we have equality in Lemma 3.3. This corresponds to finding families of graphs G with an admissible (1,2)-weight function w given by a solution to the Chinese postman problem for G, for which (G, w) has a faithful circuit cover. Using Theorems 2.2 and 2.3 we deduce:

Theorem 3.4 Let G be a 2-edge connected graph such that either

  (i) G has a nowhere-zero 4-flow, or

  (ii) G has no subgraph homeomorphic to the Petersen graph.

Then $scc(G) = cp(G)$.

Condition (i) of the above result was given independently by Zhang[Z1] and the author[J3]. Condition (ii) was given in [AGZ]. Both conditions generalise an earlier result for planar graphs obtained independently in [BJJ] and [FG]. Note that Tutte[Tu] has conjectured that any bridgeless graph which satisfies (ii) will also satisfy (i).

    Several authors [IR], [ILPR], [BJJ], [AT], [Fr], [Fa1-6], [JRT], [J3,4] have obtained upper bounds on $scc(G)$. A summary of some of the strongest results is the following.

Theorem 3.5 Let G be a bridgeless graph and let $r = \min\{|C|\}$ taken over all the even circuits C of G which have length at least six (putting $r = \infty$ if no such C exists. Then

  (a) $scc(G) \le \frac{5}{3}|E(G)|$ (see [BJJ] and [AT]).

  (b) $scc(G) \le |E(G)| + \frac{6}{5}(|V(G)| - 1)$ (see [Fa3]).

  (c) if G has a nowhere-zero 4-flow, $scc(G) \le \frac{4}{3}|E(G)|$ (see [BJJ]).

  (d) if G has a nowhere-zero 5-flow, $scc(G) \le \frac{8}{5}|E(G)|$ (see [JRT]).

  (e) if G is $2k$-edge connected for some $k > 1$, $scc(G) \le \frac{2k+2}{2k+1}|E(G)|$

                                              (see [J3]).

  (f) $scc(G) \le |E(G)| + \frac{r}{r-1}|V(G)|$ (see [Fa3]).

For the special case of cubic graphs, slightly stronger bounds are known.

Theorem 3.6 Let G be a cubic bridgeless graph of girth g. Then

  (a) $scc(G) \le \frac{44}{27}|E(G)|$ (see [Fa6]).

  (b) if $g = 6$, $scc(G) \le \frac{34}{21}|E(G)|$ (see [J4]).

  (c) if $g = 7$, $scc(G) \le \frac{19}{12}|E(G)|$ (see [Fa6]).

  (d) if $g \ge 8$, $scc(G) \le (\frac{4}{3} + \frac{2}{g})|E(G)|$ (see [J4]).

There remain the following conjectures, given in [AT] and [IR], respectively.

Conjecture 3.7 For every bridgeless graph G, $scc(G) \le \frac{7}{5}|E(G)|$.

Conjecture 3.8 For every bridgeless graph G, $scc(G) \leq |E(G)| + |V(G)| - 1$.

The upper bound on $scc(G)$ given in Conjecture 3.7 cannot be reduced because of the Petersen graph. The upper bound on $scc(G)$ given in Conjecture 3.8 cannot be reduced beyond $|E(G)| + |V(G)| - 2$ because of the complete bipartite graph $K_{3,t}$. Both conjectures hold when $scc(G) = cp(G)$ so Theorem 3.4 can be used to show they are valid if G has a nowhere-zero 4-flow or has no subgraph homeomorphic to the Petersen graph. Jamshy and Tarsi[JT2] have shown that the truth of Conjecture 3.7 will imply the truth of the circuit double cover conjecture. I refer the reader to [JRT], [JT1,2], [T], [FaR], [R] for results on shortest circuit covers of matroids.

4   Compatible circuit decompositions

Throughout this section G will be an Eulerian graph. A **system of forbidden parts** for G is a function T defined on $V(G)$ such that $T(v)$ is a partition of E_v for each $v \in V(G)$. We shall refer to the elements of $T(v)$ as the **forbidden parts** at v. We shall say that two systems of forbidden parts for G, T_1 and T_2, are **compatible** if for each vertex v of G, any two elements of $T_1(v)$ and $T_2(v)$ intersect in at most one element.

We shall call a forbidden part at v of order two a **transition** at v. If all the forbidden parts in a system of forbidden parts are transitions, then we shall call it a **transition system**. There is a natural bijection f between transition systems for G and tour decompositions for G: two edges are consecutive in some tour of $f(T)$ if and only if they form a transition of T. We use this bijection to extend the concept of compatibility to tour decompositions: a tour decomposition X for G is **compatible** with a system of forbidden parts T (or another tour decomposition Y) if and only if $f^{-1}(X)$ and T (or $f^{-1}(X)$ and $f^{-1}(Y)$) are compatible systems of forbidden parts. Given a tour decomposition X of G, I shall refer to the transitions of $f^{-1}(X)$ as **transitions** of X.

The fundamental problem on compatible circuit decompositions can be stated as:

Problem 4.1 Given a system of forbidden parts T for an Eulerian graph G, determine whether G has a circuit decomposition compatible with T.

There is a close relationship between Problems 2.1 and 4.1. Given an instance (G, T) of Problem 4.1 we can transform to an instance (H, w) of Problem 2.1, by applying the following operation to each vertex v of G (where the partition $T(v)$ of E_v is represented by the sets of edges which are incident with the 'dashed boxes' in the figure).

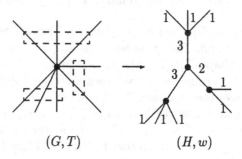

$$(G,T) \qquad\qquad (H,w)$$

On the other hand, given an instance (H, w) of Problem 2.1 we can transform to an instance (G, T) of Problem 4.1 by replacing each edge e of H by $w(e)$ parallel edges, and then making this set of parallel edges a forbidden part at v for each end vertex v of e. Note that, in these transformations, (1,2)-weight functions correspond to transition systems. The reader can thus expect to see a close correspondence between results and problems in this section and Section 2.

There is an obvious necessary condition for the existence of a circuit decomposition of G compatible with T. For any edge cut R of G and any forbidden part S of T, we must have $|R \cap S| \leq \frac{1}{2}|R|$. We shall say that a system of forbidden parts is **admissible** if it satisfies this condition. There exist examples of Eulerian graphs with admissible systems of forbidden parts which have no compatible circuit decomposition. Such an example can be obtained from K_5 by taking T to be the transition system corresponding to a tour decomposition into two 5-circuits. If one avoids K_5, however, then the above necessary condition is also sufficient. The first result along these lines was obtained by Fleischner in [F2], where he showed that if T is an admissible transition system for an Eulerian planar graph G, then G has a circuit decomposition compatible with T. This was extended to admissible systems of forbidden parts by Fleischner and Frank in [FF]. Fan and Zhang[FaZ] have recently used Theorem 2.2 to obtain the following generalisation.

Theorem 4.2 Let G be an Eulerian graph with no K_5-minor and T be an admissible system of forbidden parts for G. Then G has a circuit decomposition compatible with T.

H. Fleischner (personal communication) has constructed an infinite family of 4-connected 4-regular graphs with admissible transition systems for which there is no compatible circuit decomposition. (His construction is equivalent to using the transformation between (1,2)-weight functions and transition systems given at the beginning of this section on the cyclically 4-edge connected examples constructed using the dot product in Section 2.) Since all such examples have edge connectivity four, we are led to raise:

Conjecture 4.3 Let T be a transition system for a 6-edge connected Eulerian graph G. Then G has a circuit decomposition compatible with T.

This conjecture may even be valid under the weaker hypothesis that G has minimum degree at least four and is cyclically 6-edge connected. The truth of Conjecture 4.3 would imply the truth of the cycle double cover conjecture: it is known that the cycle double cover conjecture can be reduced to 3-connected graphs; replacing each edge of such a graph by a 2-circuit would yield a 6-edge connected Eulerian graph with the obvious transition system given by the tour decomposition into 2-circuits.

Another open problem on circuit decompositions compatible with transition systems is the following attractive conjecture made by G. Sabidussi in 1975, see [F2].

Conjecture 4.4 Let G be an Eulerian graph of minimum degree at least four and X be an Euler tour of G. Then G has a circuit decomposition compatible to X.

Theorem 4.3 implies that Sabidussi's conjecture is true for graphs with no K_5-minor. We may also deduce that it is true when all vertices of G have degree divisible by four using the following idea from [F2].

Lemma 4.5 Let G be a graph in which all vertices have degree divisible by four and let X be a tour decomposition for G in which all tours have even length. Then G has a circuit decomposition compatible with X.

Proof We may colour the edges of each tour in X with two colours, red and blue, such that consecutive edges receive different colours. The resulting red and blue subgraphs of G have all vertices of even degree so they each have a circuit decomposition, X_R and X_B, respectively. Then $X_R \cup X_B$ is a circuit decomposition compatible with X.

Note that Lemma 4.5 could also be deduced from Theorem 2.3 using the transformation between (1,2)-weight functions and transition systems given earlier in this section. Related results of Fleischner[F3] are that Sabidussi's conjecture can be reduced to the case of graphs with all vertices of degrees four and six, and that it is valid if G has all vertices of degree four apart from at most one vertex of degree six. He also shows that the truth of Sabidussi's conjecture and the truth of a conjecture of Fleischner (that every snark has a dominating circuit) would together imply the truth of the circuit double cover conjecture.

We close this section by offering one more problem, first given in [FHJ], which was inspired by the results on pairwise compatible Euler tours contained in Section 6.

Problem 4.6 Is it true that every 2-connected Eulerian graph of minimum degree at least four has two pairwise circuit decompositions compatible with each other?

5   Even circuit decompositions

Problem 5.1 Given an Eulerian graph G, determine whether G has a circuit decomposition into circuits of even length.

I shall call such a circuit decomposition an **even circuit decomposition**.

At first sight Problem 5.1 seems less restrictive than Problems 2.1 and 4.1. We shall see, however, that important special cases of these problems can be tranformed into Problem 5.1. Suppose we are given an instance (H, w) of Problem 2.1, in which H has maximum degree three and w is a (1,2)-weight function. We may assume that $|E(H)|$ is even since, if $|E(H)|$ is odd then we may subdivide an edge e_0 of H and redefine w to take the value $w(e_0)$ on the two new edges. We can then transform to an instance G of Problem 5.1, by adding a new vertex v_e for each edge e of H with $w(e) = 2$, and then joining each v_e to both end vertices of e. It can easily be seen that an even circuit decomposition of G will give rise to a faithful circuit cover of (H, w). On the other hand, since H has maximum degree three and an even number of edges, we may use [S3,Lemma 2.5] to show that a faithful circuit cover of (H, w) can be converted to an even circuit decomposition of G. Thus a good characterisation or polynomial algorithm for solving Problem 5.1 will give rise to a corresponding solution for the special case of Problem 2.1 for (1,2)-weight functions in graphs of maximum degree three. Using the transformation between Problems 2.1 and 4.1 given at the beginning of Section 4, we deduce that a similar statement holds for Problem 5.1 and the special case of Problem 4.1 for transition systems in Eulerian graphs of maximum degree six. Thus the reader can expect to see a close correspondence between results and problems of this section and Sections 2 and 4.

Even circuit decompositions were first considered by Seymour in [S3], where he pointed out that an obvious necessary condition for the existence of an even circuit decomposition of an Eulerian graph G is that each block of G should have an even number of edges. The complete graph K_5 shows that this condition is not always sufficient. Seymour showed in [S3], however, that it is both necessary and sufficient when G is planar. Zhang[Z2] has recently used Theorem 4.2 to extend this result to:

Theorem 5.2 Let G be an Eulerian graph with no K_5-minor, in which each block has an even number of edges. Then G has an even circuit decomposition.

His proof uses the following neat trick from [FF]. Choose an Euler tour in G and colour the edges alternately red and blue around the tour. Define a system of forbidden parts T for G consisting of the sets of edges at each vertex which have the same colour. Then a compatible circuit decomposition of (G,T) must necessarily consist of even cycles which alternate between red and blue edges.

One can construct an infinite family of 2-connected Eulerian graphs with an even number of edges and no even circuit decomposition using the dot-product construction of Section 2 and the transformation from Problem 2.1 described at the beginning of this section. Zhang conjectured in [Z3] that K_5 is the only 3-connected Eulerian graph with an even number of edges which does not have an even circuit decomposition. The following construction shows that Zhang's conjecture is false. We first use the dot product construction of Section 2, starting from the Petersen graph, to construct an infinite family of graphs H with the following properties:

(a) H is 3-connected and cubic;

(b) H has a (1,2)-weight function w such that the edges of H of weight two form a 1-factor of H;

(c) (H, w) has no faithful circuit cover.

We may suppose furthermore that $|E(H)|$ is even since if this is not the case then replacing a vertex of H by a triangle and redefining w in the obvious way preserves properties (a)-(c) and changes the parity of $|E(H)|$. We next construct a 3-connected Eulerian graph G by replacing each edge e of H of weight two by a graph F_e, isomorphic to K_5 minus one edge, as follows:

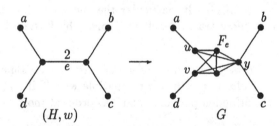

Then $|E(G)|$ has the same parity as $|E(H)|$ so is even. Suppose G has an even circuit decomposition X. If for each subgraph F_e of G, the circuit in X which contains the edge dv also contains the edge yb or yc, then X will give rise to a faithful circuit cover of (H, w). Thus for some subgraph F_e the circuit C of X which contains dv also contains ua. If C contains an even number of edges of F_e then $G - E(C)$ will have a block, $F_e - E(C)$, with an odd number of edges, attached at y. This would contradict the fact that $X - \{C\}$ is an even circuit decomposition of $G - E(C)$ and so C must contain an odd number of edges of F_e. But then the set of all circuits of X which intersect $E(F_e)$ together with $E(C) \cap E(F_e) + uv$ would be an even circuit decomposition of $F_e + uv$. This is impossible since $F_e + uv$ is isomorphic to K_5. Thus G has no even circuit decomposition.

Perhaps the conjecture becomes true if we increase the connectivity still further.

Problem 5.3 Is K_5 the only 4-connected Eulerian graph with an even number of edges which has no even circuit decomposition?

6   Compatible Euler tours

I will adopt the terminology of Section 4. In addition, given a subset S of $E_v(G)$, define the **detachment of G at v along S** to be the graph G^S with $V(G^S) = (V(G) - \{v\}) \cup \{v', v''\}$, $E(G^S) = E(G)$, $E_u(G^S) = E_u(G)$ for all $u \in V(G) - v$, $E_{v'}(G^S) = S$ and $E_{v''}(G^S) = E_v(G) - S$. Given a system of forbidden parts T for G define the **detachment of G along T** to be the graph G^T obtained by recursively detaching G along each forbidden part of T. (An elegant theory for the existence of detachments satisfying certain degree and connectivity conditions is given by Nash-Williams in [N].)

Problems concerning compatible Euler tours seem, in general, to be easier than problems concerning compatible circuit decompositions. One reason for this is because, if we form the detachment G^S in an inductive proof on G, then we can easily convert an Euler tour of G^S back to an Euler tour of G. The corresponding operation is not so easy for circuit decompositions, however, since a circuit decomposition of G^S may not give rise to a circuit decomposition of G (some elements of the circuit decomposition of G^S may be converted to tours in G). In particular the version of Problem 4.1 for Euler tours is much more tractable. It was solved by Kotzig[K1], see also Fleischner[F2].

Theorem 6.1 Let T be a system of forbidden parts for an Eulerian graph G. Then G has an Euler tour which is compatible with T if and only if for each vertex v, each forbidden part of T at v has order at most $\frac{1}{2}d(v)$.

One can even give sufficient conditions for the existence of an Euler tour which is compatible with several systems of forbidden parts. The special case when all of the systems of forbidden parts are transition systems was first considered in [J1] and then slightly extended in [JW] to give:

Theorem 6.2 Let $T_1, T_2, ..., T_m$ be transition systems for an Eulerian graph G. If each vertex of G has degree at least $2m + 4$, for m even, or at least $2m + 2$, for m odd, then G has an Euler tour which is compatible with T_i for all i, $1 \leq i \leq m$.

There exist examples which show that the degree condition of Theorem 6.2 cannot be weakened, see [J1,Fig 1] for the case when m is even.

Rather than considering several systems of forbidden parts, it seems more natural to give a set of forbidden transitions for each edge $e \in E_v$ and each

$v \in V(G)$. Formally, we define a **generalised transition system** T for G to be a set of functions $\{T_v\}_{v \in V(G)}$ such that $T_v : E_v \to 2^{E_v}$ and whenever $e_1 \in T_v(e_2)$ we have $e_2 \in T_v(e_1)$. We say that an Euler tour X is **compatible** with T if whenever $e_1 \in T_v(e_2)$ it follows that $\{e_1, e_2\}$ is not a transition of X at v for all $v \in V(G)$. The proof technique of [J1] and the result of [JW] can be used to give the following generalisation of Theorem 6.2.

Theorem 6.3 Let T be a generalised transition system for an Eulerian graph G. Suppose that for each $v \in V(G)$ we have

(i) $|T_v(e)| \leq \frac{1}{2}d(v) - 1$ when $d(v) \equiv 0 \pmod 4$ or $d(v) = 2$, and

(ii) $|T_v(e)| \leq \frac{1}{2}d(v) - 2$, otherwise.

Then G has an Euler tour compatible with T.

Proof We use induction on $|V(G)|$. Choose $v \in V(G)$ and let $|E_v| = 2m$. Define the new graph G_v with $V(G_v) = E_v$ in which two vertices e_1 and e_2 are adjacent if and only if $e_1 \notin T_v(e_2)$. Let F be the set of all transitions S at v in G such that G^S is disconnected. We may extend F to a transition system F_1 at v. Considering each transition of F_1 as an edge of G_v or $\bar{G}_v$, we may form the graph $G'_v = G_v + F_1$. Using the degree conditions (i), or (ii), it follows from results of Häggkvist[H], or [JW], respectively, that G'_v has a 1-factor L such that $F_1 \cup L$ is a Hamilton circuit of G'_v. Considering L as a transition system at v in G we may deduce from the choice of L that G^L is connected and hence Eulerian. We may now apply induction to G^L to obtain an Euler tour of G^L which is compatible to the generalised transition system T' obtained by putting $T'_u(e) = T_u(e)$ for $e \in E_u$ and $u \in V(G) - v$, and otherwise putting $T'_u(e) = \emptyset$. This gives rise to an Euler tour of G which is compatible with T.

The main motivation for obtaining Theorem 6.2 in [J1] was the following problem suggested by A.J.W. Hilton at an Open University Combinatorics Workshop in 1985.

Problem 6.4 Determine the maximum number of pairwise compatible Euler tours in a given Eulerian graph G.

A special case of this problem had previously been raised by Kotzig in [K3], where he conjectured that the complete graph K_{2k+1} has $2k-1$ pairwise compatible Euler tours. Let $pcet(G)$ represent the maximum number of pairwise compatible Euler tours of G. It seems that $pcet(G)$ is closely related to the minimum degree of G, which we will denote by $\delta(G)$. To see this, suppose that $X_1, X_2, ..., X_m$ are pairwise compatible Euler tours of G and let v be a vertex of degree $\delta(G)$. Choosing $e_1 \in E_v$ we see that e_1 must be contained in a different transition of X_i for each i, $1 \leq i \leq m$. Since there are only

$\delta(G) - 1$ edges in $E_v - e_1$, we must have $m \leq \delta(G) - 1$. Furthermore, if there exists a subset $S = \{e_1, e_2\}$ of E_v such that the detachment G^S is disconnected, then it can be seen that S cannot be a transition of X_i for all i, $1 \leq i \leq m$. In this case the above argument implies that $m \leq \delta(G) - 2$. I conjectured in [J1] that $pcet(G)$ is in fact equal to one of these two upper bounds.

Conjecture 6.5 The maximum number of pairwise compatible Euler tours in an Eulerian graph G is either $\delta(G) - 2$ or $\delta(G) - 1$.

Using Theorem 6.2 we can deduce that $pcet(G) \geq \frac{1}{2}\delta(G) - 1$, by first choosing an Euler tour X_1 and then greedily choosing an Euler tour X_i which is compatible to $X_1, X_2, ..., X_{i-1}$ for all i, $2 \leq i \leq \frac{1}{2}\delta(G) - 2$. This was slightly extended in [JW] to give:

Theorem 6.6 Let G be an Eulerian graph. Then $pcet(G) \geq \frac{1}{2}\delta(G)$.

It follows that Conjecture 6.5 is valid for $\delta(G) = 4$. It has also been verified for $\delta(G) = 6$ in [J2], and for graphs with the property that every block is a circuit (when the compatibility condition is relaxed at vertices of degree two) by Fleischner, Hilton and the author in [FHJ]. The main interest of [FHJ], however, is a reduction technique which converts the problem of determining $pcet(G)$ to one concerning 1-factorizations of complete graphs. It can be stated as

Lemma 6.7 Let G be an Eulerian graph. Suppose that for all $m \geq \frac{1}{2}\delta(G)$ and all choices of t 1-factors $F_1, F_2, ..., F_t$ of K_{2m} there exist pairwise disjoint 1-factors $L_1, L_2, ..., L_t$ of K_{2m} such that $F_i \cup L_i$ is a Hamilton circuit for all i, $1 \leq i \leq t$. Then $pcet(G) \geq t$.

This Lemma was used to obtain the above mentioned lower bounds on $pcet(G)$ in [JW] (with $t \geq m$) and [J2] (with $t = 4$ and $m \geq 3$). It follows from Lemma 6.7 that the truth of Conjecture 6.5 would be implied by the truth of

Conjecture 6.8 [FHJ] Let $F_1, F_2, ..., F_{2m-2}$ be 1-factors of K_{2m}. Then there exists a 1-factorization $L_1, L_2, ..., L_{2m-1}$ of K_{2m} such that $F_i \cup L_i$ is a Hamilton circuit for $1 \leq i \leq 2m - 2$.

This conjecture was shown to be valid in [FHJ] for the case when the 1-factors $F_1, F_2, ..., F_{2m-2}$ are identical. R. Häggkvist (personal communication) has obtained some further evidence in support of Conjecture 6.8 by showing that, if m is a power of two, then one can find a 1-factorization of K_{2m} satisfying the weaker requirement that $F_i \cap L_i = \emptyset$ for all i, $1 \leq i \leq 2m - 2$. Fleischner and the present author[FJ] obtained a similar result to Theorem 6.2 for Eulerian digraphs. This gave rise to a version of Conjecture 6.8 for complete bipartite graphs, which can be weakened to the following problem on latin squares (in which we use the term **transversal** of an $m \times m$ latin square to mean a set of m cells, no two of which lie in the same row or column).

Conjecture 6.9 Let $F_1, F_2, ... F_{m-2}$ be transversals for an $m \times m$ latin square. Then there exists an $m \times m$ latin square on $\{1, 2, ..., m\}$ such that no entry of F_i is equal to i for, $1 \leq i \leq m - 2$.

We now turn to the problem of trying to decide for which Eulerian graphs G, $pcet(G)$ takes its maximum possible value of $\delta(G) - 1$. A special case was first posed by Kotzig[K2]: characterise the 4-regular graphs which have three pairwise compatible Euler tours. The main result of [J2] gives a solution to Kotzig's problem. To describe the characterisation we shall need some further definitions. A **partial transition system** for G is a system of forbidden parts T for G such that for each vertex v of G there is at most one forbidden part in $T(v)$ which is not a transition at v, and such that $T(v) = E_v$ whenever $d(v) > \delta(G)$ (thus $T(v)$ is the trivial partition of E_V into one set in this case). Put $|T| = \sum_{v \in V(G)}(|T(v)| - 1)$. Let $c(H)$ denote the number of components of a graph H.

Theorem 6.10 Let G be an Eulerian graph. Then $pcet(G) = 3$ if and only if

(i) $\delta(G) = 4$, and

(ii) for all partial transition systems T for G we have $3(c(G^T) - 1) \leq 2|T|$.

We have already seen that if G has transition S at a vertex of degree four such that G^S is disconnected, then $pcet(G) < 3$. This can also be deduced from Theorem 6.10 by defining T to be the partial transition system for G such that $T(v) = \{S, E_v - S\}$ and $T(u) = E_u$ for all $u \in V(G) - v$. Then $|T| = 1$, $c(G^T) = c(G^S) = 2$, and $3(c(G^T) - 1) > 2|T|$.

The reader may sense some similarity between condition (ii) of Theorem 6.10 and results in matroid theory such as Edmonds' Theorem[E] on disjoint bases. Indeed, the necessity of condition (ii) was deduced by considering the transitions for G as an independence system in which a set S of transitions is independent if no two transitions of S intersect and G^S is connected. Pairwise disjoint maximum independent sets then correspond to pairwise compatible Euler tours in G. This independence system does not form a matroid, however, so the sufficiency of condition (ii) cannot be deduced from the above mentioned theorem of Edmonds. Fortunately, Bouchet[B1,2] has developed a new and powerful combinatorial structure called an **isotropic system** which, amongst other things, generalises the independence system of transitions in a 4-regular graph. I refer the reader to [B1,2] or the forthcoming monograph [B4] for a description of isotropic systems. Theorem 6.10 was deduced from a characterisation of isotropic systems having three supplementary Eulerian vectors given in [J3], which can be viewed as an analogue of Edmonds' Theorem[E] for isotropic systems.

Subsequently, Bouchet[B3] used isotropic systems to obtain a polynomial algorithm for deciding when a 4-regular graph G has $pcet(G) = 3$. The algorithm either constructs three pairwise compatible Euler tours or else constructs a partial transition system T which fails to satisfy condition (ii) of Theorem 6.10. He has recently extended this algorithm to **multimatroids** in [B5] and has suggested (personal communication) that the new algorithm could be used together with ideas from the proof of Theorem 6.10 to give a polynomial algorithm for deciding when an arbitrary Eulerian graph G has $pcet(G) = 3$.

I conjectured in [J2] that Theorem 6.10 can be generalised to hold when $pcet(G) = \delta(G) - 1 \geq 3$.

Conjecture 6.11 Let G be an Eulerian graph with $\delta(G) \geq 4$. Then $pcet(G) = \delta(G) - 1$ if and only if for all partial transition systems T for G we have

$$(\delta(G) - 1)(c(G^T) - 1) \leq (\delta(G) - 2)|T|.$$

The necessity of the condition in Conjecture 6.11 can be established in the same way as that of condition (ii) of Theorem 6.10. Unfortunately, when $\delta(G) \geq 6$, we cannot readily obtain pairwise compatible Euler tours from results on isotropic systems.

I close by mentioning one final problem on 4-regular graphs.

Problem 6.12 Given a transition system F for a 4-regular graph G, determine whether G has two Euler tours X_1 and X_2 such that F, X_1 and X_2 are pairwise compatible.

An aesthetically pleasing instance of this problem occurs when G is a plane graph and no transition of F at v is consecutive on a face boundary at v, for all vertices v. The Euler tours compatible with F are then constrained to follow the face boundaries of G (such Euler tours in a plane Eulerian graph are called **A-trails** in [F1]). A. Bouchet (personal communication) has pointed out that one may use Edmonds theorem on disjoint bases [E] to characterise the plane 4-regular Eulerian graphs which have two compatible A-trails. A direct proof of this characterisation has also been given by L.D. Andersen and the author in [AJ]. To explain this characterisation I will need one further definition. A partial transition system T for a plane Eulerian graph G is said to be **plane** if, for each vertex v, whenever $\{e_1, e_2\} \in T(v)$ it follows that $e_1 v e_2$ is consecutive on a face boundary of G.

Theorem 6.13 Let G be a plane 4-regular graph. Then G has two compatible A-trails if and only if, for all plane partial transition systems T for G, we have $2(c(G^T) - 1) \leq |T|$.

One may hope that the analogous necessary condition for Problem 6.12 (that $2(c(G^T) - 1) \leq |T|$ for all partial transition systems T for G such that whenever $\{e_1, e_2\} \in T(v)$ we have $\{e_1, e_2\} \notin F(v)$) is also sufficient. Unfortunately this is not the case, as can be seen from counterexamples given in [J2] and [AJ]. It could well be that Problem 6.12 is NP-complete.

References

[AGZ] B. Alspach, L. Goddyn and C.Q. Zhang, Graphs with the circuit cover property, *Trans. Amer. Math. Soc.*, to appear.

[AZ]   B. Alspach and C.Q. Zhang, Cycle coverings of cubic multigraphs, *Discrete Math.* to appear.

[AT]   N. Alon and M. Tarsi, Covering multigraphs by simple circuits, *SIAM J. Algebraic Discrete Methods* 6 (1985) 345-350.

[AJ]   L.D. Andersen and B. Jackson, Complementary A-trails in 4-regular plane graphs, submitted.

[AH]   K. Appel and W. Haken, Every planar map is 4-colourable, *Illinois J. Math* 21 (1977) 429-567.

[BJJ] J.C. Bermond, F. Jaeger and B. Jackson, Shortest coverings of graphs with cycles, *J. Combinatorial Theory Ser. B* 35 (1983) 297-308.

[B1]   A. Bouchet, Isotropic systems, *European J. Combinatorics* 8 (1987) 231-244.

[B2]   A. Bouchet, Graphic presentations of isotropic systems, *J. Combinatorial Theory, Ser.B* 45 (1988) 58-76.

[B3]   A. Bouchet, Compatible Euler tours and supplementary Eulerian vectors, *European J. Combinatorics*, to appear.

[B4]   A. Bouchet, *Isotropic Systems and Δ-Matroids*, to appear.

[B5]   A. Bouchet, Coverings of multimatroids by independent sets, submitted.

[BM]   J.A. Bondy and U.S.R. Murty, *Graph Theory with Applications*, Americain Elsevier, New York, 1976.

[E]   J. Edmonds, Lehman's switching game and a theorem of Tutte and Nash-Williams, *J. Res. Nat. Bur. Standards Sect. B* 69B (1965) 73-77.

[EJ]   J. Edmonds and E.L. Johnson, Matching, Euler tours and the Chinese Postman, *Math. Programming* 5 (1973) 88-124.

[El]   M.N. Ellingham, Petersen subdivisions in some regular graphs, in *Congressus Numerantum 44*, Utilitas Mathematica, Winnipeg (1984) 33-40.

[Fa1] G. Fan, Covering weighted graphs by even subgraphs, *J. Combinatorial Theory Ser. B* 49 (1990) 137-141.

[Fa2] G. Fan, Integer flows and cycle covers, *J. Combinatorial Theory Ser. B* to appear.

[Fa3] G. Fan, Covering graphs by cycles, *SIAM J. Discrete Math* to appear.

[Fa4] G. Fan, Tutte's 3-flow conjecture and short cycle covers, *J. Combinatorial Theory Ser. B* to appear.

[Fa5] G. Fan, Fulkerson's conjecture and short cycle covers, *J. Combinatorial Theory Ser. B* submitted.

[Fa6] G. Fan, Short cycle covers of cubic graphs, submitted.

[FaR] G. Fan and A. Raspaud, Covering a binary matroid by three cycles, submitted.

[FaZ] G. Fan and C.Q. Zhang, On compatible cycle decompositions of Eulerian graphs, submitted.

[F1] H. Fleischner, The importance of being Euler, *Abhandlungen aus dem Seminar der Universität Hamburg* 42 (1974).

[F2] H. Fleischner, Eulersche Linien und Kreisüberdeckungen, die vorgegebene Durchgänge in den Kanten vermeiden, *J. Combinatorial Theory Ser. B* 29 (1980) 145- 167.

[F3] H. Fleischner, Cycle decompositions, 2- coverings, and the four-colour disease, in *Progress in Graph theory* (J.A. Bondy and U.S.R. Murty eds.), Academic Press Canada 1984 233-246.

[F4] H. Fleischner, *Eulerian Graphs and Related Topics, Part 1, Vol. 1* Annals of Discrete Math. 45 (North Holland) 1990.

[F5] H. Fleischner, *Eulerian Graphs and Related Topics, Part 2, Vol. 1* Annals of Discrete Math. 50 (North Holland) 1991.

[FF] H. Fleischner and A. Frank, On circuit decompositions of planar Eulerian graphs, *J. Combinatorial Theory, Ser. B*, 50 (1990) 245-253.

[FG] M. Guan and H. Fleischner, On the cycle covering problem for planar graphs, *Ars Combinatoria* 20 (1985) 61-68.

[FHJ] H. Fleischner, A.J.W. Hilton and B. Jackson, On the maximum number of pairwise compatible Euler cycles, *J. Graph Theory* 14 (1990) 51-63.

[FJ] H. Fleischner and B. Jackson, Compatible Euler tours in Eulerian digraphs, in *Cycles and Rays* (G. Hahn et al eds.) Kluwer Academic Publishers, 1990, 95-100.

[Fr]   P. Fraisse, Cycle covering in bridgeless graphs, *J. Combinatorial Theory, Ser. B*, 39 (1985) 146-152.

[G]   L. Goddyn, Cones, lattices and Hilbert bases, submitted.

[H]   R. Häggkvist, On F-Hamiltonian graphs, in *Graph Theory and Related Topics*, (J.A. Bondy and U.S.R. Murty eds.), Academic Press, New York, 1979, 219-231.

[I]   R. Isaacs, Infinite families of non-trivial trivalent graphs which are not Tait colorable, *Amer. Math. Monthly* 82 (1975) 221-139.

[IR]   A. Itai and M. Rodeh, Covering a graph by circuits, *Automata, Languages and Programming, Lecture Notes in Computer Science* 10 (1978) 289-299.

[ILPR] A. Itai, R.J. Lipton, C.H. Papadimitriou and M. Rodeh, Covering graphs by simple circuits, *SIAM J. Computing* 10 (1981) 746-754.

[J1]   B. Jackson, Compatible Euler tours for transition systems in Eulerian graphs, *Discrete Math.* 66 (1987) 127-131.

[J2]   B. Jackson, A characterisation of graphs having three pairwise compatible Euler tours, *J. Combinatorial Theory Ser. B* 53 (1991) 80-92.

[J3]   B. Jackson, Shortest circuit covers and postman tours of graphs with a nowhere-zero 4-flow, *SIAM J. Computing* 19 (1990) 659-665.

[J4]   B. Jackson, Shortest circuit covers of cubic graphs, *J. Combinatorial Theory Ser. B* submitted.

[JW]   B. Jackson and N.C. Wormald, Cycles containing matchings and pairwise compatible Euler tours, *J.Graph Theory*, 14 (1990) 127-138.

[Ja1]   F. Jaeger, Flows and generalized coloring theorems in graphs, *J. Combinatorial Theory Ser. B* 26 (1979) 205-216.

[Ja2]   F. Jaeger, A survey of the cycle double cover conjecture, *Annals of Discrete Math.* 27 (1985) 1-12.

[JRT]   V. Jamshy, A. Raspaud and M. Tarsi, Short circuit covers for regular matroids, *J. Combinatorial Theory Ser. B* 43 (1987) 354- 357.

[JT1]   V. Jamshy and M. Tarsi, Cycle covering of binary matroids, *J. Combinatorial Theory Ser. B* 46 (1989) 154- 161.

[JT2]   V. Jamshy and M. Tarsi, Short cycle covers and the double cover conjecture *J. Combinatorial Theory Ser. B* 56 (1992) 197-204..

[K1]   A. Kotzig, Moves without forbidden transitions in graphs, *Mat. Casopis Sloven. Akad. Vied* 18, (1968) 76-80.

[K2]   A. Kotzig, Eulerian lines in finite 4-valent graphs and their transformations, in it Theory of Graphs (P. erdos and Katona eds.), 1968, 219-230.

[K3]   A. Kotzig, Problem Session, *Proc. 10th S.E. Conf. Combinatorics, Graph Theory and Computing, Congressus Numerantum XXIV* (1979) 914-915.

[N]   C.St.J.A. Nash-Williams, Detachments of graphs and generalised Euler trails, in *Surveys in Combinatorics 1985* (ed. I. Anderson), L.M.S. Lecture Notes 103, Cambridge University Press, London, 137-153.

[R]   A. Raspaud, Short cycle covers for binary matroids with a Petersen flow, submitted.

[S1]   P.D. Seymour, Sums of circuits, in *Graph Theory and Related Topics* (J.A. Bondy and U.S.R. Murty eds.) Academic Press, New York, 1978 341-355.

[S2]   P.D. Seymour, nowhere-zero 6-flows, *J. Combinatorial Theory Ser. B* 30 (1981) 130-135.

[S3]   P.D. Seymour, Even circuits in planar graphs, *J. Combinatorial Theory Ser. B* 31 (1981) 327-338.

[T]   M. Tarsi, Nowhere-zero flow and circuit covering in regular matroids, *J. Combinatorial Theory Ser. B* 39 (1985) 346-352.

[Tu]   W.T. Tutte, On the algebraic theory of graph colourings, *J. Combinatorial Theory* 1 (1966) 15-50.

[Z1]   C.Q. Zhang, Minimum cycle coverings and integer flows, *J. Graph Theory* 14 (1990) 537-546.

[Z2]   C.Q. Zhang, On even cycle decompositions of Eulerian graphs, submitted.

[Z3]   C.Q. Zhang, Cycle cover theorems and their applications, submitted.

Slicing the Hypercube

Michael E. Saks*

1 Introduction

Each real polynomial $r(x_1, x_2, \ldots, x_n)$ defines an ordered partition (Z_r, P_r, N_r) of $\mathbf{R}^n$ where Z_r (resp. P_r, N_r) is the set of points on which the polynomial is zero (resp. positive, negative). Informally, the hypersurface Z_r "slices" $\mathbf{R}^n$ into the two regions P_r and N_r (each of which may be disconnected). Much of geometry is concerned with describing the structure of these three sets in terms of properties of the polynomial r.

This paper deals with questions that arise when such partitions are restricted to the set $V^n = \{-1, 1\}^n$, sometimes called the unit hypercube. Two classes of questions are considered. The first is concerned with *sign-representation* of *boolean functions*. To each polynomial r, associate the function $f = r^\theta$ on V^n which maps $\vec{x}$ to $+1$ if $r(\vec{x}) \geq 0$ and to -1 if $r(\vec{x}) < 0$. The polynomial r is said to be a *sign-representation* of the boolean function f. For an arbitrary boolean function f one asks: what is the "simplest" polynomial r that sign-represents f? Here, the most common measures of the complexity of a polynomial are its degree and its density (number of nonzero coefficients). Of particular interest are *linear threshold functions*, boolean functions which can be sign-represented by a degree 1 (or *affine*) polynomial. Such a function defines a hyperplane which separates the points mapped to -1 from the points mapped to +1. Research in this area seeks, for example, to determine bounds on the minimum degree or density of a sign-representation of a function f in terms of other properties of f, or to determine the properties of the optimal representation of a "random" function.

The second class of questions are extremal problems about "covering" and "slicing" the hypercube by polynomial surfaces (usually hyperplanes). Typical questions here are: What is the maximum number of edges of the n-cube that can be cut by a single hyperplane? What is the minimum number of hyperplanes that slice all of the edges of the n-cube? What is the minimum number of hyperplanes with coefficients in $\{-1, +1\}$ that

---

*Dept. of Mathematics, Rutgers University, New Brunswick, NJ 08903 and Dept. of Computer Science and Engineering, UCSD, La Jolla, Ca. 92093. Research supported in part by NSF grant CCR–8911388 and AFOSR grants 89-0512 and 90-0008.

cover all of the vertices? What is the maximum integer $D(n,k)$ so that for any collection of k hyperplanes through the n-cube, there is a subcube of dimension $D(n,k)$ that is untouched by any of the hyperplanes?

Most of the work presented here was motivated by issues in computer science. Representation of boolean functions as linear threshold functions or as compositions of such functions was studied by many researchers in the context of "threshold logic" and "threshold circuits" and a number of books were written on these subjects in the 1960's and early 1970's (e.g., [48], [37], [42], [59]). Sign-representations of boolean functions by nonlinear polynomials also goes back at least to the early 1960's, sometimes under the name "perceptron" (see,e.g., [47]). More recently there has been revived interest in these areas within the fields of computational complexity and neural networks.

The intent of this paper is to survey the portion of the theory which is of intrinsic interest to combinatorial mathematicians. Questions concerned with the representation of boolean functions as *compositions* of threshold functions, i.e., by circuits of threshold gates, as opposed to questions concerned directly with sign-representations, are outside the scope of this paper. When presenting results that were motivated by computer science applications, brief remarks as to the nature of the application may be given; for more details, the interested reader is directed to the computer science literature. The survey paper by Razborov [57] on small-depth threshold circuits is of direct relevance to the present survey; there is also some connection with surveys by Parberry [53] and Orponen [52] on complexity theoretic aspects of neural networks.

The remainder of this paper is divided into two parts, corresponding to the two classes of questions identified above (sign-representations and covering/slicing problems). The two parts are, to a large extent, independent of each other; the reader interested only in the covering/slicing problems can skip to section 3 after reading section 2.1. Various research problems are identified in the text as "Conjectures", "Problems" or "Questions". The first designation is for conjectures that have appeared in other papers. The designation "Problem" refers to a specific problem on which there is previous work. The designation "Question" is used for less specific questions, or for specific questions that arose during the writing of this survey but have not been investigated, and may be trivial or uninteresting.

The research summarized in this paper is the work of a varied group of mathematicians, computer scientists, electrical engineers and operations researchers. Many of the basic results have been discovered independently by two or more sets of researchers in different fields, sometimes many years apart. Tracing and sorting out references has been somewhat tricky, and I have undoubtedly missed some relevant work in the process. I apologize for such omissions, and would appreciate receiving any information on other work in this area.

In researching this area and writing this survey I benefited enormously from the individual and collective wisdom of several colleagues. I'd like to express my thanks to Noga Alon, Martin Anthony, Jim Aspnes, Richard Beigel, Yehoshua Bruck, Mikael Goldmann, Craig Gotsman, Johan Håstad, Russell Impagliazzo, Nathan Linial, László Lovász, Ramamohan Paturi, Svatopluk Poljak, Steven Rudich, Heather Woll and Fotis Zaharoglou for their time and assistance.

2   Sign-representations of boolean functions

The research surveyed in this section is concerned with the inherent complexity of sign-representations of a boolean function as they relate to other properties of the function.

We begin, in Section 2.1 by reviewing notation and background results. In Section 2.2 we precisely define the equivalent notions of *sign-representation* of a boolean function and *sign-realization* of a *dichotomy* of the n-cube. Section 2.3 reviews elementary results on *linear threshold functions*. Section 2.4 presents a characterization theorem for the class of *threshold graphs*, graphs for which the set of characteristic vectors of their independent sets can be separated from the rest of the n-cube by a hyperplane. Section 2.5 reviews some results from Harmonic analysis which are the main tools for the next few subsections. Section 2.6 states a simple but important connection between a sign-representation of a boolean function and the Fourier transform of the function. Section 2.7 gives some basic results about the spectral support (the set of monomials with nonzero coefficients) of sign-representations. This leads into a discussion, in Section 2.8, of the threshold density of a boolean function, i.e., the minimum size of the spectral support over all sign-representations. Various results and problems concerning the minimum degree of a sign-representation of a boolean function are described Sections 2.9–2.12. Section 2.13 examines the size of the coefficients needed to sign-represent a boolean function by a polynomial of minimal degree. A notion of approximate sign-representation is discussed in Section 2.14 and another variant of sign-representation is described in Section 2.15.

2.1   Preliminaries

2.1.1   Set notation

As usual $\mathbf{R}, \mathbf{Z}$ and $\mathbf{N}$, denote, respectively, the sets of real numbers, integers and natural numbers. We use standard interval notation for real numbers, and also for integers; hence, depending on the context, $[a, b]$ denotes either the set of reals or integers between a and b. For $n \in \mathbf{N}$, $[n]$ is the set $\{1, 2, \ldots, n\}$.

For an arbitrary set X, the set of all subsets of X is denoted $\mathcal{P}(X)$. In the case that $X = [n]$, the brackets are generally omitted and we write simply $\mathcal{P}(n)$. We consider this set as partially ordered under inclusion. For $i \in \mathbf{N}$, $\mathcal{P}(X, i)$ is the set of subsets of X having cardinality i and for $J \subseteq \mathbf{N}$, $\mathcal{P}(X, J)$ is the set of all subsets Y such that $|Y| \in J$. We denote by $b(n, i)$ the cardinality of $\mathcal{P}(n, i)$, which is of course the binomial coefficient $\binom{n}{i}$. Similarly, $b(n, J) = \sum_{i \in J} b(n, i)$ is the cardinality of $\mathcal{P}(n, J)$, We make frequent use of the following standard estimates for binomial coefficients (see,e.g, [5]):

$$b(n, \lceil n/2 \rceil) = \Theta(2^n / \sqrt{n}) \tag{1}$$

and for $n \geq k \geq 0$,

$$b(n, [0, \lceil (n - k)/2 \rceil]) \leq e^{-k^2/2n} 2^n, \tag{2}$$

The latter inequality is sometimes referred to as the Chernoff bound.

For $\vec{x} \in \mathbf{R}^n$, and $J \subseteq [n]$, x_J denotes the product $\Pi_{i \in J} x_i$. For $J \subseteq [n]$, we define the *reflection* of $\vec{x}$ with respect to coordinates J, denoted $\vec{x}^{(J)}$, to be the vector with $x_i^{(J)} = -x_i$ if $i \in J$ and $x_i^{(J)} = x_i$ if $i \notin J$. The reflection by J of a subset S of $\mathbf{R}^n$, denoted $S^{(J)}$ is the set $\{\vec{x}^{(J)} : \vec{x} \in S\}$. If f is a function with domain $\mathbf{R}^n$ then the *reflection of f by J* is the function $f^{(J)}(\vec{x}) = f(\vec{x}^{(J)})$.

2.1.2   The hypercube

Let V denote the set $\{-1, 1\}$. The 2^n-element set V^n is referred to as the n-cube, and its elements are called *vertices*. The all 1's vertex is denoted by $\vec{e}$. Using the notation of the previous subsection, vertices are vectors of the form $\vec{e}^{(J)}$ and the map $J \longrightarrow \vec{e}^{(J)}$ is an order-reversing isomorphism between the partial orders $(\mathcal{P}(n), \subseteq)$ and $(V^n, \leq)$. The *weight* $w(\vec{x})$ of a vertex $\vec{x}$ is the number of -1's in it, so $w(\vec{e}^{(J)}) = |J|$. For $i \in [n]$, the set $V^n(i)$ of vertices of weight i is called the ith *level of V^n*. For $J \subseteq [0, n]$, $V^n(J)$ is the set of vertices whose weight belongs to J. Note that, by the isomorphism between V^n and $\mathcal{P}(n)$, $|V^n(J)| = b(n, J)$.

2.1.3   Boolean functions and dichotomies

A *boolean function* (sometimes called a switching function) is a function from Σ^n to Σ where Σ is a two element set. For this paper we take the set Σ to be V; this differs from the more standard convention $\Sigma = \{0, 1\}$, but is more convenient for our purposes. We will occasionally need to revert to the standard convention; in doing this we associate $x \in \{0, 1\}$ to $(-1)^x$. We let $\mathcal{B}_n$ denote the set of boolean functions on V^n.

A boolean function f specifies an ordered partition $\mathbf{W}_f = (W_f^-, W_f^+)$ of V^n into the sets that map to -1 and $+1$ respectively. Such a partition is called a *dichotomy*. Conversely, any dichotomy $\mathbf{W} = (W^-, W^+)$ specifies a unique boolean $s_\mathbf{W}$, called the *characteristic function* of $\mathbf{W}$. We refer to W^- (resp. W^+) as the negative part (resp. positive part) of the dichotomy.

We say that the function f is *balanced* if the number of negative points equals the number of positive points. The variable x_i is *relevant* for f or f *depends* on x_i if there is some $\vec{x} \in V^n$ such that $f(\vec{x}) \neq f(\vec{x}^{(i)})$. A boolean function f is said to be *monotone* if for $\vec{x}, \vec{y} \in V^n$, $\vec{x} \leq \vec{y}$ implies $f(\vec{x}) \leq f(\vec{y})$. A boolean function f is said to be *unate* if, for some $J \subseteq [n]$, $f^{(J)}$ is monotone, and such a set J is called an *orientation* for the unate function f. In general, the orientation of f is not unique, but only because f may not depend on all of its variables. A dichotomy $\mathbf{W} = (W^-, W^+)$ is *monotone* or *unate* if its characteristic function is.

Of particular importance to us is the class of *simple threshold functions*. The simple threshold function $\tau_d^n(\vec{x})$ is defined to be -1 precisely on those vertices that have at least d coordinates equal to -1. Thus, the dichotomy associated to τ_d^n is $(V^n([d, n]), V^n([0, d-1]))$. In particular, the function τ_1^n is the *OR* function, τ_n^n is the *AND* function, and for odd $n = 2k - 1$ the function t_k^n is the *simple majority* function. All simple threshold functions are monotone.

A second important class of boolean functions is the class of *parity* functions. For each $J \subseteq [n]$, the parity function $\pi_J(\vec{x})$ is equal to -1 if and only if the number of -1's in the subvector $\{x_j | j \in J\}$ is odd. Trivially we have $\pi_J(\vec{x}) = x_J = \Pi_{j \in J} x_j$; hence these functions are also called the *monomial functions*.

2.1.4   The algebra of multilinear polynomials

We consider the set $\mathcal{M}_n$ of multilinear polynomials in n formal variables $x_1, x_2, \ldots, x_n$ having coefficients in the reals, i.e., formal sums of the form:

$$r(\vec{x}) = \sum_{J \subseteq [n]} r_J x_J,$$

where $r_J \in \mathbf{R}$. Associated to each multilinear polynomial r is its *coefficient function* or *spectrum* $\hat{r}$, given by $\hat{r}(J) = r_J$ for $J \subseteq [n]$. The set of subsets J for which $\hat{r}(J) \neq 0$ is the *spectral support* of r and is denoted by $SS(r)$. The size of the spectral support is the *density* of r, denoted $\text{dns}(r)$. The degree of r, $\deg(r)$ is just the maximum cardinality of a member of $SS(r)$. A polynomial of maximum degree 1 is an *affine* polynomial; for such polynomials we typically write $a(\vec{x}) = a_0 + a_1 x_1 + a_2 x_2 + \ldots + a_n x_n$ instead of the notationally more precise $a(\vec{x}) = a_\emptyset + a_{\{1\}} x_1 + a_{\{2\}} x_2 + \ldots + a_{\{n\}} x_n$.

The set $\mathcal{M}_n$ can be viewed as an algebra over $\mathbf{R}$ with the usual addition and scalar multiplication, and multiplication given by $x_J x_I = x_{J \triangle I}$ where $J \triangle I$ is the symmetric difference. This algebra is isomorphic to the quotient $\mathbf{R}[x_1, x_2, \ldots, x_n]/I$ where $\mathbf{R}[x_1, x_2, \ldots, x_n]$ is the usual algebra of n-variate polynomials over the reals and I is the ideal generated by the set $\{x_i^2 - 1 : i \in [n]\}$.

For $\mathcal{S} \subseteq \mathcal{P}(n)$, the *spectral projection* of r onto $\mathcal{S}$ is the polynomial $(r|\mathcal{S})(\vec{x})$ given by $\sum_{J \in \mathcal{S}} r_J x_J$ and for integer $d > 0$, the *degree d projection* $r^{(d)}$ of r is given by $r^{(d)}(\vec{x}) = (r|\mathcal{P}(n, [0, d]))(\vec{x}) = \sum_{|J| \leq d} \hat{r}(J) x_J$.

2.1.5 Pseudo-boolean Functions

A *pseudo-boolean* function is a real valued function defined on V^n. The set $\mathbf{R}^{V^n}$ of all such functions is an $\mathbf{R}$-algebra with the usual addition and scalar multiplication and pointwise multiplication of functions. The set $\mathcal{B}_n$ of boolean functions is a multiplicative subgroup of $\mathbf{R}^{V^n}$, and is isomorphic to the group $\mathbf{Z}_2^{2^n}$.

Given any multilinear polynomial $r \in \mathcal{M}_n$, we may associate it to a pseudo-boolean function r^* in the obvious way. Then we have the following well known fact:

Proposition 2.1 *The map $r \longrightarrow r^*$ is an isomorphism between the algebras $\mathcal{M}_n$ and $\mathbf{R}^{V^n}$.*

In particular this implies that for each pseudo-boolean function f on V^n, there is a unique multilinear polynomial $\sum_{J \subseteq [n]} f_J x_J$ which agrees with f on all points of V^n. It is instructive to recall the explicit inductive construction of the associated polynomial: if f^- and f^+ are the boolean functions on $x_1, x_2, \ldots, x_{n-1}$ obtained by fixing x_n to -1 and $+1$ respectively and r^- and r^+ are multilinear polynomials that represent them, then $r = (1 - x_n)r^-/2 + (1 + x_n)r^+/2$ represents f.

This polynomial representation of f is called the *multilinear* or *spectral representation* of f. Henceforth we identify f with its multilinear representation, thus $\hat{f}$ is its coefficient function or spectrum, and the spectral support, density and degree of f are all well defined.

2.2 Sign representations of boolean functions

We now come to the main object of study of this part of the paper. To every real valued function $r : \mathbf{R}^n \longrightarrow \mathbf{R}$, is associated a boolean function r^θ, given by:

$$r^\theta(\vec{x}) = \begin{cases} -1 & \text{if } p(x) < 0 \\ +1 & \text{if } p(x) \geq 0 \end{cases}$$

If $f = r^\theta$, we say that r is a *sign-representation* or *threshold-representation* of f.

We will be concerned with polynomial sign-representations of boolean functions, i.e., those where the function r is a polynomial function. Obviously every boolean function f can be sign-represented by a multilinear polynomial, e.g., by its spectral representation. The question to be considered is: for a given boolean function f, what is the "simplest" multilinear polynomial that sign-represents f? Two natural measures of complexity for a multilinear polynomial are its degree and its density. For a boolean function f, the *threshold degree* and *threshold density* of f, denoted $\deg^\theta(f)$ and $\mathrm{dns}^\theta(f)$ are defined respectively to be the minimum degree and the minimum density of a multilinear polynomial that sign-represents f. (The threshold degree is called the threshold order in [62] and the strong degree in [8].) Let S be a collection of subsets of $[n]$. We say that a boolean function f is sign-representable *with spectral support* S if f can be sign-represented by a multilinear polynomial r with $SS(r) \subseteq S$. Thus the threshold density of f is equal to the minimum size of S for which f is sign-representable with spectral support S, and the threshold degree of f is the minimum d such that f is sign-representable with spectral support $\mathcal{P}(n, [0, d])$.

There are two obvious variants to the definition of sign-representation. If r is a polynomial that sign-represents f and $r(\vec{x}) \neq 0$ for all $\vec{x} \in V^n$, then r is a *strict sign-representation* or simply a *strict-representation*. In the literature, many authors require that threshold representations be strict. There is no essential difference between the two notions: if r is a (non-strict) sign-representation of f then there is always a real constant c such that $r + c$ is a strict representation. We will assume a sign-representation to be strict as needed.

A second and more substantial variant of sign-representation was introduced by Aspnes, et al. ([8]). Say that a real function r is *trivial* if it is identically zero on the n-cube and non-trivial otherwise. A real function r is a *weak sign-representation* or simply *weak-representation* of the boolean function f if (i) it is non-trivial and (ii) for all $\vec{x} \in V^n$ such that $r(\vec{x}) \neq 0$, $f(\vec{x})$ has the same sign as $r(\vec{x})$. In other words, r^θ agrees with f except on those points where r vanishes. Note that any non-trivial real function r is a weak-representation of exactly $2^{z(r)}$ distinct boolean functions, where $z(r)$ is the number of points of the cube on which r vanishes. The *weak degree*, $\deg^w(f)$ and *weak density*, $\mathrm{dns}^w(f)$ of a boolean function are, respectively, the minimum degree and density of a multilinear polynomial that weakly represents f.

Obviously, for any boolean function f we have:

$$n \geq \deg(f) \geq \deg^\theta(f) \geq \deg^w(f) \geq 0, \qquad (3)$$

and

$$2^n \geq \mathrm{dns}(f) \geq \mathrm{dns}^\theta(f) \geq \mathrm{dns}^w(f) \geq 1. \qquad (4)$$

Sign-representations, strict-representations and weak-representations of f by a multilinear polynomial r have simple geometric interpretations: r weakly-represents f means that all of the points in W_f^- lie "on or below" the surface $\{x : r(x) = 0\}$ and all of the points in W_f^+ lie "on or above" the surface; r sign-represents f if, in addition, W_f^- lies strictly below the surface; r strictly-represents f if, in addition to that, W_f^+ lies strictly above the surface. In these three cases we say that r *weakly-realizes* (resp. *sign-realizes, strictly-realizes*) the dichotomy $\mathbf{W}_f = (W_f^-, W_f^+)$. We adapt the previous terminology to the language of dichotomies: for an arbitrary vertex dichotomy $\mathbf{W} = (W^-, W^+)$, we define, for example, the *threshold degree, threshold density*, and *weak degree* of $\mathbf{W}$ to be, respectively, the threshold degree, threshold density, and weak degree of its characteristic function $s_\mathbf{W}$.

Polynomial sign-representations of boolean functions have been studied extensively over the last thirty years and many of the basic ideas and results have been rediscovered several times. There is a vast literature on the class of boolean functions of threshold degree 1, called linear threshold functions; some of the early references are given in the introduction. Representations of boolean functions as the sign of a polynomial are referred to in some of the computer science literature as *perceptrons*, which were originally introduced as a model for neural activity. Versions of this model were considered as early as the 1940's ([46]). Minsky and Papert ([47]) includes a mathematical investigation of some properties of these representations. In that literature, boolean functions are considered as functions from $\{0,1\}^n$ to $\{0,1\}$. The transformation $y \longrightarrow (y+1)/2$ from $\{-1,1\}$ and $\{0,1\}$ preserves the degree of polynomials, so the threshold degree of a function is unchanged. However, the transformation can drastically change both the support and the maximum coefficient sizes of a polynomial. Beigel, Reingold and Spielman ([12]) and Beigel ([11]) have some recent work on this subject and its relation to complexity theory.

Another early reference on functions of threshold degree higher than 1 is Cover [21]; see also the references contained within. In what follows, results from this and many of the more recent papers in the area are discussed.

2.3   Linearly realizable dichotomies

If the dichotomy $\mathbf{W} = (W^-, W^+)$ can be sign-realized by a polynomial of degree one (i.e., an *affine* function), then $\mathbf{W}$ is said to be *linearly realizable* and the function $f = s_\mathbf{W}$ is a *linear* threshold function. The canonical examples of linear threshold functions are the simple threshold functions defined in Section 2.1.3; for each k, n, τ_k^n is strictly sign-represented by the affine function $t_k^n(\vec{x}) = x_1 + x_2 + \ldots + x_n + 2k - n - 1$.

Geometrically, $\mathbf{W}$ is linearly sign-representable if there is a hyperplane that separates W^- from W^+. From standard results in convexity theory, one obtains:

Proposition 2.2 *For a dichotomy* $\mathbf{W}$ *on* V^n, *the following conditions are equivalent:*

1. **W** *is linearly realizable.*

2. *There exists a vector $\vec{a} \in \mathbf{R}^n$ such that $\vec{a} \cdot \vec{x} < \vec{a} \cdot \vec{y}$ for all $\vec{x} \in W^-$ and $\vec{y} \in W^+$.*

3. *The convex hulls of W^- and W^+ are disjoint.*

4. *For any two sequences of (not necessarily distinct) points $\vec{x}^1, \vec{x}^2, \ldots, \vec{x}^k \in W^-$ and $\vec{y}^1, \vec{y}^2, \ldots, \vec{y}^k \in W^+$, $\sum_{i=1}^k \vec{x}^i \neq \sum_{i=1}^k \vec{y}^i$.*

This last condition is referred to in the literature as *assumability*. The second condition defines a linear program which can be used to find the separating hyperplane. Many variants and extensions of the above proposition have appeared in the early literature on the subject, and algorithmic aspects of determining such a hyperplane have been investigated (see, e.g., [48],[42], [56]). The reader can easily define analogous conditions for weak-realizability.

The following necessary and sufficient condition for an affine function to weakly-realize and or to strictly-realize **W** was obtained by Dertouzos:

Theorem 2.3 [22] *Let* **W** *be a dichotomy of V^n and $\vec{w} = \sum_{(\vec{x} \in W^+)} \vec{x} - \sum_{(\vec{x} \in W^-)} \vec{x}$. A necessary and sufficient condition for the affine function $a(\vec{x}) = a_\emptyset + a_1 x_1 + \ldots + a_n x_n$ to weakly-realize* **W** *is:*

$$\sum_{i=1}^n w_i a_i + (|W^+| - |W^-|)a_\emptyset = \sum_{\vec{x} \in V^n} |a(\vec{x})|$$

Furthermore, when a does not vanish on V^n this condition is necessary and sufficient for a to strictly-realize **W**.

Later, we will see this as a special case of a more general result of Bruck (Lemma 2.12).

Finally, let a be an affine function, $f = a^\theta$, and let J be the set of $i \in [n]$ such that $a_i < 0$. If $J = \emptyset$ then f is easily seen to be a monotone boolean function. In general, the reflection $f^{(J)}$ of f is sign-represented by the function $a^{(J)}$ obtained by negating the coefficients a_i for $i \in J$. Thus $f^{(J)}$ is monotone and so we have:

Proposition 2.4 *Any linear threshold function f is a reflection of a monotone linear threshold function. Thus every linear threshold function is unate. Equivalently any linearly realizable dichotomy* **W** *is the reflection of a linearly realizable monotone dichotomy, and thus every linearly realizable dichotomy is unate.*

Various related conditions have been studied in the literature (e.g., k-monotonicity); for details see any of the books on threshold logic cited in the introduction.

The proposition implies that to characterize linearly realizable dichotomies it suffices to characterize linearly realizable monotone dichotomies. In general there does not seem to be any general characterization that is more illuminating than that given by Proposition 2.2. However, as will be seen in the next section, for certain natural classes of dichotomies, there are more interesting characterizations.

2.4   Threshold Graphs

An important and well studied class of monotone dichotomies arise from graphs in the following way. For a graph G on vertex set $[n]$, let $\mathbf{I}(G)$ denote the dichotomy on V^n with $I^+(G)$ consisting of all vectors of the form $e^{(J)}$ where J is an independent set of G. Clearly this dichotomy is monotone. Furthermore, it is easy to see (and well known) that the degree 2 polynomial $-\sum_{(i,j)\in E}(x_i - 1)(x_j - 1)$ is nonnegative precisely on vectors $e^{(J)}$ where J is independent in G, and thus is a sign-realization of the dichotomy $\mathbf{I}(G)$. A natural question is: for which graphs G is $\mathbf{I}(G)$ *linearly* realizable? Such graphs are called *threshold graphs*. Chvátal and Hammer [20] proved:

Theorem 2.5 *The following conditions on a graph G are equivalent:*

1. *G is a threshold graph.*

2. *There is a function $c : [n] \longrightarrow \mathbf{N}$ and an integer T such that for all distinct pairs i and j in $[n]$, $(i,j) \in E(G)$ if and only if $c(i)+c(j) > T$.*

3. *G has no subset of four vertices whose induced graph consists of a path, a cycle, or two disjoint edges.*

4. *The vertices of G can be totally ordered so that each vertex is adjacent to either all or none of the vertices that precede it.*

There are several other conditions known to be equivalent to the above, and these various characterizations lead to linear time algorithms for recognizing these graphs. For more details see Chapter 10 of [28].

Chvátal and Hammer also introduce a parameter $t(G)$, which has been called the *threshold dimension* of a graph, defined to be the minimum number of open halfspaces such that $I^-(G)$ is the intersection of all of these halfspaces and V^n (so $t(G) = 1$ means G is a threshold graph.) They noted that this parameter is NP-hard to compute in general. However, they obtained the following upper bound:

Theorem 2.6 *[20] For every graph G, $t(G) \leq n - \alpha(G)$, where $\alpha(G)$ is the size of the largest independent set of G. For triangle-free graphs, $t(G) = n - \alpha(G)$.*

Another related notion, that of *bithreshold graphs* was introduced and characterized by Hammer and Mahadev [33].

2.5   The harmonic analysis of boolean functions

We have noted previously that each pseudo-boolean function f has a unique spectrum $\hat{f}$. If V^n is viewed as the abelian group (isomorphic to $(\mathbf{Z}_2)^n$) with the operation $(x_1, x_2, \ldots, x_n) * (y_1, y_2, \ldots, y_n) = (x_1 * y_1, x_2 * y_2, \ldots, x_n * y_n)$, then the map $\hat{f}$ is the Fourier transform of the function f. This permits tools from harmonic analysis to be applied in the study of boolean functions.

This point of view has long been used in coding theory, where the transformation f to $\hat{f}$ corresponds to the Hadamard transform of the code, or equivalently to the representation of binary codes as weighted sums of characters in the group algebra of $\mathbf{Z}_2^n$ (see, e.g. [44]). The first explicit use of harmonic analysis in the context of boolean functions seems to be by Lechner [41]. More recently, spectral techniques have been used by Kahn, Kalai and Linial [39] to lower bound the minimum influence of variables on boolean functions, and by Linial, Mansour and Nisan [43] to give necessary conditions for the computability of a boolean function by polynomial size, bounded depth circuits, with applications to learning theory. Further applications to learning theory includes work of Furst, Jackson and Smith [26], Bellare [13] and Mansour [45]. These methods have also been used to bound the decision tree and parallel complexity of boolean functions by Brandman, et al. [16].

As we shall see in the next few sections, spectral methods also play a central role in the study of sign-representations of boolean functions, as demonstrated by the work of Bruck [17], Bruck and Smolensky [18], Aspnes, Beigel, Furst and Rudich [8], Gotsman [29] and Gotsman and Linial [30]. Before presenting this work, we review some of the background definitions and results.

Define the following inner product on the space $\mathbf{R}^{V^n}$ of pseudo-boolean functions:

$$\langle f, g \rangle = 2^{-n} \sum_{\vec{x} \in V^n} f(\vec{x}) g(\vec{x}).$$

Under this inner product, the monomial functions x_J form an orthonormal basis and the spectral representation is the representation in terms of this basis. By the orthonormality of the basis, we have $\hat{f}(J) = \langle x_J, f \rangle$. The following classical result (Parseval's equality), gives an alternative way to compute the inner product:

Proposition 2.7

$$\langle f, g \rangle = \sum_{J \subseteq [n]} \hat{f}(J) \hat{g}(J).$$

There are two natural parameterized families of norms on pseudo-boolean functions. For $p \in [1, \infty)$, define:

$$\|f\|_p = \left[2^{-n} \sum_{\vec{x} \in V^n} |f(x)|^p \right]^{1/p}$$

and

$$\||f\||_p = \left[\sum_{I \subseteq [n]} |\hat{f}(I)|^p \right]^{1/p}$$

The first norm is the *standard p-norm* and the second is the *spectral p-norm*. We also have $\|f\|_\infty$ and $\||f\||_\infty$ defined, respectively, to be the limits of $\|f\|_p$ and $\||f\||_p$ as p tends to ∞; these are equal, respectively, to the maximum over $\vec{x}$ of $|f(\vec{x})|$ and of $|\hat{f}(J)|$. It is easily shown that for any finite sequence of real numbers $b_1, b_2, \ldots, b_m$, the quantity $(\sum_{i=1}^{m} |b_i|^p)^{1/p}$ is nonincreasing in p, from which it follows that $\||f\||_p$ is nonincreasing in p.

In discussing these norms and their relationships, we call two real numbers $p, q \geq 1$ a *dual pair* if $1/p + 1/q = 1$ (where ∞ is the dual of 1). Holder's inequality for these norms is:

Lemma 2.8 *For all dual pairs p, q and pseudo-boolean functions f and g:*

$$\|fg\|_1 \leq \|f\|_p \|g\|_q$$
$$\||fg\||_1 \leq \||f\||_p \||g\||_q$$

There are also close connections between the standard and spectral norms. From Parseval's equality we obtain the identity:

$$\|f\|_2 = \sqrt{\langle f, f \rangle} = \||f\||_2$$

More generally,

Lemma 2.9 *(Haussdorff-Young inequalities [40]) For all dual pairs p, q and pseudo-boolean functions f:*

$$\|f\|_p \geq \||f\||_q \qquad 1 \leq p \leq 2 \qquad (5)$$
$$\|f\|_p \leq \||f\||_q \qquad 2 \leq p \leq \infty \qquad (6)$$

Applying the above facts to the special case of boolean (± 1-valued) functions we note first that for the standard norms, $\|f\|_p = 1$ for all p. For the spectral norms, we have $\||f\||_2 = \|f\|_2 = 1$ and more generally it is easy to show:

Proposition 2.10 *For any boolean function f on V^n, $p \in [1,2]$ and $q \in [2, \infty]$:*

$$2^{n/2} \geq |||f|||_p \geq 1 \geq |||f|||_q \geq 2^{-n/2}$$

In fact, the upper bound is tight for $p = 1$ and the lower bound is tight for $q = \infty$ for the class of *bent* functions, which are those having all spectral coefficients equal to $\pm 2^{-n/2}$ (see, e.g., [44]).

From Holder's inequality with $p = 1$ we get that for f boolean:

$$|||f|||_1 \geq \frac{1}{|||f|||_\infty}. \tag{7}$$

Finally, Gotsman [29] proved the following results about the spectrum of a random boolean function:

Theorem 2.11 *Suppose f is chosen uniformly at random from the set of all boolean functions on V^n. Then the distribution of its spectrum $\hat{f}$ satisfies:*

1. *For each $J \in \mathcal{P}(n)$, $\hat{f}(J)$ is distributed as the average of 2^n independent unbiased $\{-1, +1\}$ Bernoulli random variables, and hence converges in distribution to the normal distibution $\mathcal{N}(0, 2^{-n})$.*

2. *For distinct sets $J_1, J_2, \ldots, J_k \in \mathcal{P}(n)$ with $k = 2$ or k odd, $\hat{f}(J_1)\hat{f}(J_2) \ldots \hat{f}(J_k)$ has mean 0.*

2.6   Sign-representations and the spectrum

As noted earlier, the spectral representation of a boolean function f is also a sign-representation. It seems natural to expect that for any polynomial r that sign-represents (or weakly-represents) f, there should be some relationship between the coefficients of r and those of f. A direct connection noted by Bruck, is obtained by applying Parseval's equality (Proposition 2.7) to $\langle f, r \rangle$:

Lemma 2.12 [17] *Let f be a boolean function on V^n. Then r weakly-represents f if and only if:*

$$||r||_1 = \langle f, r \rangle = \sum_{J \subseteq [n]} \hat{r}(J)\hat{f}(J). \tag{8}$$

Furthermore r strictly-represents f if and only if (8) holds and, in addition, r does not vanish on V^n.

In the case that r is a degree 1 polynomial, this result easily reduces to Theorem 2.3.

Lemma 2.12, while elementary, is the foundation for the analysis of the weak/threshold degree/density parameters of boolean functions. As an immediate application we have:

Corollary 2.13 [17] *Let f be a boolean function and r be a strict-representation of f and suppose that r has spectral support S. Then if g is any boolean function with $(g|S) = (f|S)$ then $g = f$. In other words, f is uniquely reconstructible from its spectrum restricted to S.*

The corollary follows from lemma 2.12 since if $(g|S) = (f|S)$ then g, r satisfy equation (8) and so $g = r^\theta = f$.

2.7 Sign representations with restricted support

Recall that, for a collection S of subsets of $[n]$, a boolean function f is said to be sign-representable (resp., weakly-representable) with spectral support S if f has a sign-representation (resp. weak-representation) for which the coefficients corresponding to sets outside S are all 0. A general question is: for a given function f, for which families S of subsets does f have such a representation?

An easy first observation is:

Lemma 2.14 *Let f be a boolean function and S be a collection of subsets of $[n]$ such that $\sum_{J \notin S} |f_J| < 1$. Then the spectral projection $(f|S)$ of f onto S sign-represents f.*

This is clear since the hypothesis implies $|(f|S)(\vec{x}) - f(\vec{x})| < 1$ for all x.

Gotsman used this observation and the asymptotic normality of the spectral coefficients (Theorem 2.11) to prove:

Theorem 2.15 *Let $\epsilon \in (0, 1/2)$ and let S be any subset of $\mathcal{P}(n)$ of size at least $2^n - 2^{(1/2-\epsilon)n}$. Then, for all but a fraction $O(2^{-2n\epsilon})$ of all boolean functions on n variables, $(f|S)$ is a sign-representation of f.*

Indeed, Gotsman showed that with the given probability, the sum of the absolute values of the spectral coefficients of f outside S is less than 1, and thus lemma 2.14 applies.

Aspnes et al. obtained the following elegant "theorem of the alternative":

Theorem 2.16 [8] *For any $S \subseteq \mathcal{P}(n)$, exactly one of the following holds:*

1. *f has a weak-representation r with spectral support in S*

2. *f has a strict-representation q with spectral support in $\mathcal{P}(n) - S$.*

As noted in [8] this can be translated into a classical result in the theory of linear inequalities. View each pseudo-boolean function as a vector in $\mathbf{R}^{2^n}$ indexed by V^n. Define the $S \times V^n$ matrix A_S by $A_S(J, \vec{x}) = x_J$ and let B be the matrix obtained by multiplying column $\vec{x}$ by $f(\vec{x})$. Then r satisfies the first condition of the theorem if and only if the function $r'(\vec{x}) = r(\vec{x})f(\vec{x})$ satisfies $r' \geq 0$, $r' \neq 0$ and $r' = yB$ for some vector y indexed by $\mathcal{S}$. On the other hand, q satisfies the second condition if and only if the function $q'(\vec{x}) = q(\vec{x})f(\vec{x})$ satisfies $q' > 0$ and $Bq' = 0$ (here we use the orthogonality of the monomial functions.) The fact that exactly one of q' and r' exist is a general theorem of Stiemke [61] for systems of linear inequalities (see e.g., [19]).

We also have the following bounds on the number of boolean functions that can be sign-represented with a given spectral support. The upper bound is essentially due to Cover; the lower bound is due to Baldi.

Theorem 2.17 [21],[10] *Let* $\mathcal{S} \subseteq \mathcal{P}(n)$ *have size* k. *Then the number of boolean functions that can be sign-represented by polynomials with spectral support* $\mathcal{S}$ *is at most* $2b(2^n - 1, [0, k-1])$ *and at least* 2^k.

The proof of the upper bound uses the fact (Schläfli [58], Harding [34]) that any N hyperplanes in $\mathbf{R}^k$ through the origin, partition $\mathbf{R}^k$ into at most $2b(N - 1, [0, k - 1])$ distinct regions. Construct the matrix $A_\mathcal{S}$ as in the proof of Theorem 2.16. Then the boolean functions sign-representable with support $\mathcal{S}$ are in one-to-one correspondence with the distinct sign patterns of vectors expressible by $\vec{z}A$ where $\vec{z} \in \mathbf{R}^k$. Associating each column to a hyperplane through the origin in $\mathbf{R}^k$, this is just the number of regions defined by these hyperplanes.

The lower bound follows from the linear independence of the monomial functions. Fix a set W of k points of V^n. For each function g from W to $\{-1, +1\}$, there is a polynomial r_g with spectral support $\mathcal{S}$ whose value on W is g. Each of the 2^k boolean functions r_g^θ is distinct.

2.8   Bounds on threshold density and weak density

Recall that the threshold density of f, $\mathrm{dns}^\theta(f)$ (resp. $\mathrm{dns}^w(f)$) is the minimum density of a polynomial that sign-represents (resp. weakly-represents) f.

By inequality (4) we know $\mathrm{dns}^\theta(f) \leq \mathrm{dns}(f)$. Gotsman observed the following small improvement:

Theorem 2.18 [29] *Let* f *be a boolean function and let* T *be the spectral support of* f. *Then there is a subset* S *of* T *having size at most* $|T| - \sqrt{|T|} + 1$ *such that* $(f|S)$ *sign-represents* f. *Thus* $\mathrm{dns}^\theta(f) \leq \mathrm{dns}(f) - \sqrt{\mathrm{dns}(f)} + 1$.

To prove this, order the sets of T as $T_1, T_2, \ldots, T_{|T|}$ so that the absolute values of the spectral coefficients f_{T_i} are in nondecreasing order. Since $|||f|||_2 = 1$, the sum $\sum_{i=1}^{|T|} |f_{T_i}|$ is at most $\sqrt{|T|}$. Thus the sum of the first $\lceil \sqrt{|T|} \rceil - 1$ terms is strictly less than 1 and we may apply lemma 2.14.

Bruck obtained a lower bound on $\mathrm{dns}^w(f)$ and Bruck and Smolensky obtained an upper bound on $\mathrm{dns}^\theta(f)$ in terms of the spectral norms of f:

Theorem 2.19 [17],[18]
  For any boolean function f on V^n,

$$\frac{1}{\||f\||_\infty} \leq \mathrm{dns}^w(f) \leq \mathrm{dns}^\theta(f) \leq \lceil 2n(\||f\||_1)^2 \rceil.$$

The lower bound is obtained by considering an arbitrary sign-representation r of f with spectral support $\mathcal{S}$. We then have:

$$\||f\||_\infty \||r\||_1 \geq \sum_{J \subseteq [n]} \hat{f}(J)\hat{r}(J) = \||r\||_1 \geq \||r\||_\infty \geq \frac{\||r\||_1}{|\mathcal{S}|}.$$

The first inequality is trivial, the equality is lemma 2.12, the second inequality comes from the case $p = 1$ of the Haussdorf-Young inequality (which has an easy direct proof in this case [17]) and the final inequality uses the assumption that r has spectral support $\mathcal{S}$.

Bruck and Smolensky prove the upper bound by randomly constructing a polynomial r having the specified density and showing that, with positive probability, it sign-represents f. Let $\{\beta_J : J \subseteq [n]\}$ be the probability distribution on the set of subsets of $[n]$ given by $\beta_J = |\hat{f}(J)|/\||f\||_1$. For an integer N (to be determined), let $\tilde{J}_1, \tilde{J}_2, \ldots \tilde{J}_N$ be subsets chosen randomly and independently according to this distribution and define the random polynomial:

$$\tilde{r}(\vec{x}) = \sum_{i=1}^{N} x_{\tilde{J}_i} \mathrm{sgn}(\hat{f}(\tilde{J}_i))$$

Now, the claim is that for $N \geq 2n(\||f\||_1)^2$, there is a positive probability that $\tilde{r}$ sign-represents f, from which it follows that there must exist a sign-representation for f of density at most $\lceil 2n(\||f\||_1)^2 \rceil$. To prove the claim, it is enough to show that for each fixed $\vec{y} \in V^n$, the probability that $\tilde{r}(\vec{y})$ has the wrong sign is less than 2^{-n}. Noting that the random variable $\tilde{r}(\vec{y})$ is the sum of N independent identically distributed random variables with values in $\{-1,+1\}$, and has mean $\mu = Nf(\vec{y})/\||f\||_1$, the probability that $\tilde{r}(\vec{y})$ has the wrong sign can be bounded above (using the Chernoff bound on the tails of a sum of independent random variables [5]) by $e^{-N/(2\||f\||_1^2)}$. Choosing $N \geq 2n(\||f\||_1)^2$ gives the desired bound.

Bruck and Smolensky observed that both the upper and lower bounds given in Theorem 2.19 can be far off for particular functions. For n odd, define the boolean functions $q(\vec{x})$ and $g(\vec{x})$ by $q(\vec{x}) = \Pi_{1 \leq i < j \leq n-1} \frac{1+x_i+x_j-x_ix_j}{2}$ and let $g(\vec{x}) = (1+x_n+q(\vec{x})-x_nq(\vec{x}))/2$. Then $\||g\||_\infty = 1/2$, but $\mathrm{dns}^\theta(g) \geq \mathrm{dns}^\theta(q) \geq 1/\||q\||_\infty = 2^{(n-1)/2}$, where the last equality was shown by Bruck.

The upper bound is far off for the function function $h(\vec{x})$ on $n = 2k$ variables which is -1 on those vectors $\vec{x}$ for which the sum of the x_i is 0. This function has a sign-representation $r(\vec{x}) = (k-1) + \sum_{i \neq j} x_i x_j$ of density $n(n-1)/2 + 1$, but $|||h|||_1 \geq 2^k / k$.

Question 2.20 *Find more accurate bounds on the threshold and weak densities of f in terms of the spectrum of f.*

Remark. Bruck used the lower bound of Theorem 2.19 to show that there are functions computable by depth 2, polynomial-size, linear threshold circuits that can not be sign-represented by a polynomial of density bounded by a polynomial in n. Bruck and Smolensky used the bounds to show that there are functions in AC^0 that have superpolynomial threshold density, and that can not be computed by polynomial size majority circuits of depth 2.

The sign-representations r of f constructed in the proofs of Theorems 2.18 and 2.19 have the natural property that the spectral support of r is contained in the spectral support of f. It is not clear whether for an arbitrary function f, there is a sign-representation with minimum density that has this property.

Question 2.21 *Is the minimum density over all sign-representations for f attained by a polynomial r whose spectral support is contained in the spectral support of f?*

An affirmative answer to the above question would follow from an affirmative answer to either part of the next one:

Question 2.22 *Suppose that f has spectral support T and r is a sign-representation having spectral support S.*

1. *Must f have a sign-representation whose spectral support is contained in $S \cap T$?*

2. *Must $(r|T)$ sign-represent f?*

The second question seems rather far-fetched and there is probably an easy counterexample.

Two interesting extremal-type problems are:

Problem 2.23 *What is the maximum threshold density (resp. weak density) of any boolean function?*

Problem 2.24 *Find bounds on the threshold density and weak density that hold for almost all functions on n variables.*

We review some known results about these problems. From Theorem 2.18 we get $dns^\theta(f) \le 2^n - 2^{n/2} + 1$ for all boolean functions f. Theorem 2.19 does not imply any non-trivial bound on the maximum of $dns^\theta(f)$ since there are boolean functions for which $|||f|||_1$ is $2^{n/2}$.

In the case of weak density there is a somewhat better upper bound, obtained from Theorem 2.16 by Aspnes et al.:

Proposition 2.25 [8] *Every boolean function on n variables has weak density at most 2^{n-1}.*

As an immediate consequence of Theorem 2.17 we have:

Theorem 2.26 *The number of functions with threshold density at most k is bounded above by $2b(2^n - 1, [0, k - 1])b(2^n, k)$.*

Using this together with the standard bounds on the tails of the binomial distribution given in Section 2.1 one obtains:

Theorem 2.27 *For some constant $\delta > 0$, almost all boolean functions on n variables have threshold density at least $\delta 2^n$.*

Based on the above discussion, we can now refine Problem 2.24:

Question 2.28   *1. Is it true that for any $\epsilon > 0$, almost all boolean functions have threshold density at least $2^n(1 - \epsilon)$?*

   2. Is it true that for any $\epsilon > 0$ almost all boolean functions have weak density at least $2^{n-1}(1 - \epsilon)$?

2.9   Threshold degree and weak degree

In the next three sections, we survey various results on the threshold degree and weak degree of boolean functions. Early treatments of this topic include the work by Cover [21] and Minsky and Papert [47]. Baldi [10] considered the topic in the context of generalizations of the Hopfield neural network model. Wang and Williams [62] initiated a systematic study, considering several algorithmic questions and generalizations of concepts from linear threshold functions, including a generalization of Proposition 2.2. Bruck [17] and Gotsman [29] began a more systematic study using Harmonic analysis. Other work on threshold degrees includes papers of Aspnes, et al. [8], Gotsman and Linial [30], and Anthony [7, 6].

The first result we consider gives the weak degree and threshold degree of all monomial functions.

Lemma 2.29 [47] *If r is a weak-representation of the parity function π_J then $\hat{r}(J) \neq 0$. Thus both the weak degree and the threshold degree of π_J is $|J|$.*

This was originally proved by Minsky and Papert and is one of the earliest results concerning polynomial sign-representations. It has been rediscovered many times, and follows immediately from lemma 2.12. In particular, the functions π and $-\pi$ each have weak degree and threshold degree equal to n. In fact, as noted by Wang and Williams and also by Gotsman and Linial, both the weak degree and threshold degree of any other functions is at most $n-1$. To see this, note that if f is any boolean function, then (by Lemma 2.14) $f^{(n-1)}$ sign-represents f unless $f_{[n]} = 1$, i.e., unless f is π or $-\pi$:

Proposition 2.30 [62] [30] *The only functions with threshold degree or weak degree n are π and $-\pi$.*

For general d, it seems unlikely that there is a nice characterization of functions with threshold degree or weak degree at most d, but the following may be interesting:

Question 2.31 *Is there a useful characterization of functions with threshold degree or weak degree exactly $n-1$?*

Next, we have a result of Aspnes, et al., which establishes a tight connection between the threshold degree of an arbitrary boolean function f and the weak degree of the product of f with π:

Theorem 2.32 [8] *For any boolean function f, $\deg^w(f\pi) + \deg^\theta(f) = n$*

To see $\deg^w(f\pi) + \deg^\theta(f) \geq n$ observe that if r strictly-represents f and q weakly-represents $f\pi$ then qr weakly-represents π and so has degree at least n. For the reverse inequality, let $d = \deg^\theta(f)$ and apply Theorem 2.16 to f with $\mathcal{S}$ equal to $\mathcal{P}([n], [0, d-1])$. Then the theorem implies that f has a weak representation q in which all monomials have degree at least d, so $q\pi$ weakly-represents $f\pi$ and has degree at most $n-d$.

2.10   The number of functions with threshold degree d

The number of boolean functions on n-variables is 2^{2^n}. How many of these can be sign-represented by a polynomial of degree at most d? Let $T(n, d)$ denote the number of boolean functions on n-variables of threshold degree at most d. As an immediate consequence of the upper bound in Theorem 2.17, we have:

Theorem 2.33 *For all positive integers $n \geq d \geq 1$, we have:*

$$T(n,d) \leq 2b((2^n - 1), [1, b(n, [0, d])]).$$

The case $d = 1$ was proved independently many times; see [21] for references. For higher d, this result was essentially stated by Baldi [10], and also by Anthony [7, 6]. For fixed d, this bound is $2^{O(n^{d+1})}$. Bruck [17] and Gotsman [29] give alternative proofs of this asymptotic upper bound.

From the lower bound of Theorem 2.17, Baldi [10] noted a $2^{\Omega(n^d)}$ lower bound for fixed d; a similar bound was obtained by Anthony [7, 6]. This can be improved using:

Theorem 2.34 *For all $n \geq 1$,*

$$T(n,1) \geq 2^{n(n-1)/2}$$

$$T(n, n-1) = 2^{2^n} - 2$$

For $2 \leq d \leq n-2$:

$$T(n,d) \geq T(n-1,d)T(n-1,d-1)$$

The expression for $T(n, n-1)$ follows from Proposition 2.30. The lower bound on $T(n,1)$ has several references and has been improved slightly; see the discussion in [48]. The given bound follows directly from $N(2,1) \geq 2$ and the recurrence $N(n+1,1) > N(n,1)(2^n + 1)$ for $n \geq 2$. To prove the recurrence, we show that for each linear threshold function f on n variables, there are at least $2^{n-1}+1$ linear threshold functions g on $n+1$ variables with the property that fixing x_{n+1} to -1 yields f. Suppose f is sign-represented by an affine polynomial a. By perturbing the weights of a slightly one can obtain another affine sign-representation b of f such that $b(\vec{x})$ is distinct for all $\vec{x} \in V^n$. Now, it is easily seen that as the parameter b_{n+1} varies over $\mathbf{R}$, the linear polynomials $b(\vec{x}) + b_{n+1}(x_{n+1} + 1)/2$ sign-represent $2^{n-1} + 1$ distinct boolean functions, each having the property that when restricted to $x_{n+1} = -1$ the function represented is equal to f.

The recurrence for $T(n,d)$ for $2 \leq d \leq n-1$ appears to be new. To prove it, let f^+ and f^- be boolean functions on n variables where f^+ is sign-represented by the degree d polynomial r^+ and f^- is sign-represented by the degree $d-1$ polynomial r^-. We may assume the sign-representations are strict, i.e., neither r^+ nor r^- vanish on V^n. Let h be the function on $n + 1$ variables defined to be f^- when $x_{n+1} = -1$ and f^+ when $x_{n+1} = 1$. Then for a sufficiently large positive constant K, h is sign-represented by $Kr^-(\vec{x})(1 - x_n)/2 + r^+(\vec{x})$.

It can be shown that, for fixed d, Theorem 2.34 gives lower bounds of the form $2^{\Omega(n^{d+1})}$, so in some sense the upper and lower bounds are close.

Problem 2.35 *Define* $R(n,d) = \log_2 T(n,d)/n^{d+1}$. *By the above bounds, as n tends to ∞, this is bounded between two positive constants. Does it approach a limit, and if so, what is it?*

In particular, in the case for $d = 1$, $R(n,1)$ is known to be bounded between $1/2 + o(1)$ and $1 + o(1)$.

Problem 2.36 *Determine the smallest possible $d(n)$ such that almost all functions on n variables have threshold degree at most d.*

Let $u(n,d) = (T(n,d) - T(n,d-1))/2^{2^n}$, i.e., the proportion of boolean functions on n variables whose threshold degree is exactly d. Based on exact computations for small values of n, Wang and Williams conjectured that the function $u(n,d)$ is unimodal:

Conjecture 2.37 [62] *For positive integers n,d, $u(n,d) < u(n,d+1)$ if $d \le (n-1)/2$ and $u(n,d) > u(n,d+1)$ otherwise.*

In particular, this asserts that for fixed n, the maximum of $u(n,d)$ occurs at $d = \lfloor n/2 \rfloor$. In fact, they made a much stronger conjecture; essentially the same conjecture was made by Aspnes et al.:

Conjecture 2.38 [62] [8]

1. *As k tends to ∞, almost all functions on $2k$ variables have threshold degree exactly k, i.e., $u(2k,k) = 1 - o(1)$*

2. *As k tends to ∞, almost all functions on $2k+1$ variables have threshold degree either k or $k+1$. More precisely, $u(2k+1,k) = 1/2+o(1)$ and $u(2k+1,k+1) = 1/2 + o(1)$.*

Thus, they suggest that the threshold degree is sharply concentrated and in particular, this would answer the question posed in Problem 2.36. Alon (personal communication) noted that by applying Theorem 2.15 with S equal to the set $\mathcal{P}(n, [0,d])$ for $d = (1-\epsilon)n$, a routine calculation produces an $\epsilon > 0$ for which almost all functions have threshold degree less than $1-\epsilon$. On the other hand, Anthony ([7],[6]) and Alon (personal communication) pointed out that Theorem 2.33 for $d = \lfloor n/2 \rfloor - 1$ together with the Chernoff bound, inequality (2), implies that almost all functions have degree at least $\lfloor n/2 \rfloor$.

One can also formulate analogous questions for weak degree. It should be noted that the number of functions of weak degree 1 is at least $2^{b(n,\lfloor n/2 \rfloor)}$, since a hyperplane that passes through all of the vertices at a largest level of the n-cube weakly represents any function which is -1 above the hyperplane and +1 below it. The following result of Aspnes, et al., which follows easily from Theorem 2.32 is relevant:

Corollary 2.39 [8] *The average weak degree over all boolean functions is at most* $n/2$.

Finally, Aspnes, et al. conjectured that the weak degree is also concentrated:

Conjecture 2.40 [8] *For* n *even and tending to* ∞, *almost all boolean functions on* n *variables have weak degree* $n/2$.

2.11   The spectrum of functions of low weak degree

By Corollary 2.13, if r is any sign-representation of the boolean function f and r has spectral support S then the entire spectrum of f is determined by its spectrum on S. Indeed, it seems natural to expect that a significant fraction of the mass of the spectrum of f is concentrated on S. In this direction, we have:

Theorem 2.41 [17],[30] *Let* f *be a boolean function and* r *be a polynomial that weakly sign-represents* f. *If the spectral support of* r *is contained in* S *then:*

$$|||(f|S)|||_1 \geq 1$$

This result was first stated explicitly by Gotsman and Linial but, as noted there, is an easy consequence of results in Bruck's paper:

$$|||r|||_\infty |||(f|S)|||_1 \geq \sum_{J \in [n]} \hat{r}(J)\hat{f}(J) = ||r||_1 \geq |||r|||_\infty.$$

Here the first inequality is trivial, the equality is lemma 2.12 and the last equality is the case $p = 1$ of the Haussdorf-Young inequalities.

Specializing this Theorem to functions of weak degree d we get:

Corollary 2.42 *If* f *has weak degree less than or equal to* d *then:*

$$|||f^{(d)}|||_1 \geq 1$$

Gotsman and Linial generalized this result to obtain lower bounds on $|||f^{(d)}|||_p$ for all $p \in [1,2]$:

Theorem 2.43 *For each* $p \in [1,2]$ *there is a constant* $c_p > 0$ *such that every boolean function* f *of weak degree at most* d *satisfies:*

$$|||f^{(d)}|||_p \geq c_p^d.$$

Furthermore $c_1 = 1$ *and* $c_2 = 1/\sqrt{2}$.

The proof is obtained by applying a series of inequalities from harmonic analysis together with Lemma 2.12. The key step in the proof is the following result of Bourgain, which upper bounds $||r||_p$ in terms of $||r||_1$ and the degree of r:

Lemma 2.44 [15] *For each $p \in [1,2]$ there exists a constant c_p such that if r is a multilinear polynomial:*

$$||r||_1 \geq c_p^{\deg(r)} ||r||_p$$

Trivially $c_1 = 1$ is best possible and it was shown by Haagerup ([32]) that the best value for c_2 is $1/\sqrt{2}$.

Now, letting q be the dual of p (i.e., $1/q + 1/p = 1$) we have:

$$|||f^{(d)}|||_p \, |||r|||_q \geq \langle f, r \rangle = ||r||_1 \geq c_p^d \, ||r||_p \geq c_p^d \, |||r|||_q,$$

which gives the theorem. The first inequality is Holder's inequality (Lemma 2.8) together with the fact that r has degree d, the equality is Lemma 2.12, the next inequality is Lemma 2.44 and the last comes from Lemma 2.9.

Gotsman and Linial also observe that the function f obtained from the parity function on n variables by reversing the value at any (one) point of the cube has threshold degree at most $n - 1$ (by Proposition 2.30), but $|||f^{(n-1)}|||_2$ is exponentially small in n, and thus, up to the choice of the constant c_p, Theorem 2.43 is tight for $p = 2$.

Nevertheless, there is a possibility of improvement along the following lines. The place where the above bound is tight occurs when the norm $|||f|||_2$ of the unprojected function is itself close to 1. If $|||f|||_p$ is much larger than 1 (as is usually the case), one would expect that $|||f^{(d)}|||_p$ would also be much larger than 1. This suggests:

Problem 2.45 *Can the ratio $|||f^{(d)}|||_p/|||f|||_p$ for functions of weak degree d be bounded below by k_p^d for some constant k_p? If not, how small can the ratio be?*

2.12 The Influence of Variables in Threshold functions

The *influence* of a variable x_i on a boolean function f, denoted $\mathbf{Inf}_i(f)$, is the fraction of points $\vec{x} \in V^n$ for which $f(\vec{x}) \neq f(\vec{x}^{(i)})$, i.e., the fraction of points where negating the value of variable i negates the function value. This notion was introduced and studied by Ben-Or and Linial ([14]). An easy counting argument shows that every balanced ($|W^-| = |W^+|$) boolean function f has a variable with influence at least $1/n$. This was improved by Kahn, Kalai and Linial ([39]) to $\Omega(\log n/n)$, which is tight by an example of Ben-Or and Linial. This example is not a linear threshold function and it is natural to ask: what is the maximum $\delta(n)$ such that every balanced linear threshold function has a variable with influence $\delta(n)$? For the simple majority function, it is easy to see that all variables have influence $\Theta(n^{-1/2})$. Gotsman and Linial proved:

Theorem 2.46 [30] *For any balanced linear threshold function f, there is a variable with influence $\Omega(n^{-1/2})$.*

This is implied by:

Theorem 2.47 *For any linear threshold function f:*

$$\sum_{i=1}^{n} \mathrm{Inf}_i^2(f) \geq \frac{1}{2} - \hat{f}(0)^2,$$

which is tight for majority up to a constant factor. Theorem 2.47 is a consequence of the case $p = 2$ and $d = 1$ of Theorem 2.43 and the following connection between influences and the spectrum of unate boolean functions (which was first noted in [39]):

Proposition 2.48 *For any unate boolean function f, $\mathrm{Inf}_i(f) = |\hat{f}(\{i\})|$.*

Gotsman and Linial asked:

Problem 2.49 [30] *For each $n \geq d \geq 1$, what is the maximum $I(n,d)$ such that every boolean function of threshold degree d has a variable of influence at least $I(n,d)$?*

2.13 Coefficient size: the weight of a linear sign-representation

If r is a polynomial that is a strict-representation of f then there is an $\epsilon > 0$ such that any polynomial obtained by perturbing the weights of r by ϵ is still a strict representation. Then multiplying r by $1/\epsilon$ and rounding the coefficients to the nearest integers yields another strict-representation. This proves:

Proposition 2.50 *If f is sign-representable by a polynomial with spectral support S, then it is sign-representable by a polynomial with spectral support S having integer coefficients.*

In this section we restrict attention to polynomials with integer coefficients. We define the *weight* of such a polynomial r to be the maximum absolute value of its coefficients, i.e., $|||r|||_\infty$. If f has threshold degree d, we say that the *representation weight* of f is the minimum weight of any degree d polynomial that sign-represents f.

Problem 2.51 *For $n \geq d \geq 1$ what is the maximum representation weight of an n variable boolean function of threshold degree d?*

The case $d = 1$ is well-studied. Since there are at least $2^{n^2/2-o(n)}$ linear threshold functions, a simple counting argument implies that most of these functions must have representation weight at least $2^{n/2(1+o(1))}$. The best worst case bounds are due to Håstad.

Theorem 2.52 [36]

1. *Every linear threshold function has representation weight at most $2^{-n}(n+1)^{(n+1)/2}$.*

2. *For n a power of 2, there exists a boolean function on n variables whose representation weight is at least $\Omega(n^{n/2}c^{-n})$ for some $c > 1$.*

The first result is a small improvement on an old result (see, e.g., [48]). It follows simply by setting up the computation of weights a_i of the threshold representation as a system of linear inequalities with $a_0, a_1, \ldots, a_n$ as variables: $a(\vec{x}) \leq -1$ for all negative points $\vec{x}$ of f and $a(\vec{x}) \geq 1$, for all positive points $\vec{x}$ of f. An extreme point of the polytope defined in this way is the solution of a system of $n + 1$ equations in $(n + 1)$ unknowns and thus, using Kramer's rule, each coefficient in the solution is a ratio of determinants of two $(n+1) \times (n+1)$ matrices, and the denominators of all coefficients are the same so they can be cleared. So each coefficient can be chosen to be the determinant of an $(n+1) \times (n+1)$ matrix with entries in ± 1 and (by Hadamard's inequality) is at most $(n+1)^{(n+1)/2}$. Håstad's additional observation is that each of these determinants is necessarily divisible by 2^n, so that factor can be divided out.

Håstad's construction of a linear threshold function with large representation weight is as follows. Let $n = 2^m$. Define an ordering $J_1, J_2, \ldots, J_n$ of the subsets of $\mathcal{P}(m)$ such that the set sizes are in nondecreasing order and adjacent sets have symmetric difference of cardinality at most 2. This ordering defines a correspondence between $[n]$ and V^m: i corresponds to $\vec{e}^{(J_i)}$. With this correspondence, we may view the points of V^n as boolean functions on V^m and the desired function can be viewed as a $\{+1, -1\}$-valued operator on the set of boolean functions f on V^m. For such functions f, define the real-valued operator $R(f) = \sum_{i=1}^{n}(n+1)^i \hat{f}(J_i)$ (keeping in mind that each $\hat{f}(J)$ is a linear function of the coordinates $f(\vec{x})$ of f). Then R^θ has large representation weight; see Håstad's paper for the (rather technical) proof.

The representation weight of functions of degree greater than 1 does not seem to have been considered in the literature. The obvious lower bounds for general degree are exponential in n while the obvious upper bounds are doubly exponential in n. It would be interesting to close this gap.

2.14   Approximate sign-representations

A sign-representation r^θ of a boolean function f is an exact representation of f, in the sense that it agrees with f on all of V^n. A weaker requirement is that r^θ agree with f on "most" points of V^n. More precisely, a boolean function g is said to *approximate* f *with error* E where $E \in [0, 2^n]$, if $f(\vec{x}) = g(\vec{x})$ for at least $2^n - E$ points in V^n. We say that the polynomial r *sign-approximates* f *with error* E if r^θ approximates f with error E. The sign-approximation is *strict* if r is non-zero on all points where it agrees with f. Aspnes, et al. ([8]) obtained several results on the question: given a function f and integer $d \le n$, what is the minimum E such that f can be sign-approximated with error E by a degree d polynomial?

The key to their investigation is a connection between sign-approximations of f and weak-representations of f. This connection comes from:

Lemma 2.53 [8] *Let r be a polynomial that strictly sign-approximates f with error E and suppose that j is an integer such that $E < b(n, [0, j])$. Then there is a polynomial s of degree at most $2j$ such that rs weakly-represents f.*

The idea is to choose s so that it is 0 for all points where r^θ disagrees with f, nonnegative on all other points, and positive on at least one point. This can be done by taking s to be the square of a nonzero polynomial t of degree j that vanishes on all of the error points, and t exists by the linear independence of the monomial functions.

If f is any boolean function, the above theorem can be applied with $j = \lfloor (\deg^w(f) - k - 1)/2 \rfloor$ to obtain:

Theorem 2.54 [8] *Let f be a boolean function and r be a polynomial of degree $k < \deg^w(f)$ that is nonzero on all points of V^n. Then r^θ disagrees with f on at least $b(n, [0, \lfloor (\deg^w(f) - k - 1)/2 \rfloor])$ points of V^n.*

By Proposition 2.30, the parity function π has weak degree n, and so:

Corollary 2.55 [8] *For any polynomial r of degree k, r^θ disagrees with π on at least $b(n, [0, \lfloor (n - k - 1)/2 \rfloor])$ points of V^n.*

In fact, this is optimal: they define for each k a single variable polynomial $v_{n,k}(z)$ of degree k such that $r(\vec{x}) = v_{n,k}(x_1 + x_2 + \ldots + x_n)$ sign-approximates parity with the optimal error. For simplicity, assume that n and k are even (the other cases are similar) and choose $v_{n,k}$ so that it has roots on k consecutive odd integer points symmetric around the origin. By assigning the proper sign to the polynomial, this guarantees that f agrees with r^θ on all points in V^n having weight (number of -1's) in the range $[(n-k)/2, (n+k)/2]$, and for about 1/2 of the remaining points. A careful accounting shows that the bound of Corollary 2.55 is exactly attained.

Asymptotically, the above bound says that for $d \leq \sqrt{n}$, parity has a degree d sign-approximator with error $2^{n-1}(1 - \Theta(d/\sqrt{n}))$ and for $d \geq \sqrt{n}$, the error is at most $e^{-d^2/n}2^n$. One might guess that, since parity is the unique function of weak degree n, that it would be hardest to approximate by a degree d polynomial. In fact, this is quite far from the truth: parity has much better low degree approximations than almost all functions.

Theorem 2.56 *The number of boolean functions on n variables that can be approximated with error at most E by a degree d threshold function is bounded above by* $2b(2^n - 1, [1, b(n, [0, d])])b(2^n, [0, E])$.

This follows from Theorem 2.33 and the fact that for an arbitrary boolean function g, the number of boolean functions that it approximates with error at most E is the sum of binomial coefficients $b(2^n, [0, E])$.

This theorem can be used to obtain a trade-off between the error E and the minimum degree d needed to achieve error E for almost all functions. For example, to achieve an error of less than $\epsilon 2^n$, degree close to $n/2$ is required for almost all functions.

On the other hand, every boolean function f has a good sign-approximator of degree only slightly more than $n/2$, in fact there is an approximator r with a stronger property. Say that r *faithfully approximates* f with error E if on all but E points $\vec{x}$, we have $r(\vec{x}) = f(\vec{x})$ (that is, not just $r^\theta(\vec{x}) = f(\vec{x})$). The following was pointed out to me by Aspnes (personal communication):

Theorem 2.57 *For $K \in [1, \sqrt{n}/2]$, for every boolean function f, there is a polynomial r of degree at most $n/2 + 2K\sqrt{n}$ that faithfully approximates f with error at most $2^{n+1}e^{-2K^2}$.*

The proof is based on an argument of Smolensky [60]. For simplicity, assume that $n = 4m$ for some integer m and that $d = 2K\sqrt{n}$ is an odd integer. Write $f = f_0 + f_1$ where f_0 is the degree $n/2$ truncation of f. Next, construct a degree d polynomial $b(\vec{x})$ which faithfully approximates parity. To do this construct a polynomial $a(z)$ of degree d which is equal to $(-1)^{j/2}$ at each of the d even integers j in the interval $[d - 1, d + 1]$. Then the polynomial $b(\vec{x}) = a(\sum_i x_i)$ computes parity (not just the sign) on all points of V^n with weight in the range $[(n - d + 1)/2, (n + d - 1)/2]$. By the Chernoff bound, inequality (2), this is all but a fraction $2e^{-2K^2}$ of the points. Observe that $f_1\pi$ is of degree at most $n/2$ and so $f_2 = f_1\pi b$ is of degree at most $n/2 + d$. Since πb is equal to 1 on all but $2^{n+1}e^{-2K^2}$ of the points, f_2 agrees with f_1 on all but that number of points, and so $f_0 + f_2$ is the desired faithful approximator.

2.15 ϵ-representations of boolean functions

For $\epsilon \in [0, 1)$, a real-valued function r is an ϵ-representation of a boolean function f if $|r(\vec{x}) - f(\vec{x})| < \epsilon$ for all $\vec{x} \in V^n$. Such a representation is, of course, a strict sign-representation.

Using a construction of Newman [49] for approximating the function $|x|$ by rational functions, Paturi and Saks showed:

Theorem 2.58 [55] *Let f be a linear threshold function with representation weight u. Then for each $k \geq 1$, f can be ϵ-represented by a rational function with numerator and denominator of degree at most k where $\epsilon = O(ue^{-\sqrt{k}})$.*

Nisan and Szegedy proved:

Theorem 2.59 [50] *If f can be $1/3$-represented by a polynomial of degree d, then its spectral representation has degree at most $O(d^8)$.*

They noted that the OR function has exact degree n and can be $1/3$-represented (using a transformation of an appropriate Chebyshev polynomial) by a polynomial of degree $\sqrt{n}$. They also related both of these quantities to the decision tree complexity of f.

Paturi established tight upper and lower bounds on the minimum degree of an ϵ-representation of any symmetric boolean function:

Theorem 2.60 [54] *Let f be a symmetric function on n variables and let k be the least integer such that f is not constant on vertices of weight in the range $[n/2 - k, n/2 + k]$. Then the minimum degree of an ϵ-representation of f is in $\Theta(\sqrt{n(n+1-2k)})$.*

3   Covering and Slicing

We now move on to the second class of problems mentioned in the introduction. The first subsection presents some additional terminology and background results. The second subsection deals with covering points of the hypercube by hyperplanes and hypersurfaces. The third subsection focuses on problems about slicing edges of the hypercube by hyperplanes. In the final subsection, analogous slicing problems for higher dimensional subcubes are considered.

3.1   More preliminaries

3.1.1   Extremal problems for hypergraphs

Most of the extremal problems discussed here fit into the following standard framework. Let $\mathcal{H} = (X, \mathcal{E})$ be a hypergraph. Two fundamental parameters associated with $\mathcal{H}$ are its cover number $\tau(\mathcal{H})$, which is the minimum size of an $\mathcal{H}$-cover (a subcollection of $\mathcal{E}$ whose union is X), and its rank, denoted $\mathrm{rk}(\mathcal{H})$, which is the size of the largest member of $\mathcal{E}$. More generally, one can define, for each natural number k, its rank sequence $(\mathrm{rk}_k(\mathcal{H}) : k \in \mathbf{N})$, where $\mathrm{rk}_k(\mathcal{H})$ is the size of the largest subset of X which can be written as the union of k members of X. Trivially, we always have $\tau(\mathcal{H}) \geq |X|/\mathrm{rk}(\mathcal{H})$. and also $\tau(\mathcal{H})$ is the smallest k such that $\mathrm{rk}_k(\mathcal{H}) = |X|$.

The questions considered in this section involve the determination of the rank and cover number of various hypergraphs associated with the n-cube.

3.1.2 Partially ordered sets and the Sperner property

An *antichain* of a partially ordered set P is a subset A of pairwise incomparable elements. The family $\mathcal{A}(P)$ of antichains of P can be viewed as a hypergraph on vertex set P. The rank and cover problems for this class of hypergraphs have been extensively studied, especially for the partially ordered set V^n (which is isomorphic to $\mathcal{P}(n)$) (see, e.g., [31]). A subset of P that is the union of k antichains. is called k-*family*, and thus $\mathrm{rk}_k(\mathcal{A}(P))$ is the size of the largest k-family.

The *height* $h(p)$ of an element p in a finite partially ordered set P, is the size of the largest chain (totally ordered set) C of elements all of whose members are strictly less than p. The *height* of P, $h(P)$ is the maximum height of any of its elements. For each integer $j \in [0, h(P)]$, P_j denotes the set of elements of height h. Each of the sets P_j is an antichain, referred to as *level j* of P. A finite poset P is said to have the *Sperner property* if the largest antichain of P is one of the levels. (This definition is slightly more general than the usual one). P has the *strong Sperner property* if for each k, the largest k-family is attained by the union of the k largest levels. We have the celebrated theorem of Sperner, as generalized by Erdős:

Theorem 3.61 [25] *For all $n \geq 1$, V^n has the strong Sperner property. Hence, for each $k \geq 1$ the largest k-family of V^n has size $b(n, [[(n - k + 1)/2], \lceil (n + k - 1)/2 \rceil])$.*

There are many known sufficient conditions for P to have the strong Sperner property. For our purposes we need the following, which is a slight generalization of a condition given by Baker:

Lemma 3.62 [9] *For each element p, let $c^+(p)$ denote the number of elements at level $h(p) + 1$ that are greater than p and let $c^-(p)$ denote the number of elements at level $h(p) - 1$ that are less than p. If for each integer $k \in [0, h(P)]$, $c^+(p)$ is the same for all $p \in P_k$ and $c^-(p)$ is the same for all $p \in P_k$ then P has the strong Sperner property.*

3.2 Vertex covering

This section describes some extremal problems and results about covering points of V^n by hyperplanes (or more generally, by hypersurfaces defined by a polynomial). If a is an affine function, let $V^n(a = 0)$ denote the set of vertices of the n-cube on which a vanishes. We say that this is the set of vertices *covered* by a.

The collection of sets $\{V^n(a = 0) : a \text{ affine}\}$ defines a hypergraph on V^n. The natural extremal problems for this hypergraph are then: (i) what is its rank, i.e., what is the maximum number of points that can be covered by one hyperplane? and (ii) what is its cover number, i.e., the minimum number of hyperplanes needed to cover all vertices? Here the answer to the second question is trivially 2, and the answer to the first is obviously 2^{n-1} (although the proof is not entirely trivial; see below).

One gets more interesting problems by considering a restricted set of hyperplanes. Say that an affine function a (and the associated hyperplane) is *skew* if all of the coefficients a_i for $i \in [n]$ are nonzero. We now can ask the same two questions above for skew hyperplanes and more generally ask:

Problem 3.63 *For $n \geq k \geq 1$ what is the maximum number of vertices of V^n that can be covered by k skew hyperplanes in V^n? In particular, what is the smallest k such that all vertices can be covered by k skew hyperplanes?*

An easy lower bound of $b(n, [\lceil (n-k+1)/2 \rceil, \lceil (n+k-1)/2 \rceil])$ is obtained by taking the k middle levels of the cube. However, this is known to be tight only for the case $k = 1$:

Theorem 3.64 *The maximum number of points covered by a skew hyperplane is $b(n, \lceil n/2 \rceil)$.*

An affine function (and its associated hyperplane) is *monotone* if all of the coefficients are nonnegative. By applying an appropriate reflection, it is enough to prove the theorem for monotone skew hyperplanes, in which case $V^n(a = 0)$ is an antichain, and so we apply Sperner's theorem (the case $k = 1$ of Theorem 3.61).

Corollary 3.65 *The maximum number of points covered by an affine function for which at least j variables have nonzero coefficients is $2^{n-j}b(j, \lceil j/2 \rceil)$*

In particular the 2^{n-1} bound on general hyperplanes follows.

For $k > 1$, the reduction to Theorem 3.61 can be used to obtain the following special case:

Theorem 3.66 *If $H_1, H_2, \ldots, H_k$ are monotone skew hyperplanes then the maximum number of vertices they cover is $b(n, [\lceil (n - k + 1)/2 \rceil, \lceil (n + k - 1)/2 \rceil])$. In particular, the minimum number of monotone hyperplanes that cover all points is $n + 1$.*

More generally, this applies if the set of hyperplanes has a *common orientation*, i.e., for each $i \in [n]$, the coefficients of x_i for each hyperplane have the same sign.

In general, however, the question for $k > 1$ is open. In particular the minimum number of skew hyperplanes that cover the vertices of V^n is unknown. It is easy to cover them by n hyperplanes, but the best lower bound is the trivial bound of $2^n/b(n, \lceil n/2 \rceil)$ obtained from Theorem 3.64, which is $\Theta(\sqrt{n})$.

A partial result on the cover problem was given (implicitly) by Alon, Bergmann, Coppersmith and Odlyzko:

Theorem 3.67 [3] *Any cover of V^n by $\pm$-hyperplanes (hyperplanes defined by affine functions for which each variable has coefficient in $\{-1, +1\}$) requires at least n hyperplanes if n is even and $n + 1$ hyperplanes if n is odd.*

Some related results are also given in [3]. The following variant of the proof appears in [55]: note that for any $\pm$-affine function a, the set of vertices it covers consists of points all of even weight (number of -1's) or all of odd weight. We call a even or odd depending on which. If $A = \{a^1, a^2, \ldots, a^m\}$ is a set of $\pm$-affine functions that covers V^n, partition A into the even functions A_e and odd functions A_o. Now the function g_e obtained by multiplying together the functions of A_e, squaring the result and negating is a (non-strict) sign-representation of the parity function π_n (where multiplication is in the ring $\mathcal{M}_n$) and so, by Proposition 2.30, has degree n. Thus A_e has at least $n/2$ members, and a similar argument applies to A_o.

In another direction, Alon and Füredi considered sets of hyperplanes that *miss* at least one point. They obtained:

Theorem 3.68 [4] *For $n \geq m \geq 1$, any set of m hyperplanes that misses at least one point misses at least 2^{n-m} points.*

In particular, to cover exactly $2^n - 1$ points requires n hyperplanes. To prove this special case, consider a set of m hyperplanes that cover all but one point $\vec{z}$ and let $f(\vec{x})$ be the product of the affine functions that define the hyperplanes (again in the ring $\mathcal{M}_n$). Then $f(\vec{x}) = 0$ for all $\vec{x} \neq \vec{z}$ and $f(\vec{z}) \neq 0$. Now, since every pseudo-boolean function has a unique multilinear representation, we must have $f(\vec{x}) = K\Pi_{i=1}^n(x_i + z_i)$ where K is some constant, and we conclude that f has degree n, and thus $m \geq n$ The proof of Theorem 3.68 is now obtained from this special case by an easy induction on $n - m$.

Alon and Füredi's paper has some further results concerning hyperplane coverings of finite sets of points in $\mathbf{R}^n$.

We mention one final generalization. The only property of the sets $V^n(a = 0)$ that is used in the proof of Theorem 3.66 is that they are antichains. For general (non-monotone) affine functions a, the set $V^n(a = 0)$ is the reflection of an antichain. This suggests:

Question 3.69 *For $n \geq k \geq 1$ what is the maximum number of vertices of V^n that can be covered by k reflected antichains of V^n? In particular, what is the smallest k such that all vertices can be covered by k reflected antichains?*

Obviously, the first quantity is bounded below and the second is bounded above by the corresponding quantities for skew hyperplanes.

3.3   Slicing edges

From covering vertices we move on to slicing edges. Before discussing specific problems we need to clarify what we mean when we say that a polynomial r slices an edge. An edge of the cube is specified by a pair $\{\vec{x}, \vec{y}\}$ of vertices that differ in exactly one coordinate. An edge can be viewed geometrically as the line segment between those vertices, or combinatorially, as the pair itself. Combinatorially, we say a polynomial r slices the edge if one of $r(\vec{x})$ and $r(\vec{y})$ is positive and the other is negative. An alternative geometric definition is to say that r slices the edge if r is not identically zero on the segment $[\vec{x}, \vec{y}]$ but is zero on some interior point of the segment. In the case that r is an affine function (slicing by hyperplanes) the geometric and combinatorial definitions coincide. For higher dimensional polynomials, the second is less restrictive. In this case we distinguish the two definitions as *combinatorial slicing* and *geometric slicing*. Notice that under the geometric definition we are evaluating the polynomial r on points outside of V^n and thus polynomials that are not multilinear may be relevant. In particular, the polynomial $\sum_i x_i^2 - K$, where K is chosen slightly smaller than n, slices all edges of the cube in the geometric sense but slices no edges in the combinatorial sense.

Throughout this and later sections, unless specifically stated otherwise we adopt the combinatorial point of view. Thus an edge is a pair of elements in V^n. We let E_n denote the set of such pairs and thus (V^n, E_n) is a graph.

The problems and results for edge slicing are closely related to the vertex covering problems by skew hyperplanes considered in the previous section. Consider first: what is the maximum number, $M(n)$, of edges that can be sliced by a single hyperplane? An obvious candidate for the maximum is a hyperplane which strictly separates the two levels $V^n(\lfloor n/2 \rfloor)$ and $V^n(\lfloor n/2 \rfloor + 1)$. P.E. O'Neil proved that this is best possible:

Theorem 3.70 [51]. *For all $n \geq 1$, the maximum number of edges that can be sliced by a hyperplane is $M(n) = (n - \lfloor \frac{n}{2} \rfloor) b(n, \lfloor \frac{n}{2} \rfloor)$.*

It is easy to see that this formula gives the number of edges sliced by the given hyperplane. The upper bound will follow as a special case of a result to be presented later (Theorem 3.79).

Next we consider:

Problem 3.71 *What is the minimum size, $S(n)$, of a set of hyperplanes that slices all edges of the cube?*

Such a set of hyperplanes is called an *edge slicing set*. This problem has been asked independently several times and the earliest reference seems to be O'Neil [51]. There are two simple nonisomorphic constructions that show $S(n) \leq n$: a set of hyperplanes which separates all pairs of adjacent levels of the cube and the set of hyperplanes $\{V^n(x_i = 0) : i \in [n]\}$ which separates each pair of parallel facets. It is natural to conjecture, as many have done, that this is best possible. This is easily verified for $n \leq 3$ and was proved for $n = 4$ by Emamy-Khansary [24, 23]. However, the following unpublished example of M. Paterson shows that the conjecture fails for $n = 6$.

Example 3.72 *The set of linear functions:*

$$
\begin{aligned}
a^1 &= & x_1 &+ & x_2 &+ & x_3 &+ & 3x_4 &+ & 3x_5 &- & 4x_6, \\
a^2 &= & -2x_1 &- & 2x_2 &- & 2x_3 &+ & 3x_4 &+ & 3x_5 &- & x_6, \\
a^3 &= & 3x_1 &+ & 3x_2 &+ & 3x_3 &+ & x_4 &+ & x_5 &- & 4x_6, \\
a^4 &= & -x_1 &- & x_2 &- & x_3 &+ & 3x_4 &+ & 3x_5 &+ & 6x_6, \\
a^5 &= & 3x_1 &+ & 3x_2 &+ & 3x_3 &+ & x_4 &+ & x_5 &+ & 8x_6.
\end{aligned}
$$

is a slicing set for V^6.

Now, if $\{a^i(x_1, x_2, \ldots, x_n) : i \in I\}$ is a slicing set for V^n and $\{b^j(x_1, x_2, \ldots, x_m) : j \in J\}$ is a slicing set for V^m then $\{a^i(x_1, x_2, \ldots, x_n) : i \in I\} \cup \{b^j(x_{n+1}, x_{n+2}, \ldots, x_{n+m}) : j \in J\}$ is a slicing set for V^{n+m}. Thus:

Proposition 3.73 $S(n)$ *is subadditive, i.e.,* $S(n + m) \leq S(n) + S(m)$,

and from $S(6) \leq 5$ we get:

Corollary 3.74 *For all n,* $S(n) \leq \lceil \frac{5n}{6} \rceil$.

This is the best upper bound that seems to be known. On the other hand, the best lower bound known, observed by O'Neil, is $S(n) \geq n2^{n-1}/M(n) = \Omega(\sqrt{n})$. Thus the current state of knowledge on the problem is analogous to that for the vertex covering problem for skew hyperplanes.

Some special cases of Problem 3.71 have been settled:

Proposition 3.75 *Let A be a set of affine functions that slice all edges.*

1. *If each $a \in A$ is monotone then $|A| \geq n$.*

2. *If each $a \in A$ is a $\pm$-affine function then $|A| \geq n/2$.*

3. *If each edge is sliced an odd number of times then $|A| \geq n$.*

The first part seems to have been observed by many people. Ahlswede and Zhang [1] prove it as a consequence of a theorem in extremal set theory. A simple proof appears in [30]: For $i \in [0, n]$, let $\vec{x}^i$ be the vertex whose first i coordinates are 1. Then a monotone hyperplane can cut at most one of the edges $(\vec{x}^i, \vec{x}^{i+1})$. The second and third parts were observed by Paturi and Saks (unpublished). For the second, note that the set of edges cut by a single $\pm$-hyperplane H induce a connected bipartite graph between two subsets of V^n. There are two hyperplanes parallel to H which cover these two subsets of vertices. Thus from an edge slicing set consisting of k $\pm$-hyperplanes, we get a vertex cover consisting of $2k$ $\pm$-hyperplanes. Now apply Theorem 3.67. For the third, assume without loss of generality that no $a \in A$ covers a point of V^n and observe that the hypothesis of (3) implies that the product of the members of A sign-represents either π or $-\pi$.

It is worth noting that $\pm$-hyperplanes include all of the hyperplanes that slice a maximum number of edges.

The combinatorial definition of the set of edges sliced by a hyperplane can be extended so as to define the set of edges sliced by a boolean function f: the edge $(\vec{x}, \vec{y})$ is sliced by f if $f(\vec{x}) \neq f(\vec{y})$. The size of the set of edges sliced by f is called the *average sensitivity* of f ([30]). The problems and results above dealt with the case of edges sliced by linear threshold functions. In fact, the proof of the first part of Proposition 3.75 applies just as well to slicing by arbitrary monotone boolean functions. Now, recall that a boolean function is unate if it is the reflection of a monotone boolean function and that all linear threshold functions are unate. The bound in Theorem 3.70 extends to unate functions and it is natural to ask:

Problem 3.76 *What is the size, $S_*(n)$, of the smallest set of unate functions that slices all of the edges of the n-cube? Does $S_*(n) = S(n)$?*

Two further extensions are of interest. Gotsman and Linial [30] asked for bounds on the maximum number $M(n, k)$ of edges of the n-cube that can be sliced by k hyperplanes. Define $N(n, k)$ to be the number of edges that are between vertices on a set of $k + 1$ levels symmetric around the middle level. Then $M(n, k) \geq N(n, k)$ and, by Theorem 3.70, this is sharp for $k = 1$. On the other hand, $M(n, k) > N(n, k)$ trivially holds for k in the range $S(n)$ to $n - 1$.

Finally, Gotsman and Linial asked:

Problem 3.77 *What is the maximum number of edges that can be sliced by a multilinear polynomial of degree k?*

It is easy to construct degree k polynomials that slice $N(n, k)$ edges, and they conjecture this is best possible.

3.4   Higher dimensional slicing problems

Vertices and edges are, respectively, the 0 and 1 dimensional subcubes of the n-cube. In this section we consider slicing problems for higher dimensional subcubes. To begin with, we say that a hyperplane *slices* a subcube if at least two vertices of the subcube lie on opposite sides, and otherwise it *misses* the subcube. For $n \geq k \geq 1$, define $D(n, k)$ to be the maximum integer D such that any set of k hyperplanes in the n-cube misses a subcube of dimension D.

Problem 3.78 *For n, k, determine $D(n, k)$.*

This problem was posed by Impagliazzo, Paturi and Saks [38]. In this section we summarize their results on the problem, and consider some other related problems which generalize the edge-slicing problems in the previous section to higher-dimensional subcubes. To do this we will need some additional notation.

3.4.1   Subcubes of the n-cube

A subcube of V^n is a set of the form $\alpha_1 \times \alpha_2 \times \ldots \times \alpha_n$ where each α_i is one of the sets $\{1\}$, $\{-1\}$ or V. A subcube is denoted by an n-tuple $\alpha \in V_*^n$, where V_* is the set $\{-1, 1, *\}$. The indices i such that $\alpha_i = "*"$ are the *free* indices of α; the remaining indices are the *fixed* indices. The *basepoint* of α is the vector $\vec{b}(\alpha)$ obtained by setting all free coordinates to 1 and the *weight* of α is the weight of its basepoint, i.e., the number of coordinates in α that are -1. The *dimension* of α is the number of free indices. Note that a subcube of dimension 0 consists of a single vertex; by abuse of notation we identify the set and the unique vertex belonging to it. A subcube of dimension 1 is an *edge* and a subcube of dimension $n - 1$ is a *facet*.

The set of all subcubes of V^n is denoted $\mathcal{C}(n)$ and the set of subcubes of dimension r is denoted $\mathcal{C}^r(n)$. The set of subcubes of dimension r and weight w is denoted $\mathcal{C}^r(n, w)$; observe that this set is empty if $r + w > n$. Also, for $J \subseteq [0, n]$, $\mathcal{C}^r(n, J)$ is the set of r-dimensional subcubes whose weight belongs to the set J. The cardinalities of these various sets are denoted, respectively, by $C(n)$, $C^r(n)$, $C^r(n, w)$ and $C^r(n, J)$.

Two orderings on $\mathcal{C}(n)$ are of interest. The first is the usual containment ordering. If the subcube α is contained in the subcube β we say that α *refines* β. The second ordering is the extension to subsets of the usual "$\leq$" ordering on points: $\alpha \leq \beta$ if $\vec{x} \leq \vec{y}$ for all $\vec{x} \in \alpha$ and $\vec{y} \in \beta$.

3.4.2   Slicing r-dimensional subcubes

The parameters studied for edge slicing can be generalized: Let $M^r(n)$ be the maximum number of r-dimensional subcubes of V^n that can be sliced by a single hyperplane, and let $S^r(n)$ be the minimum size of a set of hyperplanes that slice all r-dimensional subcubes. As in the last section, we also consider the analogous functions $M_*^r(n)$ and $S_*^r(n)$ where we consider slicing by arbitrary unate functions; clearly $M_*^r(n) \geq M^r(n)$ and $S_*^r(n) \leq S^r(n)$.

As for $r = 1$, the natural candidate for achieving both $M^r(n)$ and $M_*^r(n)$ is a hyperplane that slices between levels $\lfloor n/2 \rfloor$ and $\lfloor 1 + n/2 \rfloor$ of V^n. Define $\mathcal{N}^r(n)$ to be the set of r-dimensional subcubes sliced by such a hyperplane and let $N^r(n)$ be the cardinality of $\mathcal{N}^r(n)$. Then $\mathcal{N}^r(n) = \mathcal{C}^r(n, [\lfloor n/2 \rfloor - r + 1, \lfloor n/2 \rfloor])$. We have:

Theorem 3.79 *For all* $n \geq r \geq 1$,

$$M_*^r(n) = M^r(n) = N^r(n)$$

The case $r = 1$, which is Theorem 3.70, was proved by O'Neil. The following generalizes his proof. It suffices to show $N^r(n) \geq M_*^r(n)$. So let f be an arbitrary unate function on V^n. We will show that the number of r-dimensional subcubes on which f is non-constant is at most $N^r(n)$. By applying a suitable reflection, we may assume that f is monotone. Then, as is easily checked, the set of subcubes sliced by f is an antichain in the "$\leq$" order on $\mathcal{C}^r(n)$. It now suffices to show that $\mathcal{N}^r(n)$ is the largest antichain in $\mathcal{C}^r(n)$. The obvious thing to try is to show that the product order on $\mathcal{C}^r(n)$ has the Sperner property and that $\mathcal{N}^r(n)$ is the largest level. For $r = 1$, this is true by Lemma 3.62. For $r > 1$, the hypotheses of the Lemma do not hold. To overcome this problem, we first partition the poset $\mathcal{C}^r(n)$ into r distinct posets as follows. Let J_i be the set of integers in $[0, n]$ that are congruent to $i \bmod r$ and partition $\mathcal{C}^r(n)$ into r sets $\mathcal{C}^r(n, J_i)$ for $i \in [0, r - 1]$. Then the following facts are easily checked:

1. The largest antichain of $\mathcal{C}^r(n)$ is bounded in size by the sum of the sizes of the largest antichains in $\mathcal{C}^r(n, J_i)$.

2. The product order restricted to $\mathcal{C}^r(n, J_i)$ is graded, with the jth level being the set $\mathcal{C}^r(n, jr + i)$.

3. Each $\mathcal{C}^r(n, J_i)$ satisfies Baker's condition (lemma 3.62) and so is Sperner.

4. The largest rank of $\mathcal{C}^r(n, J_i)$ is the set $\mathcal{C}^r(n, jr + i)$ where j is the largest index such that $jr + i \leq n/2$.

From these observations we conclude that the largest antichain of $\mathcal{C}^r(n)$ can not be larger than the sum of $\mathcal{C}^r(n, k)$ over k in the range $\lfloor n/2 \rfloor - r + 1$ to $\lfloor n/2 \rfloor$; this is just $N^r(n)$.

Next we consider the parameters $S^r(n)$ and $S_*^r(n)$. By slicing between levels $jr - 1$ and jr of V^n for $j \in [1, \lfloor n/r \rfloor]$, we have the following upper bound:

Proposition 3.80 *For all* $n \geq r \geq 1$:

$$S_*^r(n) \leq S^r(n) \leq \lfloor n/r \rfloor.$$

We get a lower bound from the inequality $S_*^r(n) \geq C^r(n)/M_*^r(n) = C^r(n)/N^r(n)$. This bound can be shown to be $\Omega(\max\{\frac{\sqrt{n}}{r}, 1\})$, and thus:

Proposition 3.81 *There exists a constant $K > 0$ such that for all $n \geq r \geq 1$:*

$$S^r(n) \geq S_*^r(n) \geq K \max\{\frac{\sqrt{n}}{r}, 1\}).$$

As in the case $r = 1$, the naive upper and lower bounds differ by a factor of $\sqrt{n}$ (for $r \leq \sqrt{n}$). Impagliazzo, Paturi and Saks ([38]) obtained the following improvement:

Theorem 3.82 *For all* $n \geq r \geq 1$:

$$S_*^r(n) > \sqrt{\frac{n}{(r+1)} - 1}.$$

This result was used to prove lower bounds on the sizes of threshold circuits that compute the parity function. The main step needed for that lower bound is that $D(n, k) \geq n/p(k)$ for some polynomial $p(k)$, i.e., for any collection of k linear threshold functions on n variables, there is a subcube of dimension $n/p(k)$ on which each of the functions is constant. In these terms, Theorem 3.82 is equivalent to:

Theorem 3.83 *If F is a set of unate boolean functions on V^n then there is a subcube α of dimension at least $\lfloor n/(|F|^2 + 1) \rfloor$ such that each $f \in F$ is constant on α. Thus $D(n, k) \geq \lfloor n/(k^2 + 1) \rfloor$.*

Observe that Proposition 3.81 is not sufficient to give any polynomial $p(k)$ such that $D(n, k) \geq n/p(|F|)$.

In the next subsection, we discuss the function $D(n, k)$ and sketch the proof of this theorem.

3.4.3   Finding large unsliced subcubes

The problem of determining $D(n, k)$ is contained within the problem of determining $S^r(n)$. Focusing on $D(n, k)$ suggests a different way to think about the problem: starting with a fixed set of k hyperplanes, look for a large dimensional subcube that is not sliced by any of them. It is interesting to consider as well the analogous function $D_*(n, k)$, which is the maximum d such that every set of k unate functions misses a subcube of dimension d/ Clearly $D(n,k) \geq D_*(n,k)$; the proof of Theorem 3.83 actually provides a lower bound for $D_*(n,k)$.

A trivial upper bound on both $D(n,k)$ and $D_*(n,k)$ is obtained by dividing the cube into $k+1$ groups of consecutive levels all consisting of (nearly) the same number of levels, and take one hyperplane between each consecutive pair of groups. This construction leads to an upper bound of $\lfloor n/(k+1) \rfloor$. For $k=1$ this is tight, even in the more general case of unate functions; this follows from Theorem 3.79 but there are easier direct proofs.

A simple generalization of the case $k=1$ provides the key property of unate functions which is needed to prove Theorem 3.83. This property is easy, but we need some additional notation to state it.

If α is a subcube and $J \subseteq [n]$, then $(\alpha : J)$ denotes the subcube defined by making all indices outside of J free, i.e., $(\alpha : J)_i = \alpha_i$ for $i \in J$ and $(\alpha : J)_i = $"$*$" if $i \notin J$. The subcube $(\alpha : J)$ is the *inflation* of α *relative to* J. We define the following *overlay* operation on the set of subcubes: $\alpha \triangleleft \beta$ is defined to be α_i if $\alpha_i \in V$ and β_i if $\alpha_i = $"$*$". Thus $\alpha \triangleleft \beta$ is a subcube of α but generally not of β. The operation $\triangleleft$ is associative but not necessarily commutative.

An *ordering* of a set Y of size k is a bijection Γ from $[k]$ to Y. For a fixed Γ, we refer to $\Gamma(i)$ as the ith element of Y. Also, $\Gamma(\leq i)$ denotes the set $\{\Gamma(j) : j \leq i$ and $j \in [\|Y\|]\}$ and $\Gamma(\geq i)$ denotes the set $\{\Gamma(j) : j \geq i$ and $j \in [\|Y\|]\}$.

Now we are ready to state the key property of unate functions:

Lemma 3.84 [38] *Let f be a non-constant unate function on X with orientation J and let Γ be an ordering of X. Then there is a j in $[0,n]$ such that f is identically 1 on the subcube $(\vec{e}^{(J)} : \Gamma(\leq j))$ and is identically -1 on the subcube $(-\vec{e}^{(J)} : \Gamma(\geq j))$.*

Note that the lower bound $D_*(n,1) \geq \lfloor n/2 \rfloor$ follows immediately from the lemma. To prove the lemma, first observe that, by a suitable reflection, it suffices to prove it for monotone f, i.e., $J = \emptyset$. Since f is not constant, we have $f(\vec{e}) = 1$ which means that f restricted to $(\vec{e} : \Gamma(\leq n))$ is identically 1. Let j be the least index such that f restricted to $(\vec{e} : \Gamma(\leq j))$ is identically 1. Since f is non-constant, we have $j \geq 1$. Then f restricted to $(\vec{e} : \Gamma(\leq j-1))$ is not identically 1, which implies that there is a point $\vec{z}$ that refines $(\vec{e} : \Gamma(\leq j-1))$ such that $f(\vec{z}) = -1$. Then f restricted to $(-\vec{e} : \Gamma(\geq j))$ is identically -1 by monotonicity since every point that extends $(-\vec{e} : \Gamma(\geq j))$ is less than or equal to $\vec{z}$.

We are now ready to sketch the proof of Theorem 3.83; the proof given here is a slight variant of their proof. Let $m = |F|$ and $d = \lfloor n/(m^2 + 1) \rfloor$. By restricting to a smaller subcube if necessary (fixing some variables), we may assume that $n = (m^2 + 1)d$. For each $f \in F$, let J_f be an orientation for f.

The proof uses a version of the method of *random restriction* ([27, 2, 63, 35, 5]). A random procedure P is described which, depending on the family F and some random choices, produces a subcube $\tilde{\alpha}$ of dimension d. It is then shown that with strictly positive probability, all functions in F are constant on this subcube, from which it follows that there must exist a subcube of dimension d on which all functions are constant. We adopt the following notational convention: Random variables are denoted by placing a $\tilde{}$ over the identifier. When we refer to a specific value that a random variable may assume, we denote that value by an identifier without a $\tilde{}$.

The procedure P consists of two steps. In the first step, the set $[n]$ is partitioned into $m + 1$ classes: a class $\tilde{Y}^U$ of size d and for each $f \in F$ one class $\tilde{Y}^f$ of size md. The partition is chosen uniformly at random from among all $n!/(md)!^m d!$ such partitions.

In the second step, $\tilde{\alpha}_i$ is fixed for all $i \in \cup_{f \in F} \tilde{Y}^f$. This is done as follows: for each $f \in F$, choose an ordering $\tilde{\Gamma}^f$ of Y^f uniformly at random. Select an integer $\tilde{t}_f$ at random from $\{0, 1, \ldots, md\}$. For each $j \le \tilde{t}_f$, fix the coordinate $\tilde{\alpha}_{\tilde{\Gamma}^f(j)}$ according to $\bar{e}^{(J_f)}$, and for $j > \tilde{t}_f$, fix the coordinate $\tilde{\alpha}_{\tilde{\Gamma}^f(j)}$ according to $-\bar{e}^{(J_f)}$. Finally, for $i \in \tilde{Y}^U$, set $\tilde{\alpha}_i = $"$*$".

Clearly, the subcube $\tilde{\alpha}$ has dimension d. The key property of this construction is given by:

Lemma 3.85 *For each $h \in F$, the probability that $h(\tilde{\alpha})$ is not constant is at most $d/(md + 1)$.*

It follows from this lemma that the probability that $f(\tilde{\alpha})$ is non-constant for some $f \in F$ is at most $md/(md + 1)$, and so there must be at least one subcube α of the required dimension such that $f(\alpha)$ is constant for all $f \in F$.

So it remains to prove the lemma. Fix $h \in F$. We define a modified procedure P^h which treats h in an asymmetric way. It will be easy to see that P^h produces the same distribution as P on subcubes; we then use the modified construction to verify the conclusion of the lemma.

The modified construction is as follows. In choosing the initial partition of $[n]$, do not specify the remainder block $\tilde{Y}^U$. Instead, choose $\tilde{Y}^h$ to be a set of $d(m + 1)$ elements. In the second step, the variables of $\tilde{Y}^f$ for each $f \ne h$ are fixed exactly as in P. For $\tilde{Y}^h$ we modify the variable fixing step: having chosen the ordering $\tilde{\Gamma}^h$ and $\tilde{t}_h \in \{0, \ldots, md\}$, for each $j \le \tilde{t}_h$, fix the coordinate of $\tilde{\alpha}$ in position $\tilde{\Gamma}^h(j)$ according to $\bar{e}^{(J_h)}$ (as before), and for $j > \tilde{t}_h + d$, fix the coordinate of $\tilde{\alpha}$ in position $\tilde{\Gamma}^h(j)$ according to $-\bar{e}^{(J_h)}$. The remaining coordinates, those indexed by $\{\tilde{\Gamma}^h(\tilde{t} + j) : 1 \le j \le d\}$ are set to "$*$".

It can be readily checked that the distribution on subcubes produced by the two procedures is identical. Thus, it suffices to upper bound the probability that $h(\tilde{\alpha})$ is not constant when $\tilde{\alpha}$ is constructed according to P^h. For this, condition on any fixed value of the partition of $[n]$ that is produced in the first step of P^h, on any fixed values Γ^f of the permutations $\tilde{\Gamma}^f$ for all f and on any fixed values t_f of $\tilde{t}^f$ for $f \neq h$. This conditioning determines a fixed value Y^h for the set $\tilde{Y}^h$ and fixed values α_i for $\tilde{\alpha}_i$, for each $i \notin Y^h$. The claim is that for any such conditioning, the conditional probability that h is non-constant is at most $1/m + 1$.

Let g denote the function on the variables indexed by Y^h obtained from h by the conditioning. Then g is unate, and its orientation is $J \cap Y^h$.

By Lemma 3.84 there is an index l such that the functions $g(\bar{e}^{(J_h)} : \Gamma^h(\leq l))$ and $g(-\bar{e}^{(J_h)} : \Gamma^h(\geq l))$ are both constant. Thus, unless $\tilde{t}_h$ lies in the range $[l+1, l+d]$, the fixing of the coordinates of Y^h defined by P^h will make g, and hence h, constant. Since $\tilde{t}_h$ is chosen uniformly from $[0, md]$, the probability that $\tilde{t}_h$ lies in the range $[l+1, l+d]$ is bounded above by $d/(md+1)$.

3.4.4 Simply sliced subcubes

The previous result asserts that given a "small" set of hyperplanes, there is always a "large" subcube that is not sliced by any of them. Impagliazzo, et al. also considered the following related problem. A boolean function is *simple* if either it is constant or it is equal to x_i or $-x_i$ for some variable x_i. A hyperplane is *simple* if its associated boolean threshold function is simple, i.e., either the hyperplane misses V^n entirely, or there are two maximal faces $x_i = 1$ and $x_i = -1$ that are on opposite sides of the hyperplane. Given a family of linear threshold functions (or more generally, unate functions) the problem is to find a "large" subcube α such that each hyperplane (unate function) in the family simply slices the subcube.

For a collection of functions F on V^n, let $\delta(F)$ denote the average, over all coordinates $i \in [n]$, of the number of functions which depend on variable x_i. Impagliazzo, et al. proved:

Theorem 3.86 [38] *Let F be a collection of unate functions on n variables. Then there exists a subcube α of dimension at least $n/(4\delta(F)^2 + 2)$ such that every $f \in F$ depends on at most one variable.*

The theorem was originally motivated by the problem of proving lower bounds on threshold circuits. In this case the result is used to show a superlinear lower bound on the number of edges in a bounded depth threshold circuit that computes parity.

The proof of this theorem is a probabilistic argument that is similar to, but more complicated than, the proof of Theorem 3.83; the reader can find the details in [38]. The following problem is stated in [38].

Problem 3.87 *What is the smallest exponent r such that the conclusions of Theorem 3.86 hold with $n/(4\delta^2 + 2)$ replaced by $\Omega(n/|F|^r)$?*

Various examples already discussed here show that it is impossible to take $r < 1$. Thus the best possible r is between 1 and 2.

Remark. The state of knowledge on all of the covering and slicing problems is somewhat similar: there is a significant gap between the upper and lower bounds which is similar for all of these problems. An interesting aspect of this is that for nearly all of the general bounds for covering and slicing problems by linear threshold functions, the results hold for the much more general class of unate functions. This could indicate one of two things: (i) linear threshold functions are the "optimal" unate functions for slicing and thus the corresponding "starred" and "unstarred" parameters are really all the same, or (ii) unate functions really are more powerful for slicing than threshold functions, and the weakness in the slicing lower bounds for linear threshold functions comes from the fact that the proof techniques do not take enough advantage of the special structure of linear threshold functions.

In particular, it would be very interesting if it turned out that the edges of the hypercube can be sliced by $\Theta(\sqrt{n})$ unate functions, but require $\Theta(n)$ threshold functions.

References

[1] R. Ahlswede and Z. Zhang. An identity in combinatorial extremal theory. *Advances in Mathematics*, 80(2):137–151, 1990.

[2] M. Ajtai. σ_1^1-formulae on finite structures. *Annals of Pure and Applied Logic*, 24:1–48, 1983.

[3] N. Alon, E. E. Bergmann, D. Coppersmith, and A. M. Odlyzko. Balancing sets of vectors. *IEEE Transactions on Information Theory*, 34:128–130, 1988.

[4] N. Alon and Z. Füredi. Covering the cube by affine hyperplanes. To appear in European Journal of Combinatorics, 1993.

[5] N. Alon and J. Spencer. *The Probabilistic Method.* John Wiley & Sons, 1992.

[6] M. Anthony. Classification by polynomial surfaces. Technical Report LSE-MPS-39, London School of Economics, October 1992.

[7] M. Anthony. On the number of boolean functions of a given threshold order. Technical Report LSE-MPS-36, London School of Economics, September 1992.

[8] J. Aspnes, R. Beigel, M. Furst, and S. Rudich. The expressive power of voting polynomials. In *23rd ACM Symposium on Theory of Computing*, pages 402–409, 1991.

[9] K. A. Baker. A generalization of Sperner's lemma. *J. Combinatorial Theory*, 6:224–225, 1969.

[10] P. Baldi. Neural networks, oreintations of the hypercube, and algebraic threshold functions. *IEEE Transactions on Information Theory*, 34(3):523–530, May 1988.

[11] R. Beigel. Perceptrons, PP, and the polynomial hierarchy. In *7th Annual Structure in Complexity Theory Conference*, pages 14–19, 1992.

[12] R. Beigel, N. Reingold, and D. Spielman. The perceptron strikes back. In *6th Annual Structure in Complexity Theory Conference*, pages 286–291, 1991.

[13] M. Bellare. A technique for upper bounding the spectral norm with applications to learning. In *5th Annual ACM Workshop on Computational Learning Theory*, 1992.

[14] M. Ben-Or and N. Linial. Collective coin flipping. In S. Micali, editor, *Randomness and Computation*. Academic Press, 1989.

[15] J. Bourgain. Walsh subspaces of l_p product spaces. In *In seminaire D' Analyse Fonctionelle*, pages 4.1–4.9. Ecole Polytechnique, Centre De Mathematiques, 1979-1980.

[16] Y. Brandman, A. Orlitsky, and J. Hennessy. A spectral lower bound technique for the size of decision trees and two-level and/or circuits. *IEEE Transactions on Computers*, 39(2):282–287, February 1990.

[17] J. Bruck. Harmonic analysis of polynomial threshold functions. *SIAM Journal of Discrete Mathematics*, 3(2):168–177, May 1990.

[18] J. Bruck and R. Smolensky. Polynomial threshold functions, AC^0 functions and spectral norms. *SIAM Journal of Discrete Mathematics*, 21:33–42, 1992.

[19] V. Chvátal. *Linear programming*. W. H. Freeman and Company, 1983.

[20] V. Chvátal and P. L. Hammer. Aggregation of inequalities in integer programming. *Annals of Discrete Mathematics*, 1:145–162, 1977.

[21] T. Cover. Geometrical and statistical properties of systems of linear inequalities with applications in pattern recognition. *IEEE Transactions on Electronic Computers*, EC-14(3):326–334, June 1965.

[22] M.L. Dertouzos. *Threshold Logic: A Synthesis Approach.* MIT Press, 1965.

[23] M.R. Emamy-Khansary. On the covering cuts of c^d, $d \leq 5$. Manuscript.

[24] M.R. Emamy-Khansary. On the cuts and cut number of the 4-cube. *Journal of Combinatorial Theory Series A*, 41:221–227, 1986.

[25] P. Erdös. On a lemma of Littlewood and Offord. *Bull. Amer. Math. Soc.*, 51:898–902, 1945.

[26] M. Furst, J. Jackson, and S. Smith. Improved learning of AC^0 functions. In *4th Annual ACM Workshop on Computational Learning Theory*, 1991.

[27] M. Furst, J.B. Saxe, and M. Sipser. Parity, circuits and the polynomial time hierarchy. *Mathematical Systems Theory*, 17:13–28, 1984.

[28] M.C. Golumbic. *Algorithmic Graph Theory and Perfect Graphs.* Academic Press, 1980.

[29] C. Gotsman. On boolean functions, polynomials and algebraic threshold functions. Technical Report TR-89-18, Dept. of Computer Science, Hebrew University, 1989.

[30] C. Gotsman and N. Linial. Spectral properties of threshold functions. To appear in Combinatorica, 1993.

[31] C. Greene and D. J. Kleitman. Proof techniques in the theory of finite sets. In G-C. Rota, editor, *Volume 17: Studies in Combinatorics*, pages 22–79. The Mathematical Association of America, 1978. Series: Studies in Mathematics.

[32] U. Haagerup. The best constants in the Khintchine inequality. *Studia Mathematica*, 70:231–283, 1982.

[33] P.L. Hammer and N.V.R. Mahadev. Bithreshold graphs. *SIAM Journal of Applied Mathematics*, 6:497–506, 1985.

[34] E.F. Harding. The number of partitions of a set of n points in k dimensions induced by hyperplanes. *Proceedings of the Edinburgh Mathematical Society*, 15:285–290, 1967.

[35] J. Håstad. Almost optimal lower bounds for small depth circuits. In Silvio Micali, editor, *Advances in Computing Research 5: Randomness and Computation*, pages 143–170. JAI Press, Greenwich,CT, 1989.

[36] J. Håstad. On the size of weights for threshold gates. To appear in SIAM Journal of Discrete Mathematics, 1992.

[37] S. T. Hu. *Threshold Logic*. University of California Press, Berkley and Los Angeles, California, 1965.

[38] R. Impagliazzo, M. Paturi, and M. Saks. Size-depth trade-offs for threshold circuits. Technical Report CS92-253, Dept. of CSE, University of California, San Diego, July 1992. To appear in Proceedings of 25th ACM Symposium on Theory of Computing, 1993.

[39] J. Kahn, G. Kalai, and N. Linial. The influence of variables on boolean functions. In *29th IEEE Symposium on Foundations of Computer Science*, pages 68–80, 1988.

[40] Y. Katznelson. *An Introduction to Harmonic Analysis*. Wiley, 1968.

[41] R. J. Lechner. Harmonic analysis of switching functions. In A. Mukhopadhyay, editor, *Recent Developmentsin Switching Theory*. Academic Press, 1971.

[42] P. M. Lewis and C. L. Coates. *Threshold Logic*. John Wiley & Sons, 1967.

[43] N. Linial, Y. Mansour, and N. Nisan. Constant depth circuits, fourier transforms and learnability. In *30th IEEE Symposium on Foundations of Computer Science*, pages 574–579, 1989.

[44] F. J. MacWilliams and N. J. A. Sloane. *The Theory of Error-Correcting Codes*. Noth-Holland, Amsterdam, New York, 1977.

[45] Y. Mansour. An $o(n^{\log \log n})$ learning algorithm for DNF under the uniform distribution. In *5th Annual ACM Workshop on Computational Learning Theory*, 1992.

[46] W. S. McCulloch and W. Pitts. A logical calculus for the ideas immanent in nervous activity. *Bulletin of Mathematical Biophysics*, 5:115–133, 1943.

[47] M. Minsky and S. Papert. *Perceptrons*. MIT Press, Cambridge, Mass., 1969. Expanded edition 1988.

[48] S. Muroga. *Threshold Logic and its Applications*. Wiley-Interscience, 1971.

[49] D. J. Newman. Rational approximation to $|x|$. *Michigan Math. Journal*, 11:11–14, 1964.

[50] N. Nisan and M. Szegedy. On the degree of boolean functions as real polynomials. In *24th ACM Symposium on Theory of Computing*, pages 462–467, 1992.

[51] P. O'Neil. Hyperplane cuts of an n-cube. *Discrete Mathematics*, 1:193–195, 1971.

[52] P. Orponen. *Neural networks and complexity theory*, pages 50–61. Lecture Notes in Computer Science 629. Springer-Verlag, 1992.

[53] I. Parberry. A primer on the complexity theory of neural networks. In R. B. Banerji, editor, *Formal Techniques in Artificial Intelligence: A Sourcebook*. Elsevier, 1990.

[54] R. Paturi. On the degree of polynomials that approximate symmetric boolean functions. In *24th ACM Symposium on Theory of Computing*, pages 468–474, 1992.

[55] R. Paturi and M. Saks. On threshold circuits for parity. In *31st IEEE Symposium on Foundations of Computer Science*, pages 397–404, 1990. Also, to appear as "Approximating threshold circuits by rational functions" in *Information and Computation*.

[56] U. Peled and B. Simeone. Polynomial-time algorithms for regular set-covering and threshold synthesis. *Discrete Applied Mathematics*, 12:57–69, 1985.

[57] A. A. Razborov. *On small depth threshold circuits*, pages 42–52. Lecture Notes in Computer Scince 621. Springer-Verlag, 1992.

[58] L. Schläfli. *Gesammelte Mathematische Abhandlungen I.* Verlag Birkhäuser, Basel, Switzerland, 1950. pages 209-212.

[59] C. L. Sheng. *Threshold Logic*. Academic Press, 1969.

[60] R. Smolensky. Algebraic methods in the theory of lower bounds for boolean circuit complexity. In *19th ACM Symposium on Theory of Computing*, pages 77–82, 1987.

[61] E. Stiemke. Über positive lösungen homogener linearer gleichungen. *Mathematische Annalen*, 76:340–342, 1915.

[62] C. Wang and A.C. Williams. The threshold order of a boolean function. *Discrete Applied Mathematics*, 31:51–69, 1991.

[63] A. C. Yao. Separating the polynomial-time hierarchy by oracles. In *31st IEEE Symposium on Foundations of Computer Science*, pages 1–10, 1985.

Combinatorial Designs and Cryptography

D. R. Stinson

Computer Science and Engineering Department
and Center for Communication and Information Science
University of Nebraska
Lincoln, NE 68588-0115, U.S.A.
stinson@bibd.unl.edu

Abstract

In this expository paper, we describe several applications of combinatorics (in particular, combinatorial designs) to cryptography. We look at four areas in cryptography: secrecy codes, authentication codes, secret sharing schemes, and resilient functions. In each of these areas, we find that combinatorial structures arise in a natural and essential way, and we present several examples of combinatorial characterizations of cryptographic objects.

1 Introduction

Recent years have seen numerous interesting applications of combinatorics to cryptography. In particular, combinatorial designs have played an important role in the study of such topics in cryptography as secrecy and authentication codes, secret sharing schemes, and resilient functions. The purpose of this paper is to elucidate some of these connections. This is not intended to be an exhaustive survey, but rather a sampling of some research topics in which I have a personal interest.

Some other related surveys include applications of error-correcting codes to cryptography (Sloane [60]), applications of finite geometry to crpytography (Beutelspacher [6]) and applications of combinatorial designs to computer science (Colbourn and Van Oorschot [16]).

In this introduction we will define the relevant concepts from design theory that we will have occasion to use. Beth, Jungnickel and Lenz [4] is a good general reference on design theory.

Let $1 \le t \le k < v$. An $S_\lambda(t, k, v)$ is defined to be a pair $(X, \mathcal{B})$, where X is a v−set (of *points*) and $\mathcal{B}$ is a collection of k−subsets of X (called *blocks*), such that every t−subset of points occurs in exactly λ blocks. An $S_\lambda(t, k, v)$ is called *simple* if no block occurs more than once in $\mathcal{B}$. In this paper we will restrict our attention to simple $S_\lambda(t,k,v)$. If $\lambda = 1$, then we write $S(t,k,v)$. Note that any $S(t,k,v)$ is simple.

The set of all k−subsets of a v−set is the same thing as an $S(k,k,v)$. A *large set* of $S_\lambda(t,k,v)$ is defined to be a partition of an $S(k,k,v)$ into $S_\lambda(t,k,v)$. The number of designs in the large set is $\binom{v-t}{k-t}/\lambda$.

More generally, we will require (simple) $S_\lambda(t,k,v)$ that can be partitioned into $S_{\lambda'}(t',k,v)$ for some $t' < t$. In Section 4, we will be using $S(t,k,v)$ that can be partitioned into $S(t-1,k,v)$.

An *orthogonal array* $OA_\lambda(t,k,v)$ is a $\lambda v^t \times k$ array of v symbols, such that in any t columns of the array every one of the possible v^t ordered pairs of symbols occurs in exactly λ rows. If $\lambda = 1$, then we write $OA(t,k,v)$.

A *perpendicular array* $PA_\lambda(t,k,v)$ is a $\lambda\binom{v}{t} \times k$ array of v symbols, such that each row contains k distinct symbols, and in any t columns of the array every one of the possible $\binom{v}{t}$ unordered pairs of symbols occurs in exactly λ rows. If $\lambda = 1$, then we write $PA(t,k,v)$. Observe that a $PA(1,v,v)$ is just a Latin square of order v.

An orthogonal or perpendicular array is said to be *rowwise simple* if no two rows are identical. Of course, an array with $\lambda = 1$ is rowwise simple. In this paper, we consider only rowwise simple arrays.

A *large set* of orthogonal arrays $OA_\lambda(t,k,v)$ is defined to be a set of v^{k-t}/λ rowwise simple arrays $OA_\lambda(t,k,v)$ such that every possible k−tuple of symbols occurs in exactly one of the OA's in the set. (Equivalently, the union of the OA's forms an $OA(k,k,v)$.)

2 Secrecy codes

One fundamental objective of cryptography is to enable two people, usually referred to as Alice and Bob, to communicate over an insecure channel in such a way that an observer, Oscar, cannot understand what is being said. Alice does this by encrypting the plaintext message, using a predetermined key, and sending the resulting ciphertext over the channel. Oscar, upon seeing the ciphertext, cannot determine what the plaintext was; but Bob, who knows the encryption key, can decrypt the ciphertext and reconstruct the plaintext.

This is described more formally using the following mathematical notation. A *cryptosystem* is a five-tuple $(\mathcal{P},\mathcal{C},\mathcal{K},\mathcal{E},\mathcal{D})$, where the following conditions are satisfied:

1. $\mathcal{P}$ is a finite set of possible *plaintexts*

2. $\mathcal{C}$ is a finite set of possible *ciphertexts*

3. $\mathcal{K}$, the *keyspace*, is a finite set of possible *keys*

4. For each $K \in \mathcal{K}$, there is an *encryption rule* $e_K \in \mathcal{E}$ and a corresponding *decryption rule* $d_K \in \mathcal{D}$. Each $e_K : \mathcal{P} \to \mathcal{C}$ and $d_K : \mathcal{C} \to \mathcal{P}$ such that $d_K(e_K(x)) = x$ for every plaintext $x \in \mathcal{P}$.

Alice and Bob will follow the following protocol. First, they jointly choose a random key $K \in \mathcal{K}$. This is done in secret. At a later time, suppose Alice wants to communicate a symbol $x \in \mathcal{P}$ to Bob over an insecure channel. The plaintext x is encrypted using the encryption rule e_K. Hence, Alice computes $y = e_K(x)$, and y is sent over the channel. When Bob receives y, he decrypts it using the decryption function d_K, obtaining x.

Clearly, it must be the case that each encryption function e_K is an injection (i.e. one-to-one), for otherwise, decryption could not be accomplished in an unambiguous manner.

What we will do in this section is to give a short exposition of Shannon's work on perfect secrecy, which can be found in his 1949 paper [49]. This work concerns cryptosystems that are unconditionally secure. This means that the cryptosystem cannot be broken, even with infinite computational resources. Shannon's work is probably the first example where combinatorial designs (in this case, Latin squares) play an essential role in cryptography.

One famous example of an unconditionally secure cryptosystem is the *one-time pad* of Vernam [76], which was described in the mid-1920's. Let n be an integer, and take $\mathcal{P} = \mathcal{C} = \mathcal{K} = (\mathbf{Z}_2)^n$. Each possible key is chosen with probability $1/2^n$. For $K \in (\mathbf{Z}_2)^n$, define $e_K(x)$ to be the vector sum modulo 2 of K and x (or, equivalently, the exclusive-or of the two associated bit strings).

It was conjectured for many years that the Vernam cryptosystem was unbreakable, but the first rigorous proof was given by Shannon. We shall see that the proof uses only elementary facts concerning probability. Perhaps Shannon's real insight was to establish the appropriate mathematical model in order that the proof could be accomplished.

Suppose $\mathbf{X}$ and $\mathbf{Y}$ are random variables. The *joint probability* $p(x, y)$ is the probability that $\mathbf{X}$ takes on the value x and $\mathbf{Y}$ takes on the value y. The *conditional probability* $p(x|y)$ denotes the probability that $\mathbf{X}$ takes on the value x given that $\mathbf{Y}$ takes on the value y. Joint probability can be related to conditional probability by the formula

$$p(x, y) = p(x|y)p(y).$$

Interchanging x and y, we have that

$$p(x, y) = p(y|x)p(x).$$

From these two expressions, we immediately obtain Bayes' Theorem, which states that

$$p(x|y) = \frac{p(x)p(y|x)}{p(y)}$$

provided $p(y) > 0$.

For the time being, we assume that a particular key is used for only one encryption. Let us suppose that there is a probability distribution on $\mathcal{P}$. We denote the *a priori* probability that plaintext x occurs by $p_{\mathcal{P}}(x)$. We also assume that the key K is chosen (by Alice and Bob) using some fixed probability distribution (usually a key is chosen at random, so all keys will be equiprobable, but this need not be the case). Denote the probability that key K is chosen by $p_{\mathcal{K}}(K)$. Recall that the key is chosen before Alice knows what the plaintext will be. Hence, we make the reasonable assumption that the key K and the plaintext x are independent events.

The two probability distributions on $\mathcal{P}$ and $\mathcal{K}$ induce a probability distribution on $\mathcal{C}$. Indeed, it is not hard to compute the probability $p_{\mathcal{C}}(y)$ that y is the ciphertext that is transmitted. For a key $K \in \mathcal{K}$, define

$$C(K) = \{e_K(x) : x \in \mathcal{P}\}.$$

That is, $C(K)$ represents the set of possible ciphertexts if K is the key. Then, for every $y \in \mathcal{C}$, we have that

$$p_{\mathcal{C}}(y) = \sum_{\{K: y \in C(K)\}} p_{\mathcal{K}}(K) p_{\mathcal{P}}(d_K(y)).$$

We also observe that, for any $y \in \mathcal{C}$ and $x \in \mathcal{P}$, we can compute the conditional probability $p_{\mathcal{C}}(y|x)$ (i.e. the probability that y is the ciphertext, given that x is the plaintext) to be

$$p_{\mathcal{C}}(y|x) = \sum_{\{K: x = d_K(y)\}} p_{\mathcal{K}}(K).$$

It is now possible to compute the conditional probability $p_{\mathcal{P}}(x|y)$ (i.e. the probability that x is the plaintext, given that y is the ciphertext) using Bayes' Theorem. Observe that all these computations can be performed by anyone who knows the probability distributions.

We are now ready to define the concept of perfect secrecy. Informally, perfect secrecy means that Oscar can obtain no information about the plaintext by observing the ciphertext. This idea is made precise by formulating it in terms of the probability distributions we have defined. Formally, we say that a cryptosystem has *perfect secrecy* if

$$p_{\mathcal{P}}(x|y) = p_{\mathcal{P}}(x) \text{ for all } x \in \mathcal{P}, y \in \mathcal{C}.$$

That is, the *a posteriori* probability that the plaintext is x, given that the ciphertext y is observed, is identical to the *a priori* probability that the plaintext is x.

Let us investigate perfect secrecy in general, following Shannon [49]. We make the reasonable assumption that $p_{\mathcal{C}}(y) > 0$ for all $y \in \mathcal{C}$ (for if

$p_C(y) = 0$, then ciphertext y is never used and can be omitted from C). First, we observe that, using Bayes' Theorem, the condition that $p_P(x|y) = p_P(x)$ for all $x \in P$, $y \in C$ is equivalent to $p_C(y|x) = p_C(y)$ for all $x \in P$, $y \in C$.

Fix any $x \in P$. For each $y \in C$, we have $p_C(y|x) = p_C(y) > 0$. Hence, for each $y \in C$ there must be at least one key K such that $e_K(x) = y$. It follows that $|K| \geq |C|$. In any cryptosystem, we must have $|C| \geq |P|$ since each encoding rule is an injection; hence $|K| \geq |P|$. In the boundary case $|K| = |C| = |P|$, Shannon gave a nice characterization of when perfect secrecy can be obtained.

Theorem 2.1 *[49] Suppose $(P, C, K, \mathcal{E}, \mathcal{D})$ is a cryptosystem where $|K| = |C| = |P|$. Then the cryptosystem provides perfect secrecy if and only if every key is used with equal probability $1/|K|$ and for every $x \in P$, $y \in C$, there is a unique key K such that $e_K(x) = y$.*

Proof. Suppose the given cryptosystem provides perfect secrecy. As noted above, for each $x \in P$ and $y \in C$ there must be at least one key K such that $e_K(x) = y$. Since $|K| = |C|$, if we fix an x and let y vary, we see that there must be *exactly* one such key.

Now, denote $n = |K|$. Let $P = \{x_i : 1 \leq i \leq n\}$ and fix a $y \in C$. We can name the keys $K_1, K_2, \ldots, K_n$ so that $e_{K_i}(x_i) = y$, $1 \leq i \leq n$. Using Bayes' theorem,

$$
\begin{aligned}
p_P(x_i|y) &= \frac{p_C(y|x_i) p_P(x_i)}{p_C(y)} \\
&= \frac{p_K(K_i) p_P(x_i)}{p_C(y)}.
\end{aligned}
$$

Consider the perfect secrecy condition $p_P(x_i|y) = p_P(x_i)$. ¿From this, it follows that $p_K(K_i) = p_C(y)$, for $1 \leq i \leq n$. Hence, all n keys are equiprobable, as desired.

Conversely, suppose the two hypothesized conditions are satisfied. Then we have the following:

$$
\begin{aligned}
p_C(y) &= \sum_{K \in K} p_K(K) p_P(d_K(y)) \\
&= \sum_{K \in K} \frac{1}{|K|} p_P(d_K(y)) \\
&= \frac{1}{|K|}
\end{aligned}
$$

since, for fixed y, the plaintexts $d_K(y)$ comprise a permutation of P as K varies over K, and p_P is a probability distribution.

Next, $p_C(y|x) = 1/|\mathcal{K}|$ for every x, y, since the keys are equiprobable and for every x, y there is a unique key K such that $e_K(x) = y$. Now, using Bayes' Theorem, it is trivial to compute

$$
\begin{aligned}
p_P(x|y) &= \frac{p_P(x) p_C(y|x)}{p_C(y)} \\
&= \frac{p_P(x) \frac{1}{|\mathcal{K}|}}{\frac{1}{|\mathcal{K}|}} \\
&= p_P(x),
\end{aligned}
$$

so we have perfect secrecy.   □

Suppose we define the *encryption matrix* of a cryptosystem to be the $|\mathcal{K}| \times |\mathcal{P}|$ matrix in which the entry in row K ($K \in \mathcal{K}$) and column x ($x \in \mathcal{P}$) is $e_k(x)$. Then the second condition of Theorem 2.1 can be restated in a more combinatorial way as requiring that the encryption matrix be a Latin square. It is easy to see that the encryption matrix of the Vernam Cryptosystem is a Latin square, and hence this cryptosystem achieves perfect secrecy.

2.1 t−fold secrecy

In the previous section, we assumed that each key is used to encrypt only one plaintext. It is interesting to generalize the concept of perfect secrecy to the situation where a key is used to encrypt up to t plaintexts, where $t \geq 1$ is an integer. When $t > 1$, there are several reasonable ways to define perfect t−fold secrecy (see Godlewski and Mitchell [24], for example). Here, we use the definitions from Stinson [63]. Most of the results in this section were established independently in [63] and [24].

We consider the effect of encrypting several plaintexts using the same key K. Suppose we fix an integer t and encrypt the plaintexts in X, where $|X| \leq t$, obtaining a set of ciphertexts, Y. An opponent who sees the set of ciphertexts, Y, will try to compute some information regarding the set of plaintexts, X. Our goal is that he should not be able to do so.

As before, we use probability distributions. For any $s \leq t$, suppose there is a probability distribution $p_{\binom{\mathcal{P}}{s}}$ on the set of s−subsets of $\mathcal{P}$. Also, suppose there is a probability distribution p_K on $\mathcal{K}$. Then, for each $s \leq t$, a probability distribution $p_{\binom{\mathcal{P}}{s}}$ is induced on the set of s−subsets of C. We define the following property $P(s)$:

$$
p_{\binom{\mathcal{P}}{s}}(X|Y) = p_{\binom{\mathcal{P}}{s}}(X) \text{ for every } X \subseteq \mathcal{P} \text{ and } Y \subseteq C \text{ such that } \\
|X| = |Y| = s.
$$

The following definition generalizes the concept of perfect secrecy: A cryptosystem is said to provide *perfect t−fold secrecy* provided that $P(s)$

is satisfied for $1 \leq s \leq t$. Observe that this definition coincides with the definition of perfect secrecy when $t = 1$.

Perfect t–fold secrecy means that observation of a set of at most t ciphertexts gives no information about the corresponding set of t plaintexts. Note that we are considering only a set-wise correspondence of plaintexts and ciphertexts. For example, in the case $t = 2$, it may happen that there exist plaintexts x_1, x_2 and ciphertexts y_1, y_2 such that there is no key K with $e_K(x_1) = y_1$ and $e_K(x_2) = y_2$. This in itself would not rule out the possibility that the cryptosystem has 2–fold secrecy.

Suppose we use a perpendicular array $PA_\lambda(t, k, v)$ as the encryption matrix of a cryptosystem. Choose each of the $\lambda\binom{v}{t}$ encryption rules with equal probability. Then it is not difficult to see that property $P(t)$ is satisfied, but we need property $P(s)$ to be satisfied for all $1 \leq s \leq t$. In other words, we desire that our $PA_\lambda(t, k, v)$ is also a $PA_{\lambda(s)}(s, k, v)$ for $1 \leq s \leq t$. Such a perpendicular array is called *inductive*. Simple counting shows that

$$\lambda(s) = \frac{\lambda\binom{v-s}{t-s}}{\binom{t}{s}}$$

in an inductive PA.

It is proved in [31] and [24] that any $PA_\lambda(t, k, v)$ is inductive provided $k \geq 2t - 1$. However, as noted in [8], the proof given in [31] is incomplete. Here we give a complete proof, which follows from the following lemma.

Lemma 2.2 *Suppose $t \geq s$ and $\binom{k}{t} \geq \binom{k}{s}$. Then any perpendicular array $PA_\lambda(t, k, v)$ is also a $PA_{\lambda(s)}(s, k, v)$, where $\lambda(s)$ is as above.*

Proof. Let A be a $PA_\lambda(t, k, v)$ and name the columns $1, \ldots, k$. Let Y be any subset of s symbols. For any set I of s columns, denote by $n(I)$ the number of rows in A in which the symbols in Y are contained in the columns in I. Then, for any set J of t columns, we obtain an equation

$$\sum_{I \subseteq J, |I| = s} n(I) = \lambda\binom{v-s}{t-s}.$$

In this way we get $\binom{k}{t}$ linear equations in the $\binom{k}{s}$ unknowns $n(I)$. The coefficient matrix of this system is the $\binom{k}{t} \times \binom{k}{s}$ matrix, where the rows and columns are respectively indexed by the t–subsets and s–subset of $\{1, \ldots, k\}$, and the entry in row J and column I is

$$h_{JI} = \begin{cases} 1 & \text{if } I \subseteq J \\ 0 & \text{otherwise.} \end{cases}$$

Wilson proves in [74, Remark 3] that the rows of the matrix (h_{JI}) span the real vector space of dimension $\binom{k}{s}$. This means that our system has at most one solution. Since

$$n(I) = \frac{\lambda \binom{v-s}{t-s}}{\binom{t}{s}}$$

is indeed a solution of the system, it is the only solution, and the desired result is proved.   □

Remark. If $t \geq s$, the condition $\binom{k}{t} \geq \binom{k}{s}$ is equivalent to $s + t \leq k$.

It is easy to see that $k \geq 2t - 1$ is sufficient to ensure that the hypotheses of Lemma 2.2 are satisfied for $1 \leq s \leq t$. Hence, a cryptosystem is obtained that provides perfect t-fold secrecy. The following theorem gives the converse result.

Theorem 2.3 *[63, 24] Suppose a cryptosystem provides perfect t-fold secrecy, and $|\mathcal{P}| \geq 2t - 1$. Then $|\mathcal{K}| \geq \binom{|\mathcal{P}|}{t}$, and equality occurs if and only if the encryption matrix is a perpendicular array $PA(t, |\mathcal{P}|, |\mathcal{P}|)$ and the encryption rules are used with equal probability.*

Note also that Lemma 2.2 yields necessary conditions for the existence of $PA_\lambda(t, k, v)$ with $k \geq 2t - 1$, since the quantities $\lambda(s)$ must be integers. So we obtain the following necessary numerical conditions, which were first given in [31].

Theorem 2.4 *Suppose $k \geq 2t - 1$ and a $PA_\lambda(t, k, v)$ exists. Then*

$$\lambda \binom{v-s}{t-s} \equiv 0 \bmod \binom{t}{s}$$

for $1 \leq s \leq t$.

Theorem 2.3 motivates the study of $PA(t, v, v)$ with $v \geq 2t - 1$. For $t = 1$, such an array is the same thing as a Latin square. For $t = 2$, the necessary condition from Theorem 2.4 (for $v \geq 3$) is that v is odd. A $PA(2, q, q)$ is shown to exist for any odd prime power q by Mullin *et al* in [37]. For $t = 3$, the necessary condition (for $v \geq 5$) is $v \equiv 2 \bmod 3$. However, for $t \geq 3$, the only known examples of $PA(t, v, v)$ are $PA(3, 8, 8)$ and $PA(3, 32, 32)$ (see [67, 63]). These latter two examples arise (respectively) from the 3-homogeneous groups $AGL(1, 8)$ and $A\Gamma L(1, 32)$.

If a $PA(t, v, v)$ is not known to exist, the natural approach is to look for a $PA_\lambda(t, v, v)$ with a small value of λ. In the corresponding cryptosystem, the number of encryption rules is λ times the lower bound of Theorem 2.3. The following perpendicular arrays with "small" λ are known to exist:

1. A $PA_2(2, 6, 6)$ exists [8] and a $PA_2(2, 2^n, 2^n)$ exists for any n [63].

2. A $PA_3(3,9,9)$ exists [8] and a $PA_3(3,q+1,q+1)$ exists for any prime power $q \equiv 3 \bmod 4$ [67, 63].

3. A $PA_4(4,9,9)$ and a $PA_4(4,33,33)$ both exist [67, 63].

3   Authentication codes

Whereas a secrecy code protects the secrecy of information, an authentication code protects the integrity. That is, when Alice receives a message from Bob, she wants to be sure that the message was really sent by Bob and was not tampered with.

Authentication codes were invented in 1974 by Gilbert, MacWilliams and Sloane [23], and have been extensively studied in recent years. The general theory of unconditional authentication was developed by Simmons (see e.g. [52]). Recent research has involved considerable use of combinatorial methods, which we will survey in this section. For a general survey on authentication, we recommend Simmons [58].

3.1   Authentication without secrecy

We will first discuss authentication without secrecy. As in the previous section, there are three participants: Alice, Bob, and an opponent (Oscar). Alice wants to communicate some information to Bob using a public communication channel. However, Oscar has the ability to introduce messages into the channel and/or to modify existing messages. When Oscar places a (new) message into the channel, this is called *impersonation*. When Oscar sees a message m and changes it to a message $m' \neq m$, this is called *substitution*.

We will use the following notation. An *authentication code without secrecy* is a four-tuple $(\mathcal{S}, \mathcal{A}, \mathcal{K}, \mathcal{E})$, defined as follows:

1. $\mathcal{S}$ is a finite set of *source states* (a source state is analogous to a plaintext)

2. $\mathcal{A}$ is a finite set of *authenticators*

3. $\mathcal{K}$ is a finite set of *keys*

4. For each $K \in \mathcal{K}$, there is an *authentication rule* $e_K \in \mathcal{E}$. Each $e_K : \mathcal{S} \to \mathcal{A}$.

Alice and Bob will follow the following protocol. As with a secrecy code, they jointly choose a secret key $K \in \mathcal{K}$. At a later time, suppose Alice wants to communicate a source state $s \in \mathcal{S}$ to Bob. Alice uses the authentication rule e_K to produce the authenticator $a = e_K(s)$. The *message* $m = (s, a)$ is sent over the channel. When Bob receives m, he verifies that $a = e_K(s)$ to authenticate the source state s.

When Oscar performs impersonation or substitution, his goal is to have his bogus message m' accepted as authentic by Bob, thus misleading Bob as to the state of the source. That is, if e is the authentication rule being used (which is *not* known to Oscar), then he is hoping that $m' = e(s)$ for some source state s.

We assume that there is some probability distribution on $\mathcal{S}$, which is known to all the participants. Given this probability distribution, Alice and Bob will choose a probability distribution for $\mathcal{E}$, called an *authentication strategy*. Once Alice and Bob have chosen their authentication strategy, it is possible to compute, for $i = 0, 1$, a *deception probability* denoted P_{d_i}, which is the probability that Oscar can deceive Bob by impersonation and substitution, respectively. In computing the deception probabilities, we assume that Oscar is using an optimal strategy.

P_{d_0} is calculated as follows. Oscar can compute, for each possible message $m \in \mathcal{M} = \mathcal{S} \times \mathcal{A}$, a quantity $payoff(m)$ which denotes the probability that m will be accepted as authentic by Bob. Oscar wishes to maximize his probability of deceiving Bob, so

$$P_{d_0} = \max\{payoff(m) : m \in \mathcal{M}\}.$$

It is easy to see that

$$payoff(s, a) = \sum_{\{e \in \mathcal{E} : e(s) = a\}} p_{\mathcal{E}}(e).$$

Then it follows that

$$\sum_{m \in \mathcal{M}} payoff(m) = |\mathcal{S}|.$$

Hence, there exists a particular message $m_0 \in \mathcal{M}$ such that

$$payoff(m_0) \geq \frac{|\mathcal{S}|}{|\mathcal{M}|} = \frac{1}{|\mathcal{A}|}.$$

Thus $P_{d_0} \geq 1/|\mathcal{A}|$, and $P_{d_0} = 1/|\mathcal{A}|$ if and only if $payoff(m) = 1/|\mathcal{A}|$ for every message m. We summarize this as follows.

Theorem 3.1 *[51] $P_{d_0} \geq 1/|\mathcal{A}|$. Further, $P_{d_0} = 1/|\mathcal{A}|$ if and only if*

$$\sum_{\{e \in \mathcal{E} : e(s) = a\}} p_{\mathcal{E}}(e) = \frac{1}{|\mathcal{A}|}$$

for every message (s, a).

We now turn to the computation of P_{d_1}. Suppose Oscar sees the message $m = (s, a)$ in the channel. He will substitute this message with a message $m' = (s', a')$, where $s' \neq s$. Denote by $payoff(m, m')$ the probability that the message m' will be accepted as authentic, given that m is observed in the channel. For each $m \in \mathcal{M}$, Oscar will find the message $m' = f(m)$ that maximizes $payoff(m, m')$. Then P_{d_1} is computed as the weighted average of these payoffs:

$$P_{d_1} = \sum_{m \in \mathcal{M}} p_{\mathcal{M}}(m) payoff(m, f(m)),$$

where the probability distribution $p_{\mathcal{M}}$ is induced by the probability distributions p_S and $p_{\mathcal{E}}$.

Denote by e_0 the authentication rule being used by Alice and Bob, and let $m = (s, a)$ be the message observed by Oscar in the channel. Then we have the following:

$$
\begin{aligned}
payoff(m, m') &= p(a' = e_0(s') | a = e_0(s)) \\
&= \frac{p(a' = e_0(s') \wedge a = e_0(s))}{p(a = e_0(s))} \\
&= \frac{\sum_{\{e : e(s) = a, e(s') = a'\}} p_{\mathcal{E}}(e)}{\sum_{\{e : e(s) = a\}} p_{\mathcal{E}}(e)}.
\end{aligned}
$$

Now, it follows that

$$\sum_{a' \in \mathcal{A}} payoff(m, (s', a')) = 1$$

for any $s' \neq s$. Hence, for every s', there exists an authenticator a' such that $payoff(m, (s', a')) \geq 1/|\mathcal{A}|$. Hence it follows that $P_{d_1} \geq 1/|\mathcal{A}|$, and $P_{d_1} = 1/|\mathcal{A}|$ if and only if $payoff(m, m') = 1/|\mathcal{A}|$ for every $m' = (s', a')$ with $s' \neq s$.

Combining the above discussion with Theorem 3.1, the following is straightforward.

Theorem 3.2 *[63] Suppose $(\mathcal{S}, \mathcal{A}, \mathcal{K}, \mathcal{E})$ is an authentication code without secrecy. Then $P_{d_0} = P_{d_1} = 1/|\mathcal{A}|$ if and only if*

$$\sum_{\{e : e(s) = a, e(s') = a'\}} p_{\mathcal{E}}(e) = \frac{1}{|\mathcal{A}|^2}$$

for every $s, s', a, a', s \neq s'$.

These results also lead to a lower bound on the number of authentication rules and a characterization as to when equality can occur.

Theorem 3.3 *[63] Suppose we have an authentication code without secrecy in which* $P_{d_0} = P_{d_1} = 1/|\mathcal{A}|$. *Then* $|\mathcal{E}| \geq |\mathcal{A}|^2$, *and equality occurs if and only if the authentication matrix is an orthogonal array* $OA(2, |\mathcal{S}|, |\mathcal{A}|)$ *and the authentication rules are used with equal probability.*

Proof. Suppose $P_{d_0} = P_{d_1} = 1/|\mathcal{A}|$. Let $s \neq s'$. Then, for every a, a',

$$|\{e : e(s) = a, e(s') = a'\}| \geq 1.$$

Hence, $|\mathcal{E}| \geq |\mathcal{A}|^2$.

In order that $|\mathcal{E}| = |\mathcal{A}|^2$, it must be the case that $|\{e : e(s) = a, e(s') = a'\}| = 1$ for every s, s', a, a', where $s \neq s'$; and $p_{\mathcal{E}}(e) = 1/|\mathcal{A}|^2$ for every $e \in \mathcal{E}$. The authentication matrix is clearly an orthogonal array $OA(2, |\mathcal{S}|, |\mathcal{A}|)$.

Conversely, suppose we start with an orthogonal array $OA(2, |\mathcal{S}|, |\mathcal{A}|)$. Use each row as an authentication rule with equal probability $1/|\mathcal{A}|^2$. Then we obtain a code with the stated properties from Theorem 3.2.   □

Remark. An equivalent result was shown independently by De Soete, Vedder and Walker in [19] under the assumption that authentication rules are chosen equiprobably.

Theorem 3.3 provides a nice characterization, as far as it goes. However, existence of an $OA(2, k, v)$ requires that $k \leq v + 1$. So the characterization applies only in the situation where $|\mathcal{S}| \leq |\mathcal{A}| + 1$. It is also of interest to characterize authentication codes in the situation where $|\mathcal{S}| > |\mathcal{A}| + 1$. In fact, this situation is more likely to arise in practice, for typically one would want to authenticate a "long" source state with a relatively "short" authenticator.

The second characterization involves orthogonal arrays $OA_\lambda(2, k, v)$ $\lambda > 1$. First, let us observe that we can use an $OA_\lambda(2, k, v)$ to construct an authentication code with $|\mathcal{S}| = k$, $|\mathcal{A}| = v$, and $|\mathcal{E}| = \lambda v^2$. If the encoding rules are used equiprobably, then $P_{d_0} = P_{d_1} = 1/|\mathcal{A}|$ follows from Theorem 3.2. (This construction was first given by Stinson in [61].)

Our objective is to minimize $|\mathcal{E}|$. If we are using an orthogonal array, then we want to minimize λ, given k and v. The well-known Plackett-Burman bound [38], proved in 1945, states that

$$\lambda \geq \frac{k(v-1)+1}{v^2}$$

if an $OA_\lambda(2, k, v)$ exists. The resulting authentication code will have

$$|\mathcal{E}| \geq |\mathcal{S}|(|\mathcal{A}| - 1) + 1.$$

But this does not yet rule out the possibility of constructing an authentication code with the same deception probabilities and fewer authentication rules by some other method. However, this is shown to be impossible by Stinson in [64], where the following result is proved.

Theorem 3.4 *[64] Suppose we have an authentication code without secrecy in which $P_{d_0} = P_{d_1} = 1/|\mathcal{A}|$. Then $|\mathcal{E}| \geq |\mathcal{S}|(|\mathcal{A}| - 1) + 1$, and equality occurs if and only if the authentication matrix is an orthogonal array $OA_\lambda(2, |\mathcal{S}|, |\mathcal{A}|)$ where*

$$\lambda = \frac{|\mathcal{S}|(|\mathcal{A}| - 1) + 1}{|\mathcal{A}|^2},$$

and the authentication rules are used with equal probability.

Remark. It is not difficult to construct orthogonal arrays that meet the Plackett-Burman bound. In 1946, Rao showed in [40] that there exists an $OA_{q^{d-2}}(2, (q^d - 1)/(q - 1), q)$ for any prime power q and any integer $d \geq 2$. In fact, this class of OA's is equivalent to the well-known Hamming codes discovered a few years later.

The proof of Theorem 3.4 is more difficult than that of Theorem 3.3, and uses a linear-algebraic technique.

Another orthogonal array characterization is obtained when studying the so-called entropy bounds for authentication codes. If $\mathbf{X}$ is a random variable that takes on finitely many possible values, then the *entropy* of the probability distribution p is defined to be

$$H(\mathbf{X}) = - \sum_{\{x:p(x)>0\}} p(x) \log_2 p(x).$$

Suppose also that $\mathbf{Y}$ is a random variable that takes on finitely many possible values, and for any y we have a conditional probability distribution $p(x|Y = y)$. The entropy of this probability distribution is

$$H(\mathbf{X}|y) = - \sum_{\{x:p(x|Y=y)>0\}} p(x|Y = y) \log_2 p(x|Y = y).$$

The *conditional entropy* $H(\mathbf{X}|\mathbf{Y})$ is defined to be the weighted average of the entropies $H(\mathbf{X}|y)$:

$$H(\mathbf{X}|\mathbf{Y}) = - \sum_{\{y:p(y)>0\}} \sum_{\{x:p(x|Y=y)>0\}} p(y)p(x|Y = y) \log_2 p(x|Y = y).$$

For an authentication code $(\mathcal{S}, \mathcal{A}, \mathcal{K}, \mathcal{E})$, we have probability distributions on $\mathcal{S}$ and $\mathcal{E}$. Together these induce a probability distribution on $\mathcal{M}$. Here is a combinatorial characterization of authentication codes involving entropy bounds, due to Brickell:

Theorem 3.5 *[10] $P_{d_0} \geq 2^{H(\mathbf{E}|\mathbf{M})-H(\mathbf{E})}$ and $P_{d_1} \geq 2^{-H(\mathbf{E}|\mathbf{M})}$. Further, we have equality in both inequalities if and only if the authentication matrix is an $OA(2, |\mathcal{S}|, |\mathcal{A}|)$ and encoding rules are used equiprobably.*

As with many of the other characterizations, it is easier to prove the "if" part than the "only if" part. Suppose the authentication matrix is an $OA(2, |\mathcal{S}|, |\mathcal{A}|)$ and encoding rules are used equiprobably. Then $P_{d_0} = P_{d_1} = 1/|\mathcal{A}|$. Now, $H(\mathbf{E}) = \log_2 |\mathcal{E}| = \log_2 |\mathcal{A}|^2$, and $H(\mathbf{E}|\mathbf{M}) = \log_2 |\mathcal{A}|$, so both inequalities are met with equality.

3.2   Other types of authentication codes

In the last section we have focused on authentication codes without secrecy. Many other variations have been studied in the literature. We will briefly mention some of these, and the combinatorial designs that are involved.

General authentication codes

Here we have an arbitrary message space $\mathcal{M}$, so a message does not consist of a source state with an appended authenticator. We do not care whether or not a message reveals the source state. The bounds on the deception probabilities are different from the authentication without secrecy case. It is fairly straightforward to show that

$$P_{d_0} \geq \frac{|\mathcal{S}|}{|\mathcal{M}|}$$

and

$$P_{d_1} \geq \frac{|\mathcal{S}| - 1}{|\mathcal{M}| - 1}$$

(see [51, 35]).

Codes satisfying these bounds with equality and having the minimum number of encoding rules have been characterized in terms of designs $S(2, k, v)$:

Theorem 3.6 *[64] Suppose $P_{d_0} = |\mathcal{S}|/|\mathcal{M}|$ and $P_{d_1} = (|\mathcal{S}| - 1)/(|\mathcal{M}| - 1)$. Then*

$$|\mathcal{E}| \geq \frac{|\mathcal{S}|(|\mathcal{S}| - 1)}{|\mathcal{M}|(|\mathcal{M}| - 1)},$$

and equality occurs if and only if the encryption matrix is an $S(2, |\mathcal{S}|, |\mathcal{M}|)$, encoding rules are used equiprobably, and source states occur equiprobably.

Authentication codes with secrecy

Suppose we desire perfect secrecy, as in a secrecy code, in addition to the authentication capability. The only difference between this situation and the situation of general authentication codes is that the numerical condition

$$|\mathcal{M}| - 1 \equiv 0 \bmod |\mathcal{S}|(|\mathcal{S}| - 1)$$

must also be satisfied [63, Theorem 6.4] (in fact, this condition relates to finding suitable systems of distinct representatives).

Authentication codes with splitting

The term *splitting* refers to the situation where there can be more than one message encrypting a particular source state under one encoding rule (so encryption is non-deterministic). It seems that no characterization theorems of the type we have been discussing have been proved, but several constructions for codes with splitting that employ designs have been given in the literature. The following types of designs have been used: orthogonal multi-arrays [10]; partial geometries [21]; affine resolvable $S(2, k, v)$ [21]; and mutually orthogonal F−squares [1].

t−fold secure authentication codes

As was done with secrecy codes, we could use one encoding rule for the transmission of several source states. Suppose Oscar observes i messages in the channel, which have been encrypted using the same encoding rule, and then he inserts a new message of his own choosing, hoping to have it accepted by Bob as authentic. This was termed *spoofing* of *order i* in [35]. (Observe that impersonation is spoofing of order 0 and substitution is spoofing of order 1.) The deception probability P_{d_i} is the probability that Oscar will succeed in deceiving Bob with an order i spoofing attack. A code is t−*fold secure against spoofing* if P_{d_i} is the minimum possible, for $0 \leq i \leq t$.

We can consider several different possibilities regarding the secrecy of an authentication code that is t−fold secure against spoofing:

1. authentication without secrecy

2. general authentication (no secrecy requirement)

3. t'−fold secrecy (the case that has been most studied is $t' = t + 1$).

We briefly summarize combinatorial constructions and characterizations in each of these cases.

In the case of authentication without secrecy, it can be shown that $P_{d_i} \geq 1/|\mathcal{A}|$ for $0 \leq i \leq t$ (see, for example, [63]). So a code will be t−fold secure against spoofing if $P_{d_i} = 1/|\mathcal{A}|$ for $0 \leq i \leq t$. Theorem 3.3 can be generalized in a fairly straightforward way, as follows:

Theorem 3.7 *[63] Suppose we have an authentication code without secrecy that is t−fold secure against spoofing. Then $|\mathcal{E}| \geq |\mathcal{A}|^t$, and equality occurs if and only if the authentication matrix is an orthogonal array $OA(t, |\mathcal{S}|, |\mathcal{A}|)$ and the authentication rules are used with equal probability.*

Research has also been done concerning entropy bounds. The bounds of Theorem 3.5 were generalized by Walker [73], who proved that

$$P_{d_i} \geq 2^{H(\mathbf{E}|\mathbf{M}^{i+1}) - H(\mathbf{E}|\mathbf{M}^i)}$$

for any i. It then follows immediately that

$$\frac{\sum_{i=0}^{t} P_{d_i}}{t+1} \geq |\mathcal{E}|^{-1/(t+1)}.$$

The case of equality is characterized by Mitchell, Walker and Wild in [36] in terms of orthogonal arrays $OA(t, |\mathcal{S}|, |\mathcal{A}|)$.

For general authentication codes, the deception probabilities are bounded below according to the formula

$$P_{d_i} \geq \frac{|\mathcal{S}| - i}{|\mathcal{M}| - i}$$

(this was first shown by Massey [35]). A code of this type will be t−fold secure against spoofing if

$$P_{d_i} = \frac{|\mathcal{S}| - i}{|\mathcal{M}| - i}$$

for $0 \leq i \leq t$.

The following is a combination of results from [35], [46] and [17].

Theorem 3.8 *Suppose an authentication code is t−fold secure against spoofing. Then*

$$|\mathcal{E}| \geq \frac{\binom{|\mathcal{S}|}{t+1}}{\binom{|\mathcal{M}|}{t+1}},$$

and equality occurs only if the encryption matrix is an $S(t+1, |\mathcal{S}|, |\mathcal{M}|)$. Conversely, if there exists an $S(t+1, k, v)$, then there exists an authentication code for k equiprobable source states, having v possible messages, and using $\binom{v}{t+1}/\binom{k}{t+1}$ equiprobable encoding rules, that is t−fold secure against spoofing.

Remark. In contrast to Theorem 3.4, this is not stated as an "if and only if" result, since the question of whether the source states and encoding rules are necessarily equiprobable is not yet resolved.

We should also mention that Walker's entropy bound, mentioned above, was generalized by Rosenbaum [43] to the general case (including splitting).

If we desire $(t+1)$−fold secrecy in addition to t−fold security against spoofing, the main tool is a special type of perpendicular array called an *authentication perpendicular arrays*. We have already mentioned that a $PA_\lambda(t+1, k, v)$ yields a code that provides $(t+1)$−fold secrecy; the question is how to ensure that the code will also be t−fold secure against spoofing.

The following property was defined in [67], generalizing the definition from [62] which corresponds to the case $t = 2$: an $APA_\lambda(t, k, v)$, say A, is a

$PA_\lambda(t, k, v)$ such that, for any $s \leq t - 1$ and for any $s + 1$ distinct symbols x_i $(1 \leq i \leq s + 1)$, we have that among the rows of A that contain all the symbols x_i $(1 \leq i \leq s+1)$, the s symbols x_i $(1 \leq i \leq s)$ occur in all possible subsets of s columns equally often.

This property may seem complicated, but it is precisely what is needed for a $PA_\lambda(t + 1, k, v)$ to be t-fold secure against spoofing.

As an example to illustrate the definition, we present an $APA(2, 3, 11)$ (due to van Rees) from [62]. Construct a 55×3 array A by developing the following five rows modulo 11:

$$
\begin{array}{ccc}
0 & 1 & 2 \\
0 & 9 & 7 \\
0 & 3 & 6 \\
0 & 4 & 8 \\
0 & 5 & 10 \ .
\end{array}
$$

Any unordered pair $\{x, y\}$ will occur in three rows of A. Within these three rows, x will occur once in each of the three columns, as will y.

Authentication perpendicular arrays have been studied in several papers: [62, 67, 63, 7, 8, 72].

Authentication codes with arbitration

These codes were introduced by Simmons [54]; they include a fourth participant, the arbiter, who resolves disputes between Alice and Bob. The construction of these codes uses affine geometries.

4 Secret sharing schemes

Informally, a secret sharing scheme is a method of sharing a secret key K among a finite set of participants in such a way that certain specified subsets of participants can compute the secret key K. The value K is chosen by a special participant called the *dealer*.

We will use the following notation. Let $\mathcal{P} = \{P_i : 1 \leq i \leq w\}$ be the set of participants. The dealer is denoted by D and we assume $D \notin \mathcal{P}$. $\mathcal{K}$ is *key set* (i.e. the set of all possible keys) and $\mathcal{S}$ is the *share set* (i.e. the set of all possible shares). Let Γ be a set of subsets of $\mathcal{P}$; this is denoted mathematically by the notation $\Gamma \subseteq 2^{\mathcal{P}}$. The subsets in Γ are those subsets of participants that should be able to compute the secret. Γ is called an *access structure* and the subsets in Γ are called *authorized subsets*.

When a dealer D wants to share a secret $K \in \mathcal{K}$, he will give each participant a share from $\mathcal{S}$. The shares should be distributed secretly, so no participant knows the share given to another participant. At a later time, a subset of participants will attempt to determine K from the shares they collectively hold. We will say that a scheme is a *perfect secret sharing scheme realizing* the access structure Γ provided the following two properties are satisfied:

1. If an authorized subset of participants $B \subseteq \mathcal{P}$ pool their shares, then they can determine the value of K.

2. If an unauthorized subset of participants $B \subseteq \mathcal{P}$ pool their shares, then they can determine nothing about the value of K (in an information-theoretic sense).

If an unauthorized subset can obtain partial information regarding the secret, then the scheme is *non-perfect*. We will primarily restrict our attention to perfect schemes.

Suppose that $B \in \Gamma$, $B \subseteq C \subseteq \mathcal{P}$ and the subset C wants to determine K. Since B is an authorized subset, it can already determine K. Hence, the subset C can determine K by ignoring the shares of the participants in $C \backslash B$. Stated another way, a superset of an authorized set is again an authorized set. What this says is that the access structure should satisfy the *monotone* property:

$$\text{if } B \in \Gamma \text{ and } B \subseteq C \subseteq \mathcal{P}, \text{ then } C \in \Gamma.$$

If Γ is an access structure, then $B \in \Gamma$ is a *minimal* authorized subset if $A \notin \Gamma$ whenever $A \subseteq B$, $A \neq B$. The set of minimal authorized subsets of Γ is denoted Γ_0 and is called the *basis* of Γ.

The efficiency of a secret sharing scheme is measured by the information rate. Since the secret key K comes from a finite set $\mathcal{K}$, we can think of K as being represented by a bit-string of length $\log_2 |\mathcal{K}|$, by using a binary encoding, for example. In a similar way, a share can be represented by a bit-string of length $\log_2 |\mathcal{S}|$. The *information rate* is the ratio

$$\rho = \frac{\log_2 |\mathcal{K}|}{\log_2 |\mathcal{S}|}.$$

One of the central problems in secret sharing is to construct schemes with ρ as large as possible. For a recent survey discussing research in this area, see Stinson [65].

4.1   Ideal schemes

It is easy to prove that $\rho \leq 1$ in any (perfect) scheme. Since $\rho = 1$ is the optimal situation, we refer to such a scheme as an *ideal* scheme. Some results of Brickell and Davenport [11] and Jackson and Martin [28] show some interesting and surprising connections between ideal schemes and matroids. We will not go into the details here, as it is beyond the scope of this survey, but we will describe some of the main results in an informal way. For a general reference on matroid theory, see [77].

Brickell and Davenport showed in [11] that if a matroid $\mathcal{M}$ is co-ordinatizable over a finite field, then we could take one element x of the matroid to represent the dealer, and the remaining elements to represent participants, and obtain an ideal scheme for the access structure having as a basis the circuits of $\mathcal{M}$ through x, with x deleted. This is a fairly straightforward linear-algebraic construction.

More difficult is the partial converse also proved in [11]. This converse states that the existence of an ideal scheme for an access structure gives rise to a matroid on $X = \mathcal{P} \cup \{D\}$. The circuits of this matroid consist of subsets $B \subseteq X$ with the property that there is an $x \in B$ such that the value of the information given to x (i.e a share if $x \in \mathcal{P}$, or the key, if $x = D$) can always be deduced from the information given to $B\backslash\{x\}$.

We call this matroid the *associated matroid* for the scheme $\mathcal{F}$. If we start with a co-ordinatizable matroid $\mathcal{M}$, and construct an ideal scheme from it, then $\mathcal{M}$ is the associated matroid of the scheme.

The associated matroid is defined in terms of a given ideal scheme. Hence, if we have an access structure Γ, and we want to determine if there exists an ideal scheme realizing Γ, the Brickell-Davenport result does not tell us anything. However, Jackson and Martin [28] have shown how the associated matroid can be computed as a function of the access structure Γ only (i.e. the associated matroid doesn't depend on the particular ideal scheme realizing Γ). Given an access structure Γ, it is possible to compute a pair $(X, \mathcal{C})$ such that $\mathcal{C}$ is the set of circuits of a matroid whenever an ideal scheme realizing the access structure Γ exists. This often allows us to prove the non-existence of ideal schemes realizing certain access structures. For details of the construction, see [28] (an exposition is also given in [65]).

These results do not completely characterize ideal schemes, for it may happen that an associated matroid is not co-ordinatizable over a finite field. No conclusion can be reached in this case. The only example of this situation that has been studied is the Vamos matroid, which Seymour proves is not the associated matroid of any ideal scheme [47].

Many of the most useful direct constructions for (non-ideal) secret sharing schemes use finite geometries. This is a very active area of research in secret sharing. The most powerful general method is the Simmons geometric construction. We will not discuss this method here, but we refer the reader to the following papers: Simmons [53, 55, 56, 57], Simmons, Jackson and Martin [59], and Jackson and Martin [29, 30].

4.2 Threshold schemes

A $(t, w)-$*threshold access structure* has a basis consisting of all $t-$subsets of w participants. A *threshold scheme* is a secret sharing scheme for a threshold access structure. Threshold schemes were the first type of secret sharing scheme that were constructed — by Shamir [48] using polynomial interpolation, and by Blakley [9] using finite geometries — in 1979. In fact, the Shamir scheme is an ideal threshold scheme for any $t \leq w$.

There have been several constructions given for non-perfect threshold schemes that use combinatorial designs. Examples include Steiner systems $S(t, k, v)$ [5] and generalized quadrangles [18, 20]. However, we will restrict our attention to perfect schemes for the remainder of this section.

The first result we want to mention in this section is a combinatorial characterization of ideal threshold schemes due to Jackson and Martin [28].

Theorem 4.1 *An ideal (t, w)-threshold scheme with $|\mathcal{K}| = v$ is equivalent to an $OA(t, w + 1, v)$.*

Remark. Orthogonal arrays of this type (for arbitrarily large t) can be obtained from Bush's construction in [12].

It is easy to obtain a threshold scheme from an orthogonal array. We associate w columns of the array with the w participants, and the last column (say) with the key. If the dealer wants to share key K, then he chooses a random row of the array such that K occurs in the last column, and gives out the remaining w elements in that row as the shares. Now, any t shares uniquely determine a row of the array, and then the key is revealed as the element in the last column. But $t - 1$ shares leave the key completely undetermined, since for any hypothetical value K of the key, there is a unique row containing the $t - 1$ given shares and having K in the last column.

We now turn our attention to a special type of threshold scheme which Martin has termed "anonymous." There are two ways in which this differs from the model we have been looking at thus far:

1. In an anonymous scheme, the w participants get w distinct shares.

2. The key can be computed solely as a function of t shares, without knowledge of which participant holds which share.

The second property means that the key computation can be performed by a black box that is given t shares and does not know the identities of the participants holding those shares.

Anonymous threshold schemes were first investigated in [68]. Further results can be found in [45] and [14]. The following characterization is proved in [68].

Theorem 4.2 *In an anonymous (t, w)-threshold scheme, $|\mathcal{S}| \geq |\mathcal{K}|(w - t + 1) + t - 1$. Further, equality occurs if and only if there is an $S(t, w, v)$ that can be partitioned into $S(t - 1, w, v)$.*

Let us briefly review the known results concerning the partitionable designs required by the above theorem. An $S(2, k, v)$ that can be partitioned into $S(1, k, v)$ is known as a *resolvable* $S(2, k, v)$. The necessary condition is that $v \equiv k \bmod k(k-1)$. The necessary conditions are known to be sufficient for $k = 2$ (trivial, as a resolvable $S(2, 2, v)$ is just a one-factorization of K_v); for $k = 3$ (Ray-Chaudhuri and Wilson [42] — this is the famous Kirkman schoolgirl problem); and for $k = 4$ (Hanani, Ray-Chaudhuri and Wilson [27]). The case $k = 5$ was almost completed by Zhu, Du and Zhang, with 109 possible exceptions currently unresolved (pun intended) [78]. Greig has similarly almost finished the case $k = 8$, with 95 possible exceptions [25]. For other values $k \geq 6$, not much is known regarding the spectrum of these resolvable designs beyond the asymptotic existence result of Ray-Chaudhuri and Wilson [39].

In the case $t = 3$, $w = 3$, we are asking for a large set of $S(2, 3, v)$. These exist if and only if $v \equiv 1, 3 \bmod 6$, $v \neq 7$. This is a celebrated problem that was solved mostly by Lu [32, 33], with the final exceptions done by Teirlinck [69].

In the case $t = 3$, $w = 4$, it is known for all $n \geq 2$ that there is an $S(3, 4, 4^n)$ that can be partitioned into $S(2, 4, 4^n)$. This was proved independently by Baker [2] and Zaicev, Zinoviev and Semakov [75]. Some new infinite classes of partitionable $S(3, 4, v)$ have recently been found by Teirlinck [71]; the smallest example is an $S(3, 4, 100)$ that can be partitioned into $S(2, 4, 100)$.

As an illustration of Theorem 4.1, we show how a large set of $S(2, 3, 9)$ can be used to construct an anonymous $(3, 3)$ threshold scheme with seven possible keys. The large set consists of seven disjoint $S(2, 3, 9)$. Here is one well-known way to construct the large set: Take a set of points $X = \mathbf{Z}_7 \cup \{\infty_1, \infty_2\}$, and produce seven designs $\mathcal{B}_i$ ($0 \leq i \leq 6$) by developing the following "base design" $\mathcal{B}_0$ modulo 7:

$$\begin{array}{llll}
\{\infty_1, \infty_2, 0\} & \{0, 1, 6\} & \{\infty_1, 1, 5\} & \{\infty_2, 3, 1\} \\
\{1, 2, 4\} & \{0, 2, 5\} & \{\infty_1, 2, 3\} & \{\infty_2, 6, 2\} \\
\{3, 5, 6\} & \{0, 3, 4\} & \{\infty_1, 4, 6\} & \{\infty_2, 5, 4\}
\end{array}$$

Now, the seven designs correspond to the seven possible keys; the shares are the points in X. If the dealer's secret is K ($0 \leq K \leq 6$) then he chooses a random block B from $\mathcal{B}_K$ and gives the three points in B to the three participants as their shares. Now, three shares uniquely determine the block B, which occurs in exactly one of the seven designs (since we have a large set). Hence three participants can compute K. But from two shares, say x and y, nothing can be determined, since for each possible secret K, there is a (unique) block in $\mathcal{B}_K$ that contains x and y.

5   Resilient functions

The concept of resilient functions was introduced independently in the two papers Chor *et al* [15] and Bennett, Brassard and Robert [3]. Here is the definition. Let $n \geq m \geq 1$ be integers and suppose

$$f : \{0,1\}^n \to \{0,1\}^m.$$

We will think of f as being a function that accepts n input bits and produces m output bits. Let $t \leq n$ be an integer. Suppose $(x_1, \ldots, x_n) \in \{0,1\}^n$, where the values of t arbitrary input bits are fixed by an opponent, and the remaining $n - t$ input bits are chosen independently at random. Then f is said to be $t-$*resilient* provided that every possible output $m-$tuple is equally likely to occur. More formally, the property can be stated as follows: for every $t-$subset $\{i_1, \ldots, i_t\} \subseteq \{1, \ldots, n\}$, for every choice of $z_j \in \{0,1\}$ $(1 \leq j \leq t)$, and for every $(y_1, \ldots, y_m) \in \{0,1\}^m$, we have

$$p(f(x_1, \ldots, x_n) = (y_1, \ldots, y_m)|x_{i_j} = z_j, 1 \leq j \leq t) = \frac{1}{2^m}.$$

We will refer to such a function f as an $(n, m, t)-$resilient function.

A closely related concept is that of a correlation-immune function, which is defined by Siegenthaler in [50] and further studied in [44], [26] and [13]. Let $n \geq 1$ be an integer and suppose $f : \{0,1\}^n \to \{0,1\}$. As before, suppose $(x_1, \ldots, x_n) \in \{0,1\}^n$, where the values of t arbitrary input bits are fixed by an opponent, and the remaining $n - t$ input bits are chosen independently at random. Then f is said to be *correlation-immune of order* t provided that for every $t-$subset $\{i_1, \ldots, i_t\} \subseteq \{1, \ldots, n\}$, for every choice of $z_j \in \{0,1\}$ $(1 \leq j \leq t)$, and for $y = 0, 1$, we have

$$p(f(x_1, \ldots, x_n) = y|x_{i_j} = z_j, 1 \leq j \leq t) = p(f(x_1, \ldots, x_n) = y).$$

A correlation-immune function is *balanced* if

$$p(f(x_1, \ldots, x_n) = y|x_{i_j} = z_j, 1 \leq j \leq t) = 1/2).$$

In other words, a balanced correlation-immune function is the same thing as an $(n, 1, t)-$resilient function.

Two possible applications of resilient functions are mentioned in [3] and [15]. The first application concerns the generation of shared random strings in the presence of faulty processors. The second involves renewing a partially leaked cryptographic key. Correlation-immune functions are used in stream ciphers as combining functions for running-key generators that are resistant to a correlation attack (see, for example, Rueppel [44]).

Many interesting results on resilient functions can be found in [3] and [15]. The basic problem is to maximize t given m and n; or equivalently, to maximize m given n and t. Here are some examples from [15] (all addition is modulo 2):

1. $m = 1, t = n - 1$. Define $f(x_1, \ldots, x_n) = x_1 + \ldots + x_n$.

2. $m = n - 1, t = 1$. Define $f(x_1, \ldots, x_n) = (x_1 + x_2, x_2 + x_3, \ldots, x_{n-1} + x_n)$.

3. $m = 2, n = 3h, t = 2h - 1$. Define

$$f(x_1, \ldots, x_{3h}) = (x_1 + \ldots + x_{2h}, x_{h+1} + \ldots + x_{3h}).$$

In fact, all three of these examples are optimal. It is easy to see that $n \geq m + t$, so the first two examples are optimal. The result that $t < \lfloor \frac{2n}{3} \rfloor$ if $m = 2$ is much more difficult; it is proved in [15].

Resilient functions turn out to be equivalent to certain large sets of orthogonal arrays. The following result is proved in [66].

Theorem 5.1 *An (n, m, t)-resilient function is equivalent to a large set of orthogonal arrays $OA_{2^{n-m-t}}(t, n, 2)$.*

In fact, for any $y \in \{0,1\}^m$, the inverse image $f^{-1}(y)$ yields an $OA_{2^{n-m-t}}(t, n, 2)$, and the 2^m orthogonal arrays thus obtained form a large set.

A related result for correlation-immune functions was proved earlier in [13]:

Theorem 5.2 *A correlation-immune function $f : \{0,1\}^n \to \{0,1\}$ of order t is equivalent to an orthogonal array $OA_\lambda(t, n, 2)$ for some integer λ.*

In fact, we get two orthogonal arrays: an $OA_{\lambda_0}(t, n, 2)$ from $f^{-1}(0)$ and an $OA_{\lambda_1}(t, n, 2)$ from $f^{-1}(1)$. For $i = 0, 1$, we have $\lambda_i = |f^{-1}(i)|/2^t$, and the union of the two orthogonal arrays is an $OA(k, k, n)$.

In view of Theorem 5.1, any necessary condition for the existence of an orthogonal array $OA_{2^{n-m-t}}(t, n, 2)$ is also a necessary condition for the existence of an (n, m, t)-resilient function. One classical bound for orthogonal arrays is the Rao bound [41], proved in 1947 (the Rao bound is a generalization of the Plackett-Burman bound mentioned in Section 3). We record the Rao bound as the following theorem.

Theorem 5.3 *Suppose there exists an $OA_\lambda(t, k, v)$. Then*

$$\lambda v^t \geq 1 + \sum_{i=1}^{t/2} \binom{k}{i} (v-1)^i$$

if t is even; and

$$\lambda v^t \geq 1 + \sum_{i=1}^{(t-1)/2} \binom{k}{i} (v-1)^i + \binom{k-1}{(t-1)/2} (v-1)^{(t+1)/2}$$

if t is odd.

We obtain the following corollary which gives a necessary condition for existence of an (n, m, t)−resilient function.

Corollary 5.4 *[15, 66] Suppose there exists an (n, m, t)−resilient function. Then*

$$m \leq n - \log_2 \left[\sum_{i=0}^{\lceil t/2 \rceil} \binom{n}{i} \right]$$

if t is even; and

$$m \leq n - \log_2 \left[\sum_{i=0}^{\lceil (t-1)/2 \rceil} \binom{n}{i} + \binom{n-1}{(t-1)/2} \right]$$

if t is odd.

Proof. Set $v = 2$ in Theorem 5.3 and apply Theorem 5.1.   □

Remark. For t even, the bound of Corollary 5.4 was proved in [15] from first principles. For t odd, the bound was proved in [66] and is a slight improvement over the bound in [15].

The Bush bound for orthogonal arrays with $\lambda = 1$ [12] also will provide a necessary existence condition for certain resilient functions. This bound is as follows:

Theorem 5.5 *[12] Suppose there exists an $OA(t, k, v)$, where $t > 1$. Then*

$$
\begin{aligned}
k &\leq v + t - 1 &&\text{if } v \geq t, v \text{ even} \\
k &\leq v + t - 2 &&\text{if } v \geq t \geq 3, v \text{ odd} \\
k &\leq t + 1 &&\text{if } v \leq t.
\end{aligned}
$$

As a corollary, we can obtain the following result that was proved in [3] from first principles:

Corollary 5.6 *[3] There exists an (n, m, t)−resilient function with $n = m + t$ if and only if $t = 1$ or $m = 1$.*

Proof. The cases $t = 1$ and $m = 1$ were given earlier in examples. So, suppose $n = m + t$ and $2 \leq t \leq n - 2$. Apply Theorem 5.5 with $v = 2$ to get $m + t \leq t + 1$, or $m \leq 1$, a contradiction.   □

The most important construction method for resilient functions uses (linear) binary codes. We will be using several standard results from coding theory without proof; see MacWilliams and Sloane [34] for background information on error-correcting codes. An (n, m, d) linear code is an m−dimensional subspace C of $(GF(2))^n$ such that any two vectors in C have Hamming distance at least d. Let G be an $m \times n$ matrix whose rows form a basis for C; G is called a *generating matrix* for C. The following construction for resilient functions was given in [3, 15]:

Theorem 5.7 *Let G be a generating matrix for an (n, m, d) linear code C. Define the function $f : (GF(2))^n \to (GF(2))^m$ by the rule $f(x) = xG^T$. Then f is an $(n, m, d-1)$−resilient function.*

This result can easily be seen to be true using the orthogonal array characterization. The inverse image $f^{-1}(0, \ldots, 0)$ is in fact the dual code $C^\perp$. It is well-known that $C^\perp$ is an orthogonal array $OA_{2^{n-m-d+1}}(d-1, n, 2)$ (see for example [34, p. 139]). In fact, this is obvious since any $d-1$ columns of the generating matrix for $C^\perp$ (= the parity check matrix for C) are linearly independent. Now, any other inverse image $f^{-1}(y)$ is an additive coset of $C^\perp$, and thus is also an $OA_{2^{n-m-d+1}}(d-1, n, 2)$. Hence we obtain 2^m OA's that form a large set. By Theorem 5.1, f is an $(n, m, d-1)$−resilient function.

As an example, suppose we start with the perfect binary Hamming code [34, p. 25]. This is an $(2^r - 1, 2^r - r - 1, 3)$ code. It gives rise to a $(2^r - 1, 2^r - r - 1, 2)$ resilient function; or equivalently, a large set of orthogonal arrays $OA_{2^{r-2}}(2, 2^r - 1, 2)$. These resilient functions are optimal in view of Corollary 5.4.

As another example, suppose we start with the Reed-Muller code $\mathcal{R}(1, s)$ [34, p. 376]. This is a $(2^s, s + 1, 2^{s-1})$ linear code, which yields a $(2^s, s+1, 2^{s-1}-1)$−resilient function. (Note that a $(2^s, s, 2^{s-1}-1)$−resilient function is constructed in [15]. This function corresponds to the code obtained from $\mathcal{R}(1, s)$ by deleting the row $1, 1, \ldots, 1$ from the generating matrix. So we get one more output bit than [15], while maintaining the same resiliency.)

Here is an interesting question for future research. It is conceivable that a (rowwise simple) orthogonal array might exist, but a large set (= resilient function) does not. One interesting situation where this might happen concerns the parameters $n = 3h$, $m = 2$, $t = 2h$. It was mentioned earlier that there is no resilient function with these parameters. But the proof of this fact, which is found in [15], does not seem to rule out the existence of an $OA_{2^h-2}(2h, 3h, 2)$. So this is a case where an OA might exist even though the large set does not.

In fact, there is no $OA_{2^h-2}(2h, 3h, 2)$ if $h = 2$ or $h = 3$, as can be seen by applying the Rao bound. But for $h \geq 4$, it seems that no results are known concerning this class of OA's.

Finally, we mention that Teirlinck has observed in [70] that existence of an orthogonal array $OA(t, k, v)$ (with $\lambda = 1$) implies the existence of a large set of $OA(t, k, v)$. Also, recent results of Friedman [22] show that, for certain other parameter situations, existence of an OA implies the existence of a large set.

Acknowledgements

D. R. Stinson's research is supported by NSF Grant CCR-9121051.

References

[1] M. H. G. Anthony, K. M. Martin, J. Seberry and P. Wild. Some remarks on authentication systems. *Lecture Notes in Computer Science* **453** (1990), 122-139.

[2] R. Baker. Partitioning the planes of $AG_{2m}(2)$ into 2−designs. *Discrete Math.* **15** (1976), 205-211.

[3] C. H. Bennett,G. Brassard and J.-M. Robert. Privacy amplification by public discussion. *SIAM J. Comput.* **17** (1988), 210-229.

[4] Th. Beth, D. Jungnickel and H. Lenz. *Design Theory*, Bibliographisches Institut, Zurich, 1985.

[5] A. Beutelspacher. Geometric structures as threshold schemes. In *Cryptography and Coding*, Institute of Mathematics and its Applications Conference Series **20** (1989), 255-268.

[6] A. Beutelspacher. Applications of finite geometry to cryptography. In *Geometries, Codes and Cryptography*, Springer Verlag, 1990, 161-186.

[7] J. Bierbrauer and Tran van Trung. Halving $PGL(2, 2^f)$, f odd: a series of cryptocodes. *Designs, Codes, Cryptography* **1** (1991), 141-148.

[8] J. Bierbrauer and Tran van Trung. Some highly symmetric authentication perpendicular arrays. *Designs, Codes, Cryptography* **1** (1991), 307-319.

[9] G. R. Blakley. Safeguarding cryptographic keys *AFIPS Conference Proceedings* **48** (1979), 313-317.

[10] E. F. Brickell. A few results in message authentication. *Congressus Numer.* **43** (1984), 141-154.

[11] E. F. Brickell and D. M. Davenport. On the classification of ideal secret sharing schemes *J. Cryptology* **4** (1991), 123-134.

[12] K. A. Bush. Orthogonal arrays of index unity. *Ann. Math. Stat.* **23** (1952), 426-434.

[13] P. Camion, C. Carlet, P. Charpin and N. Sendrier. On correlation-immune functions. *Lecture Notes in Computer Science* **576** (1992), 86-100.

[14] D. Chen and D. R. Stinson. Recent results on combinatorial constructions for threshold schemes. *Australasian Journal of Combinatorics* **1** (1990), 29-48.

[15] B. Chor, O. Goldreich, J. Hastad, J. Friedman, S. Rudich and R. Smolensky. The bit extraction problem or t−resilient functions. *Proc. 26th IEEE Symp. on Foundations of Computer Science* (1985), 396-407.

[16] C. J. Colbourn and P. C. Van Oorschot. Applications of combinatorial designs in computer science. *ACM Computing Surveys* **21** (1989), 223-250.

[17] M. De Soete. Some constructions for authentication − secrecy codes. *Lecture Notes in Computer Science* **330** (1988), 57-75.

[18] M. De Soete and K. Vedder. Some new classes of geometric threshold schemes. *Lecture Notes in Computer Science* **330** (1988), 389-401.

[19] M. De Soete, K. Vedder and M. Walker. Cartesian authentication schemes. *Lecture Notes in Computer Science* **434** (1990), 476-490.

[20] M. De Soete, J.-J. Quisquater and K. Vedder. A signature with shared verification scheme. *Lecture Notes in Computer Science* **435** (1990), 253-262.

[21] M. De Soete. New bounds and constructions for authentication / secrecy codes with splitting. *Journal of Cryptology* **3** (1991), 173-186.

[22] J. Friedman. On the bit extraction problem. *Proc. 33rd IEEE Symp. on Foundations of Computer Science* (1992), 314-319.

[23] E. N. Gilbert, F. J. MacWilliams and N. J. A. Sloane. Codes which detect deception. *Bell System Tech. Journal* **53** (1974), 405-424.

[24] P. Godlewski and C. Mitchell. Key minimal cryptosystems for unconditional secrecy. *Journal of Cryptology* **3** (1990), 1-25.

[25] M. Greig. Resolvable balanced incomplete block designs with a block size of 8. Preprint.

[26] X. Guo-zhen and J. L. Massey. A spectral characterization of correlation-immune functions. *IEEE Trans. Inform. Theory* **34** (1988), 569-571.

[27] H. Hanani, D. K. Ray-Chaudhuri and R. M. Wilson. On resolvable designs. *Discrete Math.* **3** (1972), 343-357.

[28] W.-A. Jackson and K. M. Martin. On ideal secret sharing schemes. Submitted to *Journal of Cryptology*.

[29] W.-A. Jackson and K. M. Martin. Cumulative arrays and geometric secret sharing schemes. Presented at AUSCRYPT 92.

[30] W.-A. Jackson and K. M. Martin. Geometric secret sharing schemes and their duals. Preprint.

[31] E. S. Kramer, D. L. Kreher, R. Rees and D. R. Stinson. On perpendicular arrays with $t \geq 3$ *Ars Combinatoria* **28** (1989), 215-223.

[32] J. X. Lu. On large sets of disjoint Steiner triple sytems I, II and III. *J. Combin. Theory A* **34** (1983), 140-182.

[33] J. X. Lu. On large sets of disjoint Steiner triple sytems IV, V and VI. *J. Combin. Theory A* **37** (1984), 136-192.

[34] F. J. MacWilliams and N. J. A. Sloane. *The Theory of Error-correcting Codes*, North-Holland, 1977.

[35] J. L. Massey. Cryptography – a selective survey. In *Digital Communications*, North-Holland, 1986, 3-21.

[36] C. Mitchell, M. Walker and P. Wild. The combinatorics of perfect authentication schemes. To appear in *SIAM J. Discrete Math.*

[37] R. C. Mullin, P. J. Schellenberg, G. H. J. van Rees and S. A. Vanstone. On the construction of perpendicular arrays. *Utilitas Math.* **15** (1979), 323-333.

[38] R. L. Plackett and J. P. Burman. The design of optimum multi-factorial experiments. *Biometrika* **33** (1945), 305-325.

[39] D. K. Ray-Chaudhuri and R. M. Wilson. The existence of resolvable block designs. In *A Survey of Contemporary Theory*, North-Holland, 1973, 361-375.

[40] C. R. Rao. Hypercubes of strength "d" leading to confounded designs in factorial experiments. *Bull. Calcutta Math. Soc.* **38** (1946), 67-78.

[41] C. R. Rao. Factorial experiments derivable from combinatorial arrangements of arrays. *J. Royal Stat. Soc.* **9** (1947), 128-139.

[42] D. K. Ray-Chaudhuri and R. M. Wilson. Solution of Kirkman's schoolgirl problem. *Amer. Math. Soc. Proc. Symp. Pure Math.* **19** (1971), 187-204.

[43] U. Rosenbaum. A lower bound on authentication after having observed a sequence of messages. To appear in *J. Cryptology*.

[44] R. Rueppel. *Analysis and Design of Stream Ciphers*, Springer Verlag, Berlin, 1986.

[45] P. J. Schellenberg and D. R. Stinson. Threshold schemes from combinatorial designs. *J. Comb. Math. and Comb. Comp.* **5** (1989), 143-160.

[46] P. Schobi. Perfect authentication systems for data sources with arbitrary statistics. Presented at EUROCRYPT 86.

[47] P. D. Seymour. On secret-sharing matroids. *Journal of Combin. Theory B* **56** (1992), 69-73.

[48] A. Shamir. How to share a secret. *Commun. of the ACM* **22** (1979) 612–613.

[49] C. E. Shannon. Communication theory of secrecy systems. *Bell Systems Technical Journal* **28** (1949), 656-715.

[50] T. Siegenthaler. Correlation immunity of nonlinear combining functions for cryptographic applications. *IEEE Trans. Inform. Theory* **30** (1984), 776-780.

[51] G. J. Simmons. Message authentication: a game on hypergraphs. *Congressus Numerantium* **45** (1984), 161-192.

[52] G. J. Simmons. Authentication theory / coding theory. *Lecture Notes in Computer Science* **196** (1985), 411-432.

[53] G. J. Simmons. Robust shared secret schemes or 'how to be sure you have the right answer even though you don't know the question'. *Congressus Numerantium* **68** (1989), 215-248.

[54] G. J. Simmons. A cartesian product construction for unconditionally secure authentication codes that permit arbitration. *Journal of Cryptology* **2** (1990), 77-104.

[55] G. J. Simmons. How to (really) share a secret. *Lecture Notes in Computer Science* **403** (1990), 390-448.

[56] G. J. Simmons. Prepositioned shared secret and/or shared control schemes. *Lecture Notes in Computer Science* **434** (1990), 436-467.

[57] G. J. Simmons. An introduction to shared secret and/or shared control schemes and their application. In *Contemporary Cryptology, The Science of Information Integrity*, IEEE Press, 1992, 441-497.

[58] G. J. Simmons. A survey of information authentication. In *Contemporary Cryptology, The Science of Information Integrity*, IEEE Press, 1992, 379-419.

[59] G. J. Simmons, W. Jackson and K. Martin. The geometry of shared secret schemes. *Bulletin of the ICA* 1 (1991), 71-88.

[60] N. J. A. Sloane. Error-correcting codes and cryptography. In *The Mathematical Gardner*, Prindle, Weber and Schmidt, 1981, 346-382.

[61] D. R. Stinson. Some constructions and bounds for authentication codes. *Journal of Cryptology* 1 (1988), 37-51.

[62] D. R. Stinson. A construction for authentication / secrecy codes from certain combinatorial designs. *Journal of Cryptology* 1 (1988), 119-127.

[63] D. R. Stinson. The combinatorics of authentication and secrecy codes. *Journal of Cryptology* 2 (1990), 23-49.

[64] D. R. Stinson. Combinatorial characterizations of authentication codes. *Designs, Codes and Cryptography* 2 (1992), 175-187.

[65] D. R. Stinson. An explication of secret sharing schemes. *Designs, Codes and Cryptography* 2 (1992), 357-390.

[66] D. R. Stinson. Resilient functions and large sets of orthogonal arrays. To appear in *Congressus Numer.*

[67] D. R. Stinson and L. Teirlinck. A construction for authentication / secrecy codes from 3−homogeneous permutation groups. *Europ. J. Combin.* 11 (1990), 73-79.

[68] D. R. Stinson and S. A. Vanstone. A combinatorial approach to threshold schemes. *SIAM J. on Discrete Math.* 1 (1988), 230-237.

[69] L. Teirlinck. A completion of Lu's determination of the spectrum for large sets of disjoint Steiner triple sytems. *J. Combin. Theory A* 57 (1991), 302-305.

[70] L. Teirlinck. Large sets of disjoint designs and related structures. In *Contemporary Design Theory – A Collection of Surveys*, John Wiley & Sons, New York, 1992, 561-592.

[71] L. Teirlinck. Some new 2−resolvable Steiner quadruple systems. Submitted to *Designs, Codes, Cryptography*.

[72] Tran van Trung. On the construction of authentication and secrecy codes. To appear in *Designs, Codes, Cryptography*.

[73] M. Walker. Information-theoretic bounds for authentication schemes. *Journal of Cryptology* **2** (1990), 131-143.

[74] R. M. Wilson. The necessary conditions for t−designs are sufficient for something. *Utilitas Math.* **4** (1973), 207-215.

[75] G. V. Zaicev, V. A. Zinoviev and N. V. Semakov. Interrelation of Preparata and Hamming codes and extension of Hamming codes to new double error-correcting codes. In *Proc. 2nd Inter. Symp. Inform. Theory*, Akademiai Kiado, Budapest, 1973, 257-263.

[76] G. S. Vernam. Cipher printing telegraph systems for secret wire and radio telegraphic communications. *J. Am. Inst. Elec. Eng.* **55** (1926), 109-115.

[77] D. J. A. Welsh. *Matroid Theory*, Academic Press, 1976.

[78] L. Zhu, B. Du and X. Zhang. A few more RBIBDs with $k = 5$ and $\lambda = 1$. *Discrete Math.* **97** (1991), 409-417.